The Oceanic Thermohaline Circulation: An Introduction

ATMOSPHERIC AND OCEANOGRAPHIC SCIENCES LIBRARY

VOLUME 39

The titles published in this series are listed at the end of this volume.

The Oceanic Thermohaline Circulation: An Introduction

Hendrik M. van Aken

Royal Netherlands Institute for Sea Research
The Netherlands

Springer

Hendrik M. van Aken
Royal Netherlands Institute for Sea Research
Department of Physical Oceanography
1790 AB Den Burg/Texel
The Netherlands
email: aken@nioz.nl

Library of Congress Control Number: 2006933292

ISBN-10: 0-387-36637-7
ISBN-13: 978-0-387-36637-1
eISBN-13: 978-0-387-48039-8

Printed on acid-free paper.

9 8 7 6 5 4 3 2 1

springer.com

Dedication

In memory of my father, Reijer van Aken

Preface

The ocean is an important part of the global climate system, as was already stated in 1798 by Benjamin Thompson, Count Rumford. The founder of modern oceanography, Matthew F. Maury, also stressed the importance of the ocean circulation for the climate by influencing the winds, the air temperature, and the hydrological cycle in his book on the physical geography of the sea published in 1855. Ongoing research over the past 150 years has confirmed these viewpoints and has identified key processes for the links between ocean circulation and climate.

With the increasing probability of climate change due to the anthropogenic emission of greenhouse gases into the atmosphere, the study of the ocean circulation has gained more than only academic interest. A good understanding of the ocean's role in the climate system is essential in order to assess the present climate status and to forecast climate changes that may result from the still-increasing emission of greenhouse gases. The globally overturning thermohaline circulation, driven by global-scale gradients in temperature and salinity, is assumed to contribute essentially to the climatic feedback mechanisms that are involved in anthropogenic climate change as well as in natural climate variability. At low latitudes the surface water is heated by interaction with the atmosphere, while at high latitudes the surface water is cooled. These cold water masses then descend into the deep abyssal basins. This deep-reaching circulation system also is involved in the storage of carbon dioxide in the ocean. While carbon dioxide is converted to organic material by primary production of phytoplankton in the near-surface photic zone, the microbial ecosystem in the deep ocean converts organic material back to dissolved inorganic carbon. When the cold deepwater from the abyssal ocean basins returns to the surface of the ocean heat and carbon dioxide again are exchanged with the atmosphere.

Nowadays numerical models of the oceanic circulation are used as part of global climate models. With these models possible future developments of the climate, which may have a profound influence on people and society all around the world, are simulated. The International Panel on Climate Change of the World Meteorological Organization and the United Nations Environmental Program also discuss the climate status of the oceans and the role of the oceans in climate in their regularly published Climate Change Scientific Assessment Reports.

In discussions with colleagues and students I often have explained the evidence for the course of the thermohaline circulation, possible changes of this circulation, and its main dynamic properties. Although nowadays the importance of the thermohaline circulation is well recognized, oceanographic textbooks generally do not present the hydrographic and dynamic aspects of the thermohaline circulation comprehensively. In many publications, especially in the fields of marine geology, chemistry, and biology, the understanding of the thermohaline circulation appears to be quite old-fashioned. Physical oceanographic papers are often rather one-sided, missing either the hydrographic, thermodynamic or hydrodynamic characteristics of this important global current system. Therefore, I decided to write this textbook, which is devoted completely to the thermohaline circulation. This book deals with a description of the thermohaline circulation, based on recent ocean observations, and with important theoretical aspects of the dynamics of the thermohaline circulations and its coupling to the atmosphere. It is intended to be used in courses for students in physical oceanography, climatology, geography and environmental sciences, paleoceanography, marine biology, and marine chemistry.

A mathematical physical approach can be very useful for the understanding of many quantitative and qualitative aspects of the thermohaline circulation. To make this book accessible for students with a variety of backgrounds, I have tried to keep the mathematics used in this book quite simple. Only some basic knowledge of differential and integral reckoning is required.

I owe thanks to several people for the support I got when preparing this book. First of all, I thank my wife Marjan, who allowed me to spend many evenings and weekends in my study. My colleagues at the Royal Netherlands Institute for Sea Research at Texel shared their expertise on several aspects of the thermohaline circulation. Leo Maas supported me in understanding the mathematical aspects of dynamic and feedback models, and was a critical reader of the manuscript. Fred Jansen introduced me to the world of paleoceanography, and Gerald Herndl and Geraldine Kramer taught me about the importance of the deep microbial ecosystem for the remineralization of organic material in the water mass of the abyssal branch of the thermohaline circulation. Wim Mook advised me with regard to the use of isotopes as tracers of the ocean circulation. Will de Ruijter of the Institute for Marine and Atmospheric Research of Utrecht University invited me to give seminars for students at his institute which form an important basis for this book. Discussions with students during the courses and seminars that I taught and during research cruises with Dutch, Indonesian, and German research vessels guided me with the choice of subjects for this textbook on the thermohaline circulation.

Finally I want to thank my late father, who in my youth stimulated my interest in science and the sea. He always encouraged and supported me in the changing choices for my studies, first engineering at the Royal Dutch Naval College, then physics and mathematics at Leiden University, and finally physical oceanography and meteorology at Utrecht University. This book is dedicated to him.

I thank Peter Wadhams, Arnold Gordon, and John Marshall, who allowed me to use figures from their publications in this book. I acknowledge all scientists who submitted their data to the public data bases that I have used to prepare the many figures, illustrating the diverse aspects of the oceanic thermohaline circulation.

Hendrik M. van Aken
Texel
October 2006

Contents

List of Abbreviations

AABW	Antarctic Bottom Water
AAIW	Antarctic Intermediate Water
ACC	Antarctic Circumpolar Current
AgC	Agulhas Current
AgRf	Agulhas Retroflection
AIW	Arctic Intermediate Water
AMS	Accelerator mass spectrometers
AOU	Apparent oxygen utilization
BOUNCE	Boundary Current Experiment
BP	Years before present (present = 1950)
CBDW	Canadian Basin Deep Water
CDW	Circumpolar Deep Water
CFC	Chlorofluorocarbon
CGFZ	Charlie–Gibbs Fracture Zone
CIRES	Cooperative Institute for Research in Environmental Sciences
DNBC	Deep northern boundary current
DSBC	Deep southern boundary current
DSDP	Deep Sea Drilling Project
DSOW	Denmark Strait Overflow Water
DWBC	Deep western boundary current
EBDW	Eurasian Basin Deep Water
ECC	Equatorial Counter Current
ECMWF	European Centre for Medium-Range Weather Forecasting
EGC	East Greenland Current
EMC	East Madagascar Current
EOS	Equation of state
ERBE	Earth Radiation Budget Experiment
GCM	Global climate model
GEOSECS	Geochemical Ocean Section Study
GISP	Greenland Ice Sheet Project
GNAIW	Glacial North Atlantic Intermediate Water
GRIP	Greenland Ice Core Project
GSDW	Greenland Sea Deep Water
HOAPS	Hamburg ocean atmosphere parameters and fluxes from satellite data
IDW	Indian Deep Water
IPCC	Intergovernmental Panel on Climate Change
IPTS-68	International practical temperature scale
ISOW	Iceland–Scotland Overflow Water

ISW	Ice Shelf Water
ITFW	Indonesian Trough-Flow Water
ITS-90	International temperature scale 1990
KNMI	Koninklijk Nederlands Meteorologisch Instituut (Royal Netherlands Meteorological Institute)
LC	Labrador Current
LCDW	Lower Circumpolar Deep Water
LDW	Lower Deep Water
LGM	Last Glacial Maximum
LSW	Labrador Sea Water
MAR	Mid-Atlantic Ridge
MARE	Mixing of Agulhas Rings Experiment
MC	Mozambique Current
MOC	Meridional overturning circulation
MSOW	Mediterranean Sea Outflow Water
NAC	North Atlantic Current
NADW	North Atlantic Deep Water
NCAR	National Center for Atmospheric Research
NCEP	National Centers for Environmental Prediction
NCW	Northern Component Water
NEADW	North East Atlantic Deep Water
NH	Northern Hemisphere
NIDW	North Indian Deep Water
NOAA	National Oceanographic and Atmospheric Administration
NSDW	Norwegian Sea Deep Water
ODP	Ocean Drilling Program
OGCM	Ocean general circulation model
OMPA	Optimal multiparameter analysis
PAIW	Pacific Arctic Intermediate Water
PDW	Pacific Deep Water
PSS-78	Practical salinity scale 1978
rAtIW	Return Atlantic Intermediate Water
RGR	Rio Grande Rise
RSOW	Red Sea Outflow Water
SAAMW	Sub-Antarctic Mode Water
SAT	Surface air temperature
SC	Somali Current
SCW	Southern Component Water
SEC	South Equatorial Current
SH	Southern Hemisphere
SI	Système International d'Unités (International System of Units)
SIOC	South Indian Ocean Current

SMOW	Standard Mean Ocean Water
SSS	Sea surface salinity
SST	Sea surface temperature
STMW	Subtropical Mode Water
SUN	Symbols, units, and nomenclature in physics
THC	Thermohaline circulation
TIC	Total inorganic carbon
TOA	Top of the atmosphere
UCDW	Upper Circumpolar Deep Water
WDW	Warm Deep Water
WGC	West Greenland Current
WOCE	World Ocean Circulation Experiment
WR	Walvis Ridge
WSBW	Weddell Sea Bottom Water
WSW	Western Shelf Water
WW	(Weddell Sea) Winter Water
YD	Younger Dryas

1. Introduction

1.1. Climate and climate variations

The equatorial zone of the earth receives more energy per surface unit from the sun in the form of short-wave radiation than the polar regions because of the spherical form of the earth. The resulting temperature difference between the equator and the poles leads to meridional heat transport by the atmospheric and the oceanic circulation, both parts of the climate system. In its turn this heat transport mitigates the extreme cold and heat, caused by the differential heating by the sun and is responsible for a moderate global climate and a habitable earth.

It has been well established that the global climate system, of which the ocean circulation is an integral part, exhibits (natural) variability on a multitude of timescales. The ocean circulation pattern will change if the climate changes, while changes in the ocean circulation may induce climate changes. In first order the ocean will follow the changes in the atmospheric climate passively, since the ocean circulation and heat and salinity distributions are responses to the boundary conditions at the sea surface, set by the atmosphere. However, in higher order the ocean also may play an important role in amplifying or damping climate variations via several feedback mechanisms, depending on changed sea surface conditions like sea surface temperature, ice cover, etc. For climate simulation and forecasting of climate change a thorough knowledge of the ocean's role in climate is required (IPCC 2001). This is not a simple task. Despite the best efforts of thousands of people and the investments of hundreds of millions of dollars, there remain serious uncertainties about the present state of the ocean, its surface boundary conditions, and the extent to which and how it is changing (Wunsch 2003). This book deals with that part of the oceanic circulation that is assumed by many to supply the most important climatic feedbacks: the global thermohaline circulation (THC).

At present the strongest climate variation cycle is the seasonal cycle, directly driven by the variation of the solar radiation which is mainly due to the obliquity of the rotation axis of the earth relative to the plane of its orbit around the sun. Longer period climate variations have characteristic timescales ranging from a few years to at least several hundred-thousands of years. A well-known climate variation on relatively short timescales, in which the ocean plays a significant role, is the so-called El Niño phenomenon, associated with anomalous warming of the eastern tropical Pacific every 3 to 5 years. This phenomenon has worldwide consequences, e.g., the precipitation pattern over southern Africa and droughts in Indonesia and Australia. It shows that changes in the ocean circulation, combined with a feedback due to wind drift and the heat exchange with the atmosphere, can influence the global climate. Recent climate phenomena with a timescale of several hundreds of years were the Mediaeval Climate Optimum and the Little Ice Age which strongly influenced nature and society. On the long-term side of the climate spectrum, with a typical timescale of hundred-thousand years are the ice ages or glacials, connected with the formation of huge continental ice caps. During the last glacial period the climate was colder than today, but also oscillated rather irregularly between cold and relatively warm periods: the interstadials or Dansgaard–Oescher events (Dansgaard et al. 1982). After the end of the last glacial a significant cooling event occurred during the Younger Dryas, a period of 1,300 years, that abruptly terminated at ~11,640 year before present (BP).

Increased emissions of the greenhouse gases carbon dioxide, methane, and nitrous oxide, due to human activities, may lead to an anthropogenic global warming of the climate. The global mean surface temperature of the atmosphere has increased over the 20th century by about 0.6°C and is expected to lead to an anthropogenic increase in the globally averaged atmospheric surface temperature, estimated to be between 1.4 and 5.8°C from 1990 to 2100 (IPCC 2001). A consequence of global warming may be a rise in sea level which will inundate many of the present-day population centers. But a slowing down of the deep ocean circulation also is predicted in many climate forecasts.

In the discussions of climate variability and climate change on timescales of several years to thousands of years the effects of the ocean circulation in the climate system play an important role because of the thermal inertia of the ocean's waters. The uppermost few meters of the ocean have the same heat capacity as the entire overlying atmosphere. The ocean therefore can easily drive or damp climatic temperature changes in the atmosphere. The THC, thought to be driven by meridional differences in temperature and salinity, is often considered as the main oceanic climate process. Widespread consequences are ascribed to the shutdown or acceleration of the THC, an oceanographic *deus ex machina* for climate change science (Munk and Wunsch 1998; Wunsch 2002). For the general public the possible shutdown of the THC played a dramatic role in the 2004 blockbuster motion picture "The Day after Tomorrow" by Roland Emmerich. In that movie the shutdown of the THC causes a massive snowstorm to pound New Delhi; tornadoes rip Los Angeles; hail the size of grapefruits batters Tokyo; and in New York City the temperature swings from sweltering to freezing in one day. This disaster scenario, sketched in the movie, is impossible, but the patterns described by the movie have a distant basis in real concepts, studied by climate scientists and oceanographers. In this book I want to discuss: what is the THC; what are its main characteristics and dynamics; and what is its interaction with global climate?

1.2. The ocean and climate

The specific heat for dry air at constant pressure, C_p, is of the order of 1000 $Jkg^{-1}{}^{o}C^{-1}$. This value is definitely lower than the C_p of seawater, approximately 4000 $Jkg^{-1}{}^{o}C^{-1}$. Moreover the density of air near the sea surface amounts to 1.25 kg/m^3, while the ocean water has a density ρ of slightly over 1000 kg/m^3. Because of these differences the heat capacity of the atmosphere over 1 m^2 of the ocean's surface is about equal to the heat capacity per m^2 of the upper 3 m of the ocean. The specific heat of soil and rock is even lower than that of air, nearly 800 $Jkg^{-1}C^{-1}$. With a typical soil density of 3000 kg/m^3, the heat capacity of the uppermost meter of the earth's soil, where most of the continental seasonal heat storage is located, is about 2.4 $MJm^{-2}{}^{o}C^{-1}$. Seasonal heat storage in the ocean reaches a typical depth of about 100 m. The ocean surface layer with a thickness of 100 m has a heat capacity of 400 $MJm^{-2}{}^{o}C^{-1}$, two orders of magnitude larger than the typical heat capacity of the continental soil for seasonal heat storage. The large specific heat of seawater and the enormous mass of the ocean make the oceans a prime moderating factor for rapid climate variations. It certainly will prevent the ocean from cooling by over 10°C over a matter of days to weeks, as suggested in the movie "The Day after Tomorrow."

Because of its relatively low heat capacity the atmosphere lacks a thermal memory surpassing a few weeks. For climatic variability with timescales of seasons to hundreds of years the oceanic storage and transport of heat is a dominant factor. A standard explanation for interannual to decadal climate variability is that high-frequency stochastic variations in the atmosphere (the weather) are "integrated" by a slower-reacting ocean. The ocean responds as a stochastically driven oscillator, leading to low-frequency climate variability (Hasselmann 1976). On the timescale of the seasonal cycle (1 year) the transport of temperature anomalies by ocean currents is limited, and the heat stored in summer will be released to the atmosphere in winter in approximately the same geographic region. On this timescale the Hasselmann model may be applied, superimposed on the seasonal cycle. Climate variability caused by this mechanism does not have a preferred timescale or a preferred spatial pattern. The ocean's thermal inertia mainly converts the white noise of the weather forcing into a red spectrum response of the ocean temperature where larger timescales dominate. However, for the observed interdecadal to centennial climate variations preferential spatial and temporal patterns have been found. Therefore, the Hasselmann mechanism cannot fully explain the climate variability on these larger timescales (Te Raa 2003). For climate variations with longer timescales the oceanic circulation may alter the location where heat from a thermal anomaly in the ocean is released to the atmosphere, compared to the location where that heat was stored in the ocean, generating feedback loops in the climate system with preferential patterns and timescales.

Heat, stored in the ocean, can be transported by ocean currents which have discernable climate effects on interannual and longer timescales. Two different circulation modes are generally discriminated in oceanic circulation studies, each with their specific dynamics and timescales. The more or less regular winds over the oceans with easterly trade winds at low latitudes, prevailing westerlies at moderate latitudes and polar easterlies at arctic and antarctic latitudes drive a regular current system in the upper layers (~1000 to 1500 m) of the ocean. The main features of this system are the oceanwide anticyclonic gyres in the subtropics, cyclonic sub-arctic gyres in the North Atlantic and North Pacific oceans, and the Antarctic Circumpolar Current (ACC) in the Southern Ocean. In the equatorial zone a complicated system of surface and subsurface zonal currents is found. The northern Indian Ocean has a deviating current pattern because of the seasonally alternating monsoon winds. In the subtropical and sub-arctic gyres we find narrow and fast western boundary currents and a more diffuse recirculation further east The theory for the main dynamics of this wind-driven circulation is reasonably well understood (Pedlosky 1996). The poleward transport of warmer water and equatorward transport of colder water by the oceanic gyres maintain a meridional heat transport in the ocean. The thickness of the upper ocean layer in which this wind-driven heat transport occurs [O(1 km)] and the typical transport rates of the gyres introduce an inertia in the climate system with a characteristic timescale of the order of decades or tens of years (Nauw 2003).

Even larger timescales for climatic variability, up to hundreds of years or even a few thousand years can be reached by storage and transport of heat whereby the whole oceanic water column with a mean depth of 3700 m is involved. Here the THC is the dominating process. Zonally averaged the THC is a meridional overturning circulation (MOC). Regions of severe surface heat loss to the atmosphere at high latitudes are loosely associated with a downward mass flux of cold water which subsequently spreads at depth to lower latitudes. The downward mass flux is replenished by a poleward transport of warmer water in the upper layers of the ocean. For reasons of mass conservation the deep flow of cold water has to be connected with the upper layers by regions with upward moving deepwater (upwelling). The concept of the THC as a closed global overturning circulation with its main source of cold water in the North Atlantic Ocean and interhemispheric and interocean mass exchange has been sketched by Arnold Gordon (1986).

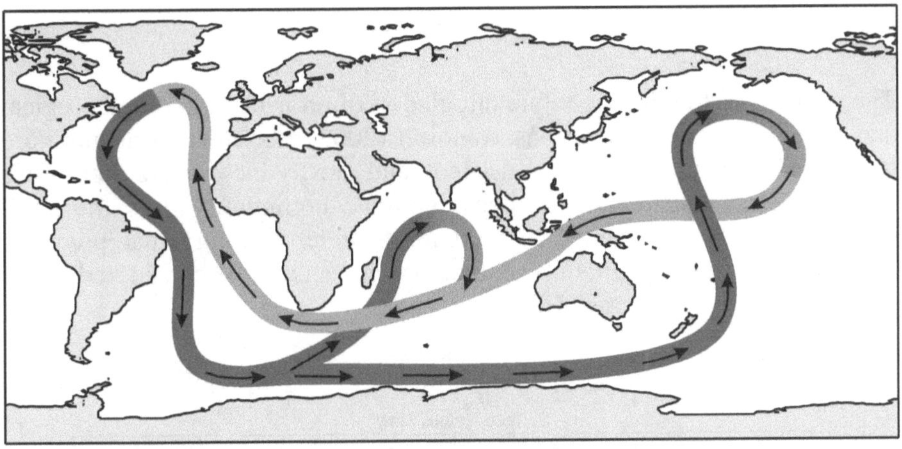

Fig. 1.1. A schematic picture of the great oceanic conveyor belt with the main source of cold deepwater in the North Atlantic Ocean, introduced by Broecker (1987). As stressed by Broecker (1991, p. 79) "other circulation 'loops' exist in the ocean and mixing occurs among the waters traveling along these intersecting pathways." But although reality is much more complex, this simple scheme drew the attention of many for the thermohaline circulation and its possible importance for the understanding and prediction of climate change.

This sketch was strongly simplified and popularized in the scheme of the "great oceanic conveyor belt" by Walter Broecker (1987, 1991) with emphasis on the North Atlantic Ocean as source of deepwater (Fig. 1.1). The downward mass flux at high latitudes and the subsequent lateral spreading of deepwater are assumed to be connected with thermal and evaporative forcing at the sea surface by the global atmosphere. The oceanic conveyor belt often is thought to operate like a heat engine, similar to the atmosphere, although a directly convectively driven overturning mass flux is presently thought to be impossible (Wunsch 2002). Contrary to the wind-driven circulation, a profound theoretical understanding of the THC is lacking, and no full general agreement has been reached yet even on the description of some major pathways in this circulation. Note also that in first order the direct effect of the THC is redistribution of heat; it does not change the global mean sea surface temperature. Any effects on the global mean temperature might occur by feedback mechanisms like changing sea ice and clouds which alter the albedo (reflection of short-wave sunlight), or atmospheric CO_2 and water vapor that change the atmosphere's opacity to long-wave radiation.

1.3. What is the THC?

The concept of the thermohaline circulation often is used in climatological and oceanographic publications. Wunsch (2002) has cited from literature seven different, sometimes inconsistent, and mostly incomplete definitions for the THC. Apparently the oceanographic community uses the THC concept widely, but there is no common understanding what precisely should be the THC circulation scheme or what should be considered as the main driving force for the THC.

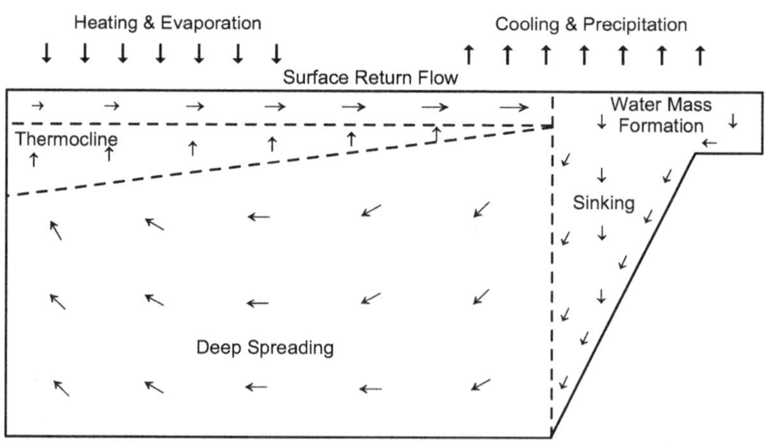

Fig. 1.2. Sketch of the overturning thermohaline circulation, adapted from Wyrtki (1961) by the inclusion of a shallow shelf sea.

It is not considered useful to come here with yet another definition of the concept of the global THC. Instead, extending the THC characteristics summed up by Wyrtki (1961), some aspects of the concept of the THC that will be discussed in this book will be emphasized (Fig. 1.2).

- *First*: Due to air–sea interaction in the source regions of deep ocean water, mainly cooling in winter, surface water will gain density (lose buoyancy). Air–sea interaction (cooling, evaporation, and precipitation) already modifies the warm surface currents flowing from the tropics to the source region of deepwater before they arrive there. The resulting decline and even reversal of the static stability of the density stratification in the source region will drive deep-reaching local convective mixing. In deep source regions a thick, homogeneous water mass is formed, while in source regions in shallow shelf seas the whole water column is homogenized.
- *Second*: The newly formed homogeneous high-density water mass descends or subducts into the abyssal ocean in a region with an average downward motion, where the cold water mass loses the possibility of direct contact with the atmosphere. The subduction region is not necessarily the region where the convective mixing and water mass formation takes place.
- *Third*: Hereafter the water spreads horizontally away from its subduction region. The hydrographic properties of this abyssal water mass (temperature, salinity, and natural and man-made tracers) are characteristic for its source region and formation process. The water can be followed along its path while it is (gradually) modified by mixing with surrounding water and biogeochemical aging. Adjustment of the abyssal density field will take place so that the deep flow can be described with the so-called geostrophic equilibrium. The spatial structure of this geostrophic adjustment depends on the locations where the vertical motion takes place, either subduction or upwelling.
- *Fourth*: In some regions of the ocean upwelling of the deepwater to shallower levels will take place whereby, due to turbulent mixing with shallower water, temperature, salinity, and density adapt to the overlying layers of the ocean.
- *Fifth*: The upwelled water returns to the regions where the deep convective mixing takes place. This return flow is not necessarily a simple direct link as suggested by the conveyor belt scheme shown in Figs. 1.1 and 1.2, and occurs for a large part or even completely in the wind-driven upper ocean.

In the last 100 years an increasing mass of so-called hydrographic observations has led to a progressive understanding of the main pathways and dominant processes of the global oceanic THC. The developing theoretical concepts of the ocean circulation have resulted in analytical models of several aspects of the THC. With the availability of electronic computers numerical models of the THC have been developed with different grades of complexity. Experiments with simplified THC models have strongly contributed to our understanding of the variability of climate, coupled to the THC. Advanced numerical ocean general circulation models form an essential part of the present state-of-the-art global climate models that are used for climate analysis and prediction.

The ocean circulation is a physical system, described by conservation equations for heat, salt, momentum, etc. Solutions of these equations depend on boundary equations for heat, freshwater, and momentum at the sea surface and the ocean floor. The boundary conditions at the sea surface are strongly determined by the state of the atmosphere over the ocean. Differences in seawater temperature and salinity, via the equation of state for seawater, are connected with density differences, resulting in pressure forces that determine the water motion. Since changes in the sea surface temperature in their turn will alter the exchange of heat and freshwater (evaporation) with the atmosphere, feedback mechanisms emerge, which may lead to either climate stability or variability of the climate.

To a certain degree the oceanic motion, driven by the stress, exerted by the winds over the ocean (the wind-driven circulation) can be separated from that part of the oceanic circulation that is mainly determined by the density differences, resulting from regional differences in the exchange of heat and freshwater between atmosphere and ocean (the THC). That is not a trivial distinction. Wind-driven ocean circulation may play an essential role for the formation phase of cold deepwater during the so-called preconditioning phase. The shallow surface return flow of the THC will be subject to forcing by the wind stress, thereby coupling the THC to the wind-driven circulation. Finally, the turbulent mixing in the ocean's interior, a process that appears to be essential for the maintenance of the THC, probably depends for a large part on the energy input from variable winds into the ocean.

The THC is a circulation process in the real ocean. That implies that its properties are strongly determined by the actual topography, including continents, undersea ridges, and deep passages. Also the global distribution of wind stress, surface heat, and freshwater fluxes and their seasonal and interannual variation influence the course and intensity of the THC. The THC is a global phenomenon with interhemispheric and interocean exchanges as is sketched in the conveyor belt scheme of Fig. 1.1. Because of the typical vertical stratification of temperature and salinity this mass exchange also involves a large-scale exchange of heat and freshwater between hemispheres and oceans. Better understanding of the THC will lead to a better understanding of climate and may contribute to a more accurate prediction of natural and man-made climate change.

1.4. Some historical notes

One of the first indications that water with low temperatures, originating from higher latitudes, can be found at depth in the subtropics, was a report by Captain Ellis of a British slave trader, the Earl of Halifax (Ellis 1751). With a bucket sea gauge, specially designed by Dr. Hales, rector of Teddington (Hales 1751), Ellis measured a temperature of 53°F (11.7°C) at a depth of 1190 m near the Azores. Modern measurements suggest that this reading was several degrees Celsius too warm due to warming of the water in the bucket during ascent to the surface (Warren 1981). From then onward subsurface temperature measurements were carried out during the 18th and early 19th century, which although inaccurate established that the ocean at depth generally was much colder than the near-surface waters (Krümmel 1907). This fact, trivial for modern oceanographers, has had profound implications for our understanding of the oceanic circulation. It is indicative of an overturning oceanic circulation which brings cold high-latitude water to the low latitudes. In the second half of the 19th century the minimum–maximum thermometer, devised by James Six in 1785, was fitted with an outer glass tube that protected the mercury bulb against pressure effects. This instrument, known as the Miller–Casella thermometer, became the classic instrument for subsurface temperature measurements, used during the famous worldwide oceanographic surveys by the British Challenger and the German Gazelle. At the end of the 19th century it had become an established fact that most of the ocean waters had low temperatures, derived from cooling at high latitudes. In the early 20th century accurate reversing deep-sea thermometers, since 1878, produced by Negretti en Zambra, came into general use. Already during the Challenger expedition from 1872 to 1876 prototypes of this instrument were used for testing purposes. In the 1970s they were rapidly replaced by electronic temperature sensors.

The concept of an oceanic flow, driven by density differences, was first proposed by the Italian Luigi Ferdinando, Count de Marsigli (1658–1730) with reference to the flow of water through the Bosporus between Asia and Europe. He attributed the existence of an undercurrent from the Mediterranean Sea to the Black Sea to the high density of the Mediterranean water due to an evaporation excess in the latter ocean basin. This hypothesis was substantiated with hydraulic experiments in a tank, filled with water with different densities, separated by a vertical wall. When holes in the wall were opened, the high-density water passed through the lower holes, while the lighter water passed through the upper holes in the opposite direction. Sir Benjamin Thompson, Count Rumford (1753–1814), born in Woburn, Massachusetts, but living in Europe, did careful experiments with regard to heat transfer in water and convective motion in relation with the density of water. He found that salt water, contrary to freshwater, has its highest density at the freezing point. He concluded that "cooled particles of *salt water* descend as soon as they have parted with their Heat, and in moving downwards force other warmer particles to move upwards; and in consequence this continual succession of warm particles which come to the surface of the sea, a vast deal of Heat is communicated to the air." (Thompson 1798, reprinted in 1968, p. 206). From the scarce and inaccurate subsurface temperature measurements which existed around 1800, Thompson derived that there exists a deep meridional temperature gradient, confirming that the deepwater is heated when it spreads from polar to lower latitudes. He assumed that "as its specific gravity is greater than that of water at the same depth in warmer latitudes, it will immediately spread towards the equator, and it must necessarily produce a current at the surface in an opposite direction" (Thompson 1798, reprinted in 1968, p. 209). The warming of the deepwater when spreading toward the equator was assumed to come ultimately from the sun at the warm latitudes. The low temperatures of deepwater in the subtropics were seen as a proof of the deep equatorward motion, while the Gulf Stream was identified as one of the return currents. On the resulting overturning current system Thompson noted that "The vast extent of the ocean, and its great depth, but still more numerous its currents, and the power of water to absorb a vast quantity of Heat, render it particularly well adapted as an equalizer of Heat" (p. 193). These remarkable and colorful insights into the large-scale interaction between the oceans and the atmosphere, the formation of deepwater, and the moderating effect that the oceans have on climate largely have stood unchanged for two centuries (Weaver et al. 1999).

In the 19th century the ideas of Thompson, Count Rumford, were not widely accepted, and it was generally assumed that the temperature of the deep ocean was 4°C, the temperature where freshwater attains its maximum density. Carpenter (1871) again argued that the density of seawater increased with decreasing temperature until the freezing point was reached. Similar to Thompson he advocated that the deep overturning circulation of the Atlantic Ocean was driven by the temperature difference of deepwater between the polar ocean and the equatorial regions. Polar cold, rather than equatorial heat, was assumed to be the *primum mobile* (first mover) of this circulation. He supported his arguments with a model experiment with a long narrow tank of water that was warmed with a hot metal plate at the surface at one side and a block of ice at the other side. Contrary to Thompson, however, Carpenter proposed that not the Gulf Stream but a warm current along the British Isles and Norway toward Spitsbergen formed the warm return flow for this overturning system. Carpenter also noted the consequences of the THC for the climate in western Europe. To substantiate his ideas with a sounder observational basis, he proposed "the systematic prosecution of observations of the temperature and motion of different strata of the ocean as part of the regular duty of the British navy, ..., in the interest of all nations" (Carpenter 1871, p. 87). An accurate description of the temperature distribution in the world ocean was lacking till then, and most descriptions of the THC were based on a far-reaching extrapolation of hydraulic model results and (often ill-founded) theoretical insights, as well as a few inaccurate observations. The following year, on December 21, 1872, the British naval ship HMS Challenger left the dock to survey the world ocean during a 4-year worldwide oceanographic expedition. Many deep temperature measurements were carried out that were interpreted as evidence for flushing of the deep Atlantic and Pacific oceans from antarctic rather than arctic latitudes. The return of that antarctic water from the northern portions of the Atlantic and the Pacific oceans was assumed to be carried by evaporation which was hurried down through the atmosphere to the zone of low barometric pressure in the southern hemisphere (Wyville Thomson 1877).

The Norwegian oceanographer and polar traveler Fridtjof Nansen was able to show from observations, collected in the early 20th century, that in the North Atlantic Ocean cold water also entered the Atlantic Basin over the shallow Greenland–Scotland Ridge, through the Denmark Strait between Greenland and Iceland as well as between Iceland and Scotland (Nansen 1912).

Only with the German Atlantic expedition with the steamship Meteor (Fig. 1.3) a complete and accurate description of the distribution of temperature, salinity, and dissolved oxygen in the deep water masses of the Atlantic Ocean became available. This allowed the German oceanographer Georg Wüst to develop water mass analysis methods in order to explore the spreading of deep water masses involved in the THC. He was able to discriminate different deep water types, originating from both arctic and antarctic latitudes, as well as from the Mediterranean Sea (Wüst 1933, 1935). His results still largely agree with the modern ideas about the Atlantic part of the THC.

Fig. 1.3. A bow view of the steamship Meteor during the German Atlantic Expedition 1925–1927. Source: Spiess (1928).

One serious problem remained, regarding the driving of the THC by density differences. This was demonstrated already nearly a century ago by the Swedish oceanographer Sandström (1908), who performed a simple laboratory experiment. In a tank filled with water he introduced a thin heating source and a cooling source. He found that when the heating was at a lower level than the cooling, a vigorous overturning circulation was excited between the two levels. If, on the other hand, they were at the same level at the water surface as in the real ocean, the ultimate stationary circulation was very weak and was confined to a thin near-surface layer. Thus, there would exist a significant flow only in a thin layer just below the surface, driven by molecular heat conduction. The rest of the ocean would be filled with stagnant cold water. Understanding of the role of turbulent mixing in the maintenance of density gradients was required to solve the apparent paradox, suggested by the experiments of Sandström.

1.5. The following chapters

The THC and its properties, kinematics, dynamics, and variability are presented in the following 11 chapters. In Chapter 2 the geography of the ocean basins is treated, with emphasis on the main topographic features, continents, ridges, and basins, that influence and guide the THC. Chapter 3 is devoted to the physical characteristics of seawater that are relevant to the THC, e.g., temperature, salinity, density, and specific heat. This chapter also contains a simple outline of the ocean dynamics used in following chapters. Chapter 4 has a dual purpose. The tracer methods for the hydrographic analysis of the THC in the deep ocean are introduced, while their use is illustrated with the description of the water masses in the deep Atlantic Ocean. These water masses are involved in the cold deep branch of the THC. This hydrographic description is continued in Chapter 5 for the water masses in the deep Southern, Indian and Pacific oceans. In Chapter 6 the hydrographic evidence for the return flow to the North Atlantic is presented, following the routes from the Pacific Ocean via the Bering Strait, the Indonesian through-flow, the Agulhas leakage, and via Drake Passage. The formation of high-density water masses in the arctic and antarctic seas and their descent into the abyssal ocean are discussed in Chapter 7. In Chapter 8 the dynamics that determine the horizontal structure of the (deep) flow in the abyssal basins are introduced, focused on the Stommel–Arons model for the abyssal circulation and observational evidence for this structure is discussed. The existence of deep boundary currents is an essential result of that theory, confirmed by observations. Attention also is given to the wind-driven upwelling. The role of turbulent mixing in determining the vertical hydrographic structure is explained in Chapter 9, as well as the energy requirements to maintain the turbulent mixing, following the abyssal recipes formulated by Walter Munk. Chapter 10 on energetics of the THC is devoted to the transport of heat and freshwater by the THC, including the air–sea exchange. Attention is given to the thermodynamic properties of the THC as a heat engine, leading to Sandström's theorem for the overturning circulation. This theorem states that the THC cannot function without an additional energy source that maintains the turbulent mixing in the ocean. Simple one- and two-dimensional analytical models for the THC are presented in Chapter 11, including the boundary conditions for such models that may lead to different types of feedback via the atmosphere, generating variability in the THC. The possible existence of multiple equilibria in the THC is a result of some of these simple models. Finally in Chapter 12 studies of climate variability involving changes in the THC are introduced, as well as paleoceanographic methods that are used to study the variability of the THC in the past. Changes in proxies for several hydrographic parameters show that during glacials the THC differed from the present situation, although the interpretation of such proxy data appears not to be straightforward.

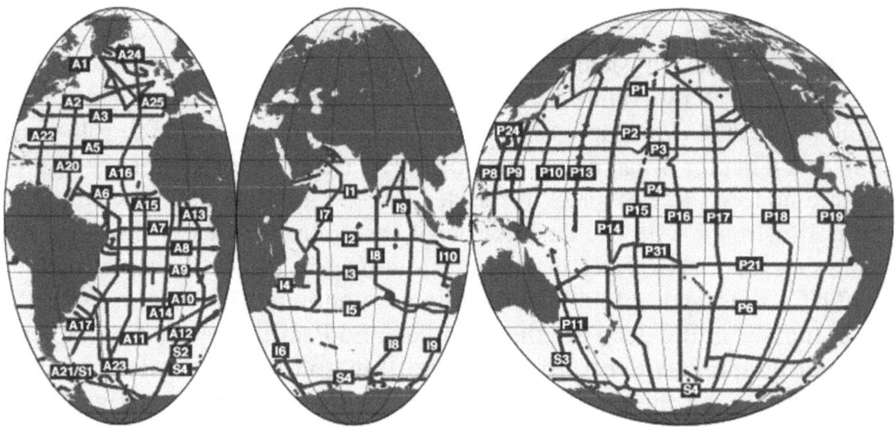

Fig. 1.4. Hydrographic sections for the WOCE Hydrographic Program one-time survey. Additional to these sections there are repeatedly surveyed hydrographic sections, XBT sections, current meter mooring arrays, satellite observations, surface drifters, subsurface floats, and tidal stations. Source: http://woceatlas.tamu.edu/Sites/html/atlas/SOA_WOCE.html.

In this book many examples of hydrographic observations are presented as an illustration of the characteristics of the THC. Most of these data were collected during the hydrographic program of the World Ocean Circulation Experiment (WOCE) in the 1990s (Fig. 1.4). The data from the hydrographic surveys were downloaded from the website of the former WOCE Hydrographic Program Office at the Scripps Institution of Oceanography, presently the CLIVAR and Carbon Hydrographic Data Office (http://cchdo.ucsd.edu/). The current meter data were collected from the site of the Buoy Group of the Oregon State University (http://kepler.oce.orst.edu/), while the data on the drift of subsurface floats were downloaded from the WOCE Subsurface Float Data Assembly Center at the Woods Hole Oceanographic Institution (http://wfdac.whoi.edu/). Other data sources are mentioned in the text or in the figure captions. The reader is encouraged to visit these sites and download oceanographic data for practice and further analysis.

2. The ocean basins

2.1. The bottom topography of the oceans

The oceans cover over 70% of the earth with a relatively thin layer of water. The earth is a slightly flattened sphere with an effective radius of 6371 km. On this sphere topographic relief is present which forms continents and deep basins; the latter contain the oceans. Within the oceans the currents in the upper 1000 to 1500 m are mainly driven by wind stress at the sea surface. The deep cold branch of the THC, below the wind-driven layer, is for a large part constrained and guided by the topography of the ocean basins and their interconnections.

Much information on the earth's relief can be found in global gridded relief data sets, available from the US NOAA National Geophysical Data Center (http://www.ngdc.noaa.gov/mgg/global/global.html). Examples are the ETOPO-5 earth's relief data set with a spatial resolution of 5'×5' (Anonymous 1988) or more recently the ETOPO-2 data set (Smith and Sandwell 1997; Jakobsson et al. 1999) with a horizontal resolution of 2'×2'.

The topographic range of the earth's relief, derived from the ETOPO-5 data set, is about 20 km (Fig. 2.1). The topographic range of the ocean bottom alone is of course smaller, about 11 km, or 0.17% of the earth's radius. The distribution of the relief height of the earth is bimodal, with a high cluster representing the continents including the shelf seas and a low cluster that represents the ocean basins. The mean depth of the ocean basins amounts to 3700 m, only 0.06% of the earth radius. The "aspect ratio" (ratio of the characteristic vertical and horizontal scales) of the ocean has a value of less than 0.1%. For illustration, the earth shrunken to the size of an orange with a diameter of 10 cm would have an ocean with a mean depth of only 29 μm. So the ocean is extremely flat, and therefore the motion of the ocean is often assumed to be two-dimensional in a horizontal plane. But despite the topographic flatness of the ocean, the vertically overturning circulation of the THC plays a major role on the stage of the global climate.

Whereas the ocean occupies 72% of the earth surface ($3.7\ 10^{14}\ m^2$) only 80% of the ocean is deeper than 2500 m ($2.9\ 10^{14}\ m^2$). The rest of the ocean is shelf sea or continental slope. Only this deep 80% of the ocean can participate in the deep circulation of the THC. The total volume of this deep part of the ocean amounts to $1.3\ 10^{18}\ m^3$.

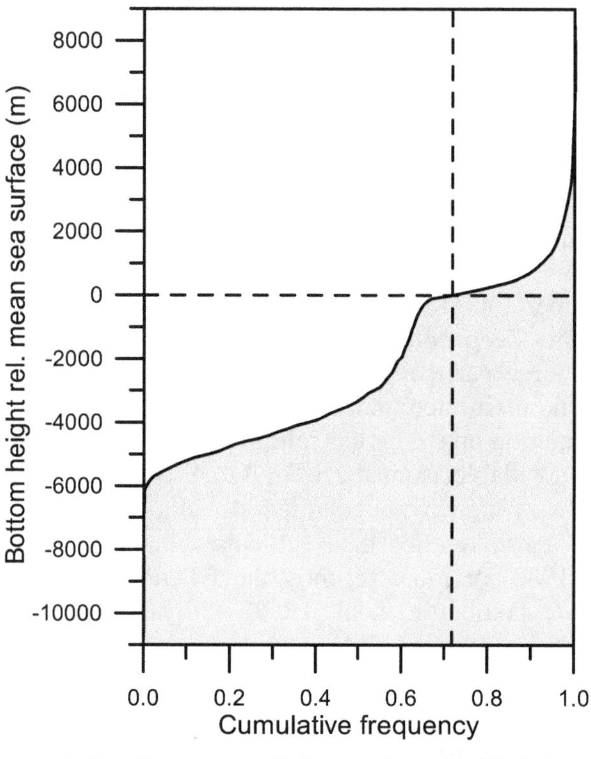

Fig. 2.1. The cumulative frequency of the earth's relief height, relative to the mean sea surface. This frequency distribution has been derived from the ETOPO-5 global relief data set.

2.2. Basins and ridges

The ocean floor displays a rich variation in topography, from isolated sea-mounts to the global system of mid-ocean ridges. The topography of the ocean basins or the bathymetry (Fig. 2.2) constrains several aspects of the ocean circulation.

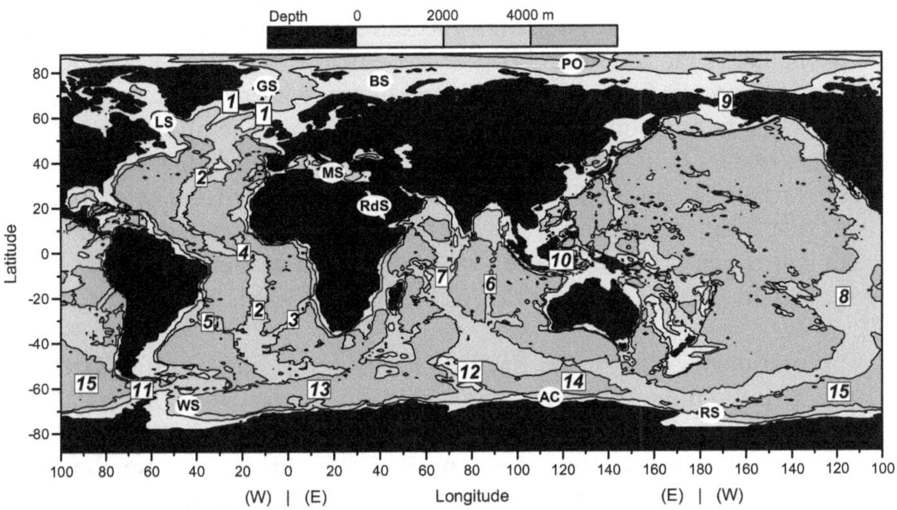

Fig. 2.2. The bathymetry of the world ocean derived from the ETOPO-5 data set. The acronyms and numbers refer to formation regions of deepwater and topographic features, discussed in the text.

The topography of the ocean basins guides and sometimes prevents the motion of deepwater, thereby controlling the THC. The main formation regions of cold deepwater in the northern hemisphere are the arctic seas, Greenland Sea (GS), Barents Sea (BS) adjacent to the Polar Ocean (PO) (or Arctic Ocean) and the Labrador Sea (LS). The Weddell Sea (WS), Ross Sea (RS), and the sea off the Adelie coast (AC) are the main formation centers in the southern hemisphere. In the semi enclosed Mediterranean Sea (MS) and Red Sea (RdS) very saline water is formed due to the local evaporation excess. Traces of that saline water are found in the upper layers of the deepwater in, respectively, the Atlantic Ocean and Indian Ocean. Between the major formation region of very dense water in the arctic seas and the main basins of the Atlantic Ocean the shallow Greenland–Scotland Ridge [1] limits the southward flow of newly formed deepwater. This deepwater flows into the Atlantic Ocean over sills on the ridge with depths between 600 and 850 m. The presence of the continents as zonal boundaries of the deep ocean basins allows the existence of deep western boundary currents that are the main agents for the meridional spreading of deepwater in the ocean. The Mid-Atlantic Ridge [2] divides the Atlantic Ocean into an eastern and a western basin. Exchange of deep and bottom water across this ridge is possible at a limited number of deep fracture zones. The eastern Atlantic basins below 3000 m are isolated from the Southern Ocean by the Walvis Ridge [3] between the Mid-Atlantic

Ridge and southwest Africa. The cold bottom water of antarctic origin cannot pass this ridge. It enters the Eastern Atlantic Basin mainly through the Romanche Fracture Zone [4] near the equator. The Rio Grande Rise at ~30ºS [5] limits the meridional exchange in the western South Atlantic Basin. Valleys with depths of 4000 to 4400 m in this ridge allow the passage of deeper waters than those that pass across the Walvis Ridge.

The Indian Ocean also is divided by a complicated series of oceanic ridges that limit and guide the flow of deepwater. The main division of this ocean is in an eastern basin, a central basin, and a western basin, separated by the Ninety-East Ridge at 90ºE [6] and the Mid-Indian Ridge near 70ºE [7].

The dominant ridge in the Pacific Ocean is the East Pacific Rise near 110ºW [8]. Coming from Antarctica this ridge runs north-north-east toward Central America. West of this ridge the huge Central Pacific Basin is found, with depths well over 5000 m, while to the east the smaller Peru and Chile basins are located. The Pacific Ocean is separated from the Polar Ocean by the shallow Bering Strait [9]. The relatively shallow passages in the East Indonesian waters [10] restrict exchange of water between the Pacific and Indian Ocean to the uppermost 1000 m.

The main obstacle for the zonal flow of bottom water in the Southern Ocean is the horseshoelike Scotia Ridge system, about 2000 km east of the Drake Passage [11], which connects South America and the Antarctic Peninsula. This ridge is generally shallower than about 1500 m and contains numerous islands, although a few passages exist that are over 3000 m deep. It curves eastward from the actual passage near 60 to 70°W to 27°W. At the latitude band of Drake Passage the only other relatively shallow topographic feature is the Kerguelen Ridge [12].

The southern Ocean is divided into three principal subbasins: the Weddell and Enderby Abyssal Plains or Atlantic–Indian Basin [13], the Australian–Antarctic Basin [14], and the Amundsen–Mornington–Bellingshausen Abyssal Plain or Pacific Antarctic Basin [15] separated from each other by ridges that are less than 4000 m deep. The fact that in the Southern Ocean a latitude belt exists without a zonal boundary shallower than about 1500 m may have profound influences for the THC. It can connect the northward directed wind-driven transport in the surface layer with a southward transport below 1500 m.

The irregular bottom topography with depth variations at a large range of horizontal scales poses a challenge for the numerical simulation of the oceanic circulation. Numerical simulation models of the oceanic circulation have a limited spatial resolution (horizontal and vertical), and the representation of the bottom topography can be crude. Even if the ocean bathymetry is perfectly known at all scales, the interpolation of that perfectly known bathymetry on the computational grid of the model may not necessarily lead to the best possible flow field. Especially the sills in narrow channels that restrict the exchanges between ocean basins are generally not well resolved in the state of the art models of the THC that are used for the simulation of climate variability. Examples are the deep channels through the Greenland–Scotland Ridge in Denmark Strait and the Faroe Bank Channel that have horizontal scales of only a few kilometers. Special parameterizations are required to simulate a realistic transport through such narrow gaps.

3. Pressure, temperature, salinity, and some thermohaline dynamics

3.1. Pressure

Pressure is the force exerted on a unit surface that is oriented perpendicular to the direction of that force. According to the international system for weights and measures (the SI system) the unit for pressure is Pascal, with symbol Pa, equivalent to N/m^2 (SUN 1985). The pressure in sea is one of the thermodynamic variables of seawater, determining together with temperature and salinity a range of seawater properties, e.g., density, specific heat, sound velocity, etc. (Fofonoff and Millard 1983). The net force, exerted by a pressure gradient on a water parcel, also is one of the main driving forces of the THC. The pressure used in oceanography is generally the sea pressure, P, which is the actual pressure minus one standard atmosphere (= 101 325 Pa). When the oceanic motions are restricted to low frequencies (no surface waves, internal waves, or convective motion), vertical accelerations can be ignored and the vertical momentum equation is to a very high accuracy approximated by the hydrostatic equilibrium:

$$\frac{\partial P}{\partial z} = -\rho g \quad . \tag{3.1}$$

Here z is the vertical coordinate (upward positive), ρ is the density of seawater and g is the gravitational acceleration. With $\rho \approx 1000\ \mathrm{kg/m^3}$ and $g \approx 10\ \mathrm{m/s^2}$ the pressure in sea increases with about 10,000 Pa for each depth increment of 1 m. In oceanography the use of the non-SI unit decibar (symbol dbar) is allowed. A pressure increase of 1 dbar (10^4 Pa) corresponds numerically with a depth increase of about 1 m. At 1000 dbar the numerical value of the depth in m is only about 0.8% less than the value of the pressure in dbar, increasing to 2.2% less at 7500 dbar because of the compression of the water at high pressure (Fig. 3.1). This is one of the reasons why in oceanography pressure in dbar often is used as a vertical coordinate. Note, however, that a faulty use of dbar instead of Pa in dynamic calculations easily leads to errors of the order of several powers of 10. Some published thermodynamic equations for seawater require a pressure input in dbar, some in bar (10^5 Pa)!

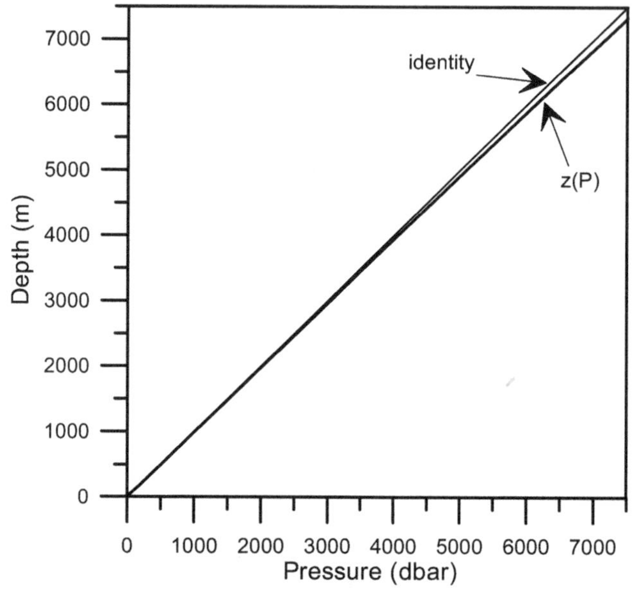

Fig. 3.1. Depth z as a function of sea pressure P (thick line), compared with the identity $x = y$ (thin line).

Assuming a sea surface air pressure close to one standard atmosphere, integration of Eq. (3.1) from the sea surface at $z = \zeta$ to a depth $z = -D$ gives the sea pressure P at a depth D below the surface:

$$P(-D) = -\int_{\zeta}^{-D} \rho g dz = \int_{-D}^{\zeta} \rho g dz \quad . \tag{3.2}$$

The horizontal gradient of $P(-D)$ in the x direction then can be written as

$$\frac{\partial P}{\partial x} = \int_{-D}^{\zeta} \frac{\partial \rho}{\partial x} g dz + \rho(\zeta) g \frac{\partial \zeta}{\partial x} \quad . \tag{3.3}$$

The second term of the right-hand side of Eq. (3.3) gives the contribution of a sloping sea surface to the pressure gradient, while the first term represents the contribution of horizontal density gradients to the horizontal pressure gradients. Since the density of seawater depends on temperature and salinity, horizontal differences in temperature and salinity therefore contribute to a horizontal thermohaline, or more common in oceanography, a baroclinic pressure gradient. The depth independent part of the pressure gradient is referred to as the barotropic pressure gradient.

3.2. Temperature

A thermodynamic variable for seawater that has been measured for many years is the temperature *T*. Already in the mid-18th century scientists supplied ship captains with special thermometers to measure subsurface temperatures (Hales 1751). In 1861, the Royal Netherlands Meteorological Institute published an oceanographic book devoted to results of scientific research with the sea thermometer (KNMI 1861). It is presently common practice in oceanography to express a temperature in terms of the difference of the (absolute) temperature from 273.15 K, the ice point. A thermodynamic temperature, expressed in this way, is known as the Celsius temperature with unit symbol °C. The temperature of the ocean is mainly in the range of –2 to +35°C. A typical temperature profile for the ocean is shown in Fig. 3.2 (thin line). Below 1500 dbar the temperature range is generally quite small. The layer between approximately 300 and 1500 dbar, where the vertical temperature gradient is relatively large, is referred to as the main thermocline. Seasonal variations of the seawater temperature, including the formation and erosion of a seasonal thermocline, are mainly restricted to the upper 300 dbar. Due to volcanic processes at mid-ocean ridges locally very small isolated volumes of much warmer water at large depths can be found, the so-called hot brines.

The determination of the thermodynamic temperature using an absolute scale is difficult. Those measurements are usually made by national standards laboratories. Such absolute measurements have been used to define a practical temperature scale based on the temperature of a few fixed points and interpolating devices (thermometers) that are calibrated at the fixed points. The practical temperature scale was revised in 1887, 1927, 1948, and 1968 as more accurate determinations of absolute temperature became accepted. The most recent scale is the International Temperature Scale of 1990 (ITS-90), replacing the International Practical Temperature Scale of 1968 (IPTS-68). The important calibration points in this scale for oceanographic applications are the triple point of water (0.0100°C) and the melting point of Gallium (29.7646°C). All oceanographic temperature measurements, performed after 1990, should be reported according to the ITS-90 scale. However, the algorithms for most other thermodynamic parameters used in oceanography have been developed before 1990 and require an input of temperature in the IPTS-68 scale, T_{68}. The relation between T_{90} and T_{68} in the oceanic temperature range (–2 to 35°C) is

$$T_{68} = 1.00024\, T_{90} \quad . \tag{3.4}$$

The difference between T_{68} and T_{90} only becomes significant above 10°C, given the accuracy of oceanographic temperature measurements (~0.002°C).

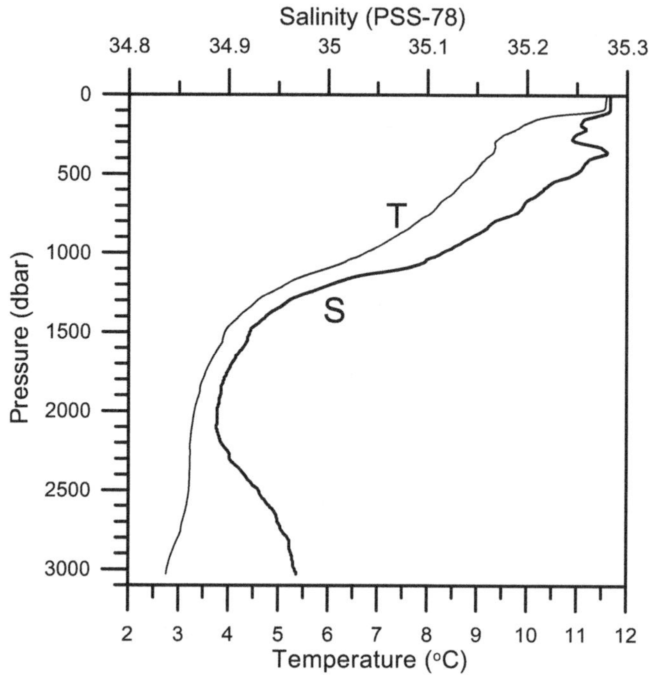

Fig. 3.2. Vertical profiles of temperature (thin line) and salinity (thick line), observed in the North Atlantic Ocean (Iceland Basin). Note that the pressure is used as vertical coordinate.

3.3. Salinity

The main constituents of dissolved matter or sea salt in seawater (~55.3% chlorine, 30.8% sodium, 7.7% sulfate, 3.7% magnesium, 1.2% calcium, and 1.1% potassium) are found with constant mutual ion ratios because of their long residence times in the ocean, compared to the typical stirring time of the world ocean of a few thousand years. Only calcium shows small deviations from this constancy because it is used for the formation of calcium carbonate shells. The near-constant mutual ratios make it possible to replace the separate mixing ratios of the individual constituents by a single thermodynamic variable, the salinity. At the simplest level, salinity is the total mass of dissolved material in one mass unit of seawater. This basic definition of salinity turned out not to be a useful definition because the total dissolved material is almost impossible to measure in practice. For example, how do we measure volatile material like gases? Nor can we evaporate seawater to dryness because chlorides are lost in the last stages of drying. A fully detailed chemical analysis is still the only practical method to determine the absolute salinity, but is too time consuming for routine use. Therefore salinity (symbol S) was defined in 1902 as a linear function of the halogen content (chlorinity), determined by titration with silver nitrate. Hereby it was assumed that all other halogens were replaced by chlorine ions. The numerical value of this bulk mixing ratio of sea salt gave the total amount of solid materials in grams dissolved in 1 kg of seawater expressed in the dimensionless ‰, or per mille.

In the 1960s the chemical analysis for the determination of the salinity was progressively replaced by measurements of the electrical conductivity of seawater. In 1978, a new definition for (practical) salinity was introduced, using only electrical conductivity, temperature, and pressure, thereby breaking the direct link with chlorinity or dissolved substances. However, the dimensionless practical salinity algorithm was developed to produce a numerical value very close to the absolute salinity expressed in ‰ or g/kg. We refer to this salinity scale as the Practical Salinity Scale of 1978 (PSS-78) which is expressed as a nonlinear function of conductivity ratio, temperature, and pressure (Fofonoff and Millard 1983). The conductivity reference standard for a practical salinity of 35.000 is a solution in pure water of 32.4356 g potassium chloride in 1 kg solution, of which the electrical conductivity is measured at a temperature of 15°C and a pressure of 1 standard atmosphere. The conductivity of that reference solution is 4.2914 S/m (Siemens per meter). The typical value of S in the ocean varies from about 33 to 37. An example of an open ocean salinity profile is given in Fig. 3.2 (thick line). In subtropical semienclosed seas like the Mediterranean Sea and the Red Sea S may reach even higher values because of the large evaporation excess in such seas and their limited exchange with the open ocean. The outflow of large rivers like the Amazon is characterized by a river plume with low surface salinities.

The factor 1000, introduced in the definition of salinity by the use of the ‰, should be considered carefully when making freshwater budgets from salinity observations. The mass of a cubic meter of seawater equals the density ρ. The total amount of sea salt dissolved in this cubic meter of seawater is $S\rho/1000$ kg. The remaining fraction of seawater represents the freshwater concentration in seawater in kg/m^3 that amounts to $\rho(1-S/1000)$. A precipitation excess over the ocean will increase the freshwater fraction in the surface water, thereby decreasing the salinity S. An evaporation excess withdraws water from the ocean and will increase the sea surface salinity.

3.4. Density

The term density (symbol ρ) has been chosen to represent the physical quantity mass of a substance per volume. Meridional density gradients are often assumed to drive the overturning THC. Increase of density at the sea surface due to heat loss to the atmosphere or evaporation may generate a static unstable stratification, with the denser water on top. In such a situation so-called convective turbulent mixing will occur, until the density in the upper part of the water column has become homogeneous. Convective mixing is assumed to play a key role in the formation of the cold deep water masses involved in the cold branch of the THC.

The equation of state (EOS) relates the density of a pure fluid to the other thermodynamic variables pressure and temperature. For a solution additional thermodynamic variables are required, presenting the mass ratio of each dissolved substance. Because of the near-constant ratio of the different ions in sea salt the bulk mass ratio salinity can perform that task of the mass ratios of the different ions. In the early 20th century the Danish oceanographer Martin Knudsen (1901) published tables for an EOS for seawater. Because of the technical problems involved in the determination of an absolute density of a concentrated solution like seawater, the Knudsen Equation of State related temperature and salinity of a water sample to the dimensionless specific gravity or relative density of the sample $d(T,S,P)$, not the density $\rho(T,S,P)$. The specific gravity is the density, divided by the density ρ_{max} of pure water at 4°C, the temperature of maximum freshwater density, at the standard atmospheric pressure ($d = \rho/\rho_{max}$). Since $\rho \approx 1000$ kg/m^3 and ρ_{max} equals 999.975 kg/m^3, $1000 \times d$ will be about 0.025 higher than the numerical value of the density ρ in kg/m^3. This difference has emerged from newer and better determinations of ρ_{max}, that originally was assumed by Knudsen (1901) to equal exactly 1000 kg/m^3. In the older literature from before 1980 there exists confusion on this subject, and d was often interpreted as density in g/cm^3 instead of dimensionless specific gravity.

With the introduction of the electrically measured practical salinity instead of the salinity derived from chlorinity it became necessary to introduce a new equation of state for seawater (the EOS-1980) in the form

$$\rho = \rho(T, S, P) = \frac{\rho_w(T) + \Delta\rho(S, T)}{1 - P / K(S, T, P)} \quad . \tag{3.5}$$

This EOS-1980 has been devised in three steps, each empirically approximated by a polynomial expression of the thermodynamic parameters (Fofonoff and Millard 1983). First a temperature-dependent density ρ_w for pure zero salinity Standard Mean Ocean Water (SMOW) with a well-defined isotopic composition is calculated. Then a (temperature-dependent) salinity correction $\Delta\rho$ is applied, and finally a pressure correction is added with a secant bulk modulus K which also depended on salinity, temperature, and pressure. Note here that because of its definition before the introduction of the ITS-90 temperature scale, the EOS-1980 requires a temperature input according to the IPTS-68 scale.

Since the value of the specific gravity of seawater often consisted of five or six significant digits after the decimal point (e.g., 1.028723), Martin Knudsen introduced the anomaly of specific gravity (or conventional specific gravity), σ, for his 1901 EOS. By writing $\sigma = (d-1)\times 10^3$ he obtained a number with only three significant digits after the decimal point (e.g., 28.723). Although σ is a dimensionless physical quantity, it often was reported in undefined "σ units". After the introduction of the EOS-1980 many oceanographers continued the use of the symbol σ, now defined as $\sigma = \rho - 1000$ (kg/m^3). Unfortunately the numerical value of this new dimensional σ differs systematically 0.025 kg/m^3 from Knudsen's dimensionless σ because the latter was based on the specific gravity, not on the density of seawater. With the introduction of the SI system in oceanography, it therefore was proposed to use a symbol different from σ, preferentially γ, for the dimensional density anomaly or density excess defined as (SUN 1985)

$$\gamma(T, S, P) = \rho(T, S, P) - 1000 \text{ kg/m}^3 \quad . \tag{3.6}$$

The nonlinear relation between the density of seawater and the thermodynamic variables temperature, salinity, and pressure is illustrated in Fig. 3.3 where lines of constant density anomaly, γ (isopycnals), are shown in a T–S diagram for two different pressures. This figure shows that while the haline contraction coefficient β,

$$\beta = \frac{1}{\rho}\frac{\partial \rho}{\partial S}\bigg|_{T,P} , \tag{3.7}$$

hardly changes with temperature or salinity, the thermal expansion coefficient α,

$$\alpha = -\frac{1}{\rho}\frac{\partial \rho}{\partial T}\bigg|_{S,P} , \tag{3.8}$$

increases with increasing temperature and pressure. The temperature dependence of α implies that in the cold source regions of the THC salinity variations are relatively more important than in the subtropical and tropical surface waters. Contrary to freshwater, seawater with a salinity in the oceanic range does not show a density maximum at a temperature above its freezing point.

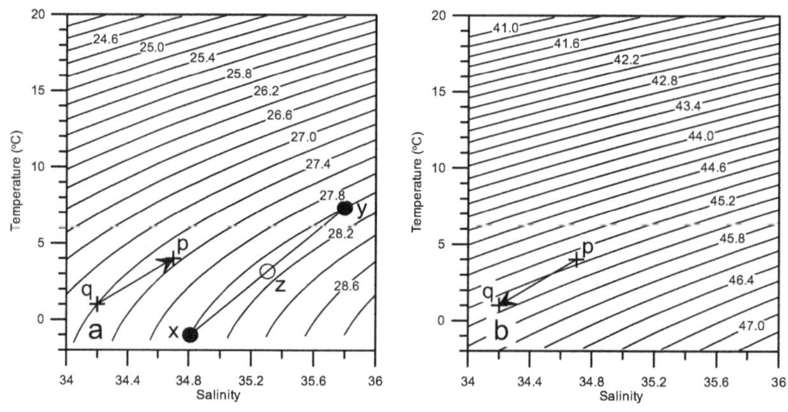

Fig. 3.3. Lines of equal density anomaly γ (kg/m³) as a function of temperature and salinity for zero sea pressure (a) and for a pressure of 4000 dbar (b). In both (a) and (b) identical *T–S* points p and q defined in the text are shown (crosses) with a connecting arrow, pointing to increasing density. These points illustrate the pressure dependence of a density difference. The points x and y in (a) are *T–S* points with equal density anomaly γ(*T*, *S*, 0) = 28.0 kg/m³. The mixing products of x and y will be located on the line between x and y. The point z is produced by mixing of equal amounts of water with x and y temperature and salinity.

The change of slope of the isopycnals with pressure in the *T–S* diagram, due to the pressure dependence of the thermal expansion coefficient, illustrates that with changing pressure the static stability of a water column may change. Take as an example two water parcels, p and q, with temperature and salinity $T_p = 4.0°C$, $S_p = 34.7$ and $T_q = 1.0°C$, $S_q = 34.2$, respectively (the crosses in Fig. 3.3). At a pressure of 0 dbar the density of water parcel p is 0.15 kg/m^3 larger than of parcel q, while at a pressure of 4000 dbar the density of water parcel p is 0.16 kg/m^3 smaller than of parcel q. This effect makes it possible that vertical migration of a stably stratified part of the water column by, e.g., internal waves changes the static stability. For marginally stable layers this may lead to static instability and resulting thermobaric convection (Garwood et al. 1994).

The temperature dependence of the thermal expansion coefficient may lead to another phenomenon, named caballing. When two adjacent water parcels x and y with identical density but different *T*,*S* properties (the black dots in Fig. 3.3a) mix, the resulting mixture will be found at a nearly straight line between these points in a T,S diagram. Due to the nonlinearity of the EOS for seawater the resulting mixture in this situation always will have a density higher than that of the original water parcels. In the example, shown in Fig. 3.3a, the water parcel z, produced by mixing of equal amounts of parcel x and y, has a $\gamma(T, S, 0)$ of 28.113 kg/m^3 compared to $\gamma(T, S, 0) = 28.000$ kg/m^3 of the original water parcels x and y. In most circumstances mixing leads to an increase in density due to the temperature dependence of the thermal expansion coefficient for seawater. The higher density water parcel z, produced by mixing, then will sink relative to its source water masses x and y, maintaining a diapycnal (= cross isopycnal) advection. Caballing by mixing of adjacent water masses with differing *T*, *S* properties is hypothesized to contribute significantly to the sinking of high-density water in the high northern latitudes of the arctic seas and polar ocean. There, cold and relatively fresh polar water is found adjacent to warmer more saline Atlantic water. In isopycnal general ocean circulation models for these regions this process should be modeled explicitly, which implies that a nonlinear EOS should be used.

3.5. Adiabatic compression, potential temperature, and potential density

If a water parcel moves vertically from pressure level p_1 to p_2 without exchange of heat and salt with the surrounding water (adiabatic motion) the density of the water parcel will change due to expansion or compression. The salinity of the water parcel will not change due to this density change, since this parameter represents a mixing ratio, not a concentration. But the temperature will change in this adiabatic process, since the surrounding water performs work on the water parcel with increasing pressure and density, while with decreasing pressure and density the water parcel performs work on its surroundings. This effect is described by the adiabatic lapse rate Γ, defined as the change of temperature per unit pressure for an adiabatic change of pressure of a parcel of seawater. From thermodynamic considerations Γ can be expressed as

$$\Gamma(T,S,P) = \left.\frac{\partial T}{\partial P}\right|_{\text{ad}} = \frac{T_a\,\partial V/\partial T}{C_p} \quad , \tag{3.9}$$

where $T_a = T + 273.15$ is the absolute temperature, $V = 1/\rho$ the specific volume, $\partial V/\partial T$ the thermal expansion in $m^3kg^{-1}{}^\circ C^{-1}$, and C_p the specific heat of seawater in $Jkg^{-1}{}^\circ C^{-1}$. In the salinity range for oceanic water Γ is positive with a typical value of 1 to $2\times10^{-4}{}^\circ$C/dbar. An empirical expression has been derived to relate Γ to T, S, and P (Fofonoff and Millard 1983).

In order to discriminate between temperature change due to advection and mixing of heat and temperature change due to adiabatic pressure changes a corrected temperature parameter, the potential temperature θ, has been introduced. This parameter is defined as the temperature that a parcel of seawater would obtain if moved adiabatically from the in situ pressure, P, to a reference pressure, P_r. The adiabatic condition not only excludes heat exchange with the surrounding water but also salt exchange. Most often the surface sea pressure of 0 dbar is used as reference pressure, but in general P_r may be larger or smaller than P. With this definition θ can be calculated as

$$\theta(T,S,P,P_r) = T(P) + \int_P^{P_r} \Gamma(T',S,P')\,dP' \quad , \tag{3.10}$$

where the integration is performed along an adiabath, that is $\partial T/\partial P|_{\text{ad}} = \Gamma$. This equation cannot be solved directly and some iterative approximation method is required. Several algorithms and empirical equations exist for the calculation of θ. The difference between T and θ is shown in Fig. 3.4a.

Similar to the potential temperature a potential density ρ_{pt} can be defined as the density a parcel of seawater would have if moved adiabatically from the in situ pressure, P, to a reference pressure, P_r. The potential density simply can be calculated from the equation of state with the potential temperature relative to the pressure level P_r, θ_{Pr}, as being $\rho_\theta = \rho(\theta_{Pr}, S, P_r)$. The symbol for the corresponding potential density anomaly is γ_θ. Following the tradition established in the past for the use of the specific gravity anomaly σ, we can add another subscript (e.g., reference pressure $P_r/1000$) to the potential density anomaly γ if the reference pressure P_r differs from zero. Then the variable γ_2 is the potential density anomaly relative to a reference pressure of 2000 dbar. The difference between the in situ density and the potential density is illustrated in Fig. 3.4b.

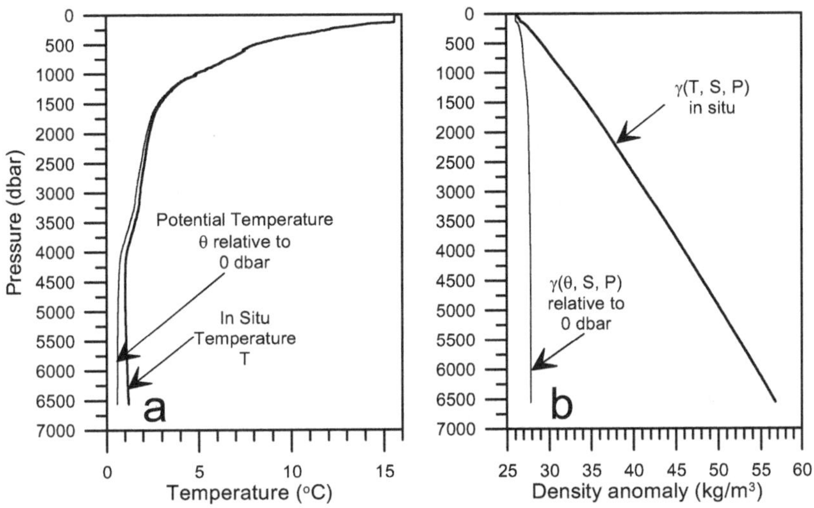

Fig. 3.4. Profiles of the in situ and potential temperature (a) and density (b) anomaly for a hydrographic station, measured during the international World Ocean Circulation Experiment (WOCE) in the Pacific Ocean at 32°30'S, 177°40'W. The in situ temperature and density anomaly have been drawn with thick lines, the potential temperature and density with thin lines.

From Fig. 3.4 it can be seen clearly that the difference between the potential temperature or density and the in situ values is significant. The in situ temperature in this figure has its minimum value at a pressure of 4730 dbar, while the lowest potential temperature is found near the maximum observed pressure of 6555 dbar. The compressibility of seawater causes more than a doubling of the in situ density anomaly between the sea surface and the bottom (from 26.1 to 56.7 kg/m^3), while the potential density anomaly increases with only 6% (1.67 kg/m^3) over this pressure interval.

3.6. Freezing point and specific heat

Seawater freezes at a temperature lower than the freezing point of pure water, and its freezing point T_f at standard atmospheric pressure according to the IPTS-68 scale depends on salinity according to the expression

$$T_f(S) = a_0 S + a_1 S^{3/2} + a_2 S^2 \qquad (3.11)$$

(Fofonoff and Millard 1983). The pressure dependence of the freezing point is similar to that for pure water, decreasing 0.075°C for a pressure increase of 100 dbar. When we compare the freezing point of seawater as a function of salinity with the temperature at which the density maximum is found it appears that at salinities above about 24.7 the water starts freezing before the density maximum according to the EOS has been reached (Fig. 3.5). This implies that for all ocean waters ($S > 33$) the maximum density of fluid seawater is found at the freezing point.

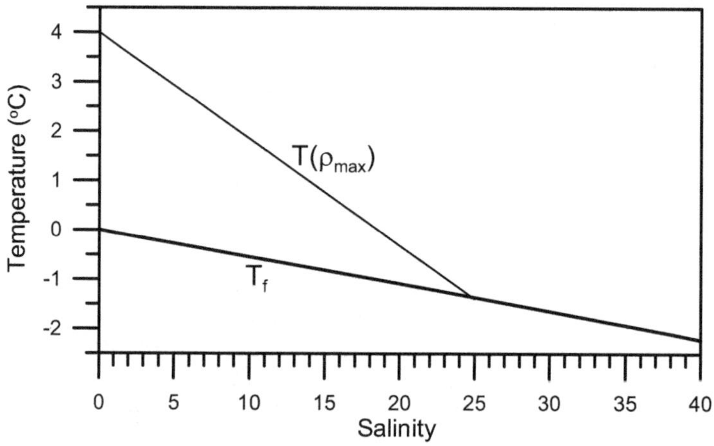

Fig. 3.5. Plot of the freezing point of seawater (T_f) and the temperature of maximum seawater density according to the EOS-80 ($T_{(\rho max)}$), both at zero sea pressure, as functions of salinity.

This effect has consequences for the relatively slow way in which the salty ocean freezes over, compared with a freshwater lake. Surface cooling will increase the density of the surface water, leading to convective mixing and cooling of a large part of the water column. In a freshwater lake further surface cooling, after a temperature of 4.0°C has been reached, will decrease the surface density and suppress convective mixing. Only a relatively shallow surface layer will be cooled below 4°C. In the ocean convective mixing can continue until ice formation sets in at the sea surface. Even then convection may continue, since with sea ice formation only part of the sea salt is incorporated into the ice matrix. This leads to the rejection of extra saltwater (brine) from the newly formed sea ice that will increase the salinity and water density of the seawater under the ice and support some more convective mixing.

The specific heat of seawater at constant pressure, C_p, can be expressed as the specific heat of pure water with a temperature-dependent salinity correction and a temperature-dependent pressure correction followed by a final correction for temperature, salinity, and pressure combined (Fofonoff and Millard 1983). The dependence of C_p on variations of P, T, and S is illustrated in Fig. 3.6.

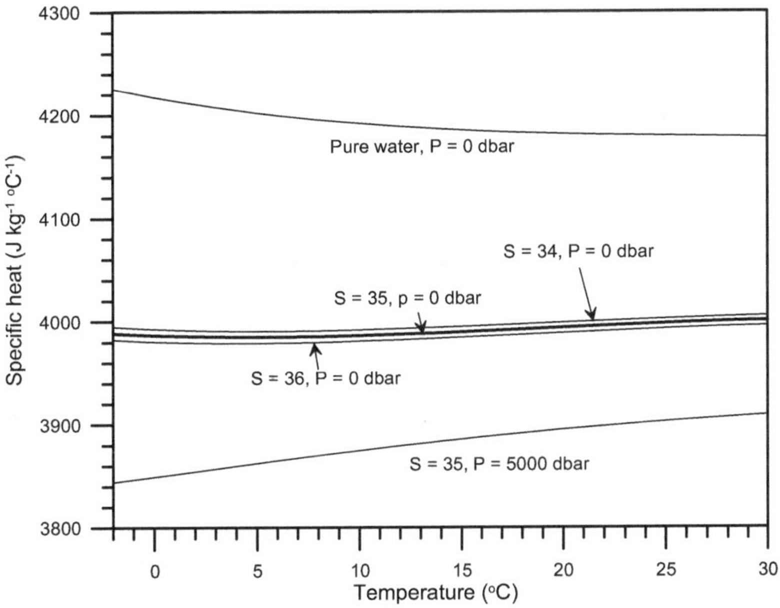

Fig. 3.6. The specific heat C_p of (sea)water in J kg^{-1}°C^{-1}, as a function of temperature for different salinities and pressures.

The specific heat for seawater within the oceanic salinity range, O(4000 J kg^{-1}°C^{-1}), is about 5% lower than the C_p of pure water. Within that salinity range and the oceanic temperature range C_p at atmospheric pressure varies only a few tenths of a percent. At a pressure of 5000 dbar, however, C_p is nearly 3% lower than at the sea surface.

3.7. Pressure gradient forces

When a horizontal component of the pressure gradient at a level z, in the x direction exists, it will exert a net force on a unit water volume directed from the high to the low pressure, contributing to a horizontal acceleration (= force per mass unit) a_x in the x direction, presented by

$$a_x = -\frac{1}{\rho}\frac{\partial P}{\partial x}\bigg|_z \quad . \tag{3.12}$$

This equation can be rewritten when we change the vertical coordinate from z to P as suggested in Section 3.1. In order to illustrate the consequences of this change for the formulation of the horizontal pressure gradient we will use the scheme shown in Fig. 3.7.

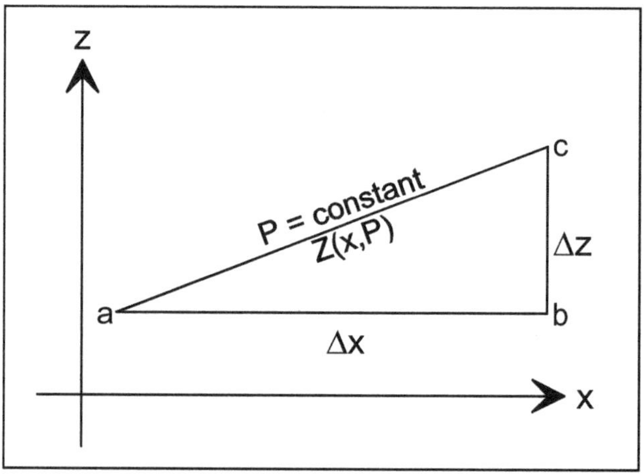

Fig. 3.7. Scheme to illustrate the consequences of the change of the vertical coordinate from z to P for the formulation of the horizontal pressure gradient.

With z as vertical coordinate, the pressure P is a function of the independent variables x, y, and z. If P replaces z as vertical coordinate, the depth of an isobaric surface will be a function of the independent variables x, y, and P. If we follow, in an x–z section, the pressure change from a point a on the P isobar, horizontally over a distance Δx via b and then vertically Δz upward to point c again on the P isobar (Fig. 3.7), we find the following equality:

$$\frac{\partial P}{\partial x}\Delta x + \frac{\partial P}{\partial z}\Delta z = 0 \quad . \tag{3.13}$$

The ratio $\Delta z/\Delta x$ equals the slope of the P isobar in the limit of vanishing Δx. Combining Eqs. (3.13) and (3.1) then leads to

$$a_x = -g\frac{\partial z}{\partial x}\bigg|_P \quad . \tag{3.14}$$

So the horizontal acceleration due to the presence of a pressure gradient is proportional to the slope of the isopycnals. The vertical derivative of a_x with P as vertical coordinate, derived from Eq. (3.14) with the use of Eq. (3.1) then equals the horizontal gradient of the specific volume V:

$$\frac{\partial a_x}{\partial P} = \frac{\partial}{\partial x}\left(\frac{1}{\rho}\right) = \frac{\partial V}{\partial x} \quad . \tag{3.15}$$

Equation (3.15) shows that the horizontal acceleration due to a pressure gradient changes with depth or pressure if a horizontal gradient of density or specific volume due to temperature or salinity gradients is present. This is the physical concept behind the intuitively formulated idea that meridional thermohaline differences in density should drive an overturning THC. If, for example, the x coordinate is directed northward in the North Atlantic Ocean, while at high latitudes the water column is colder, with higher ρ and lower V than at lower latitudes, $\partial a_x/\partial P$ will be negative, implying a downward increasing southward acceleration a_x which forces cold and deep arctic water southwards. Often oceanographers assume that at a certain pressure level the horizontal pressure gradient and the velocity are zero, the "level of no motion." Then the pressure forces can be computed from a measured density distribution by integrating Eq. (3.15) vertically from that level of no motion, the so-called dynamic method (Fomin 1964). In the surface layer a_x may be directed northward due to the presence of a slope of the sea surface ζ according to Eq. (3.3). The American oceanographer Tom Rossby (1999) suggested that similarly the 0.5 m sea level difference between the tropical Pacific and the North Atlantic Ocean is the main driving agent of the upper branch of the THC that brings warm and relatively freshwater in the near-surface branch of the Atlantic–Pacific THC back into the North Atlantic.

3.8. Geostrophic and near-geostrophic flow

When an unopposed horizontal component of the pressure gradient accelerates the water, a stationary flow will not be possible. For a stationary current system to exist, additional forces are required that balance the pressure gradient. The most important of these forces in the large-scale ocean circulation is the Coriolis force, an apparent force due to the rotation of the earth (Stommel and Moore 1989). The horizontal component of this apparent force is directed perpendicular to the horizontal velocity component. Effects of vertical velocity components can be ignored here. The x and y components of acceleration due to the Coriolis force are, respectively,

$$a_x = fv \quad \text{and} \quad a_y = -fu \quad , \tag{3.16}$$

where $f = 2\Omega\sin(\phi)$ is the Coriolis parameter, with Ω representing the angular frequency of the earth's rotation ($7.29\ 10^{-5}\ s^{-1}$) and ϕ being the geographic latitude, while u and v are, respectively, the horizontal x and y components of the current velocity. The Coriolis parameter f represents the local vertical component of the earth's rotation. Since ϕ and $\sin(\phi)$ change sign at the equator the direction of Coriolis force relative to the velocity vector differs between hemispheres, being to the right on the northern hemisphere and to the left on the southern hemisphere relative to the horizontal velocity vector (Fig. 3.8).

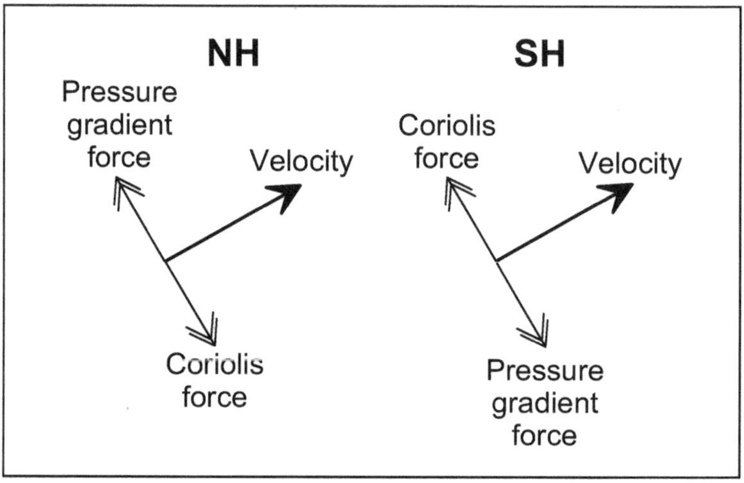

Fig. 3.8. Scheme of the direction of the Coriolis force and the pressure gradient force relative to the horizontal current velocity in geostrophic balance for the northern hemisphere (NH) and the southern hemisphere (SH).

For most of the ocean's interior the large-scale horizontal currents can be described accurately by a balance between the Coriolis force and the pressure gradient, where friction and acceleration terms can be ignored. This is called the geostrophic balance:

$$fv = \frac{1}{\rho}\frac{\partial P}{\partial x}\bigg|_z \quad , \quad fu = -\frac{1}{\rho}\frac{\partial P}{\partial y}\bigg|_z \quad \text{or} \tag{3.17a}$$

$$fv = g\frac{\partial z}{\partial x}\bigg|_P \quad , \quad fu = -g\frac{\partial z}{\partial y}\bigg|_P \quad . \tag{3.17b}$$

This force balance leads to a current, named geostrophic current, in a direction perpendicular to the pressure gradient force, to the right on the northern hemisphere and to the left on the southern hemisphere (Fig. 3.8). Geostrophic currents therefore will follow the isobars.

The current velocity, pressure, and density fields in the ocean, when not in balance, have a tendency to adapt mutually to fulfill the geostrophic balance, the so-called geostrophic adjustment. The application of this balance in most parts of the oceans presents an apparent problem for the THC. The currents and water transport will be perpendicular to the large-scale meridional pressure gradient generated by the density difference between cold polar latitude waters and the warmer water at subtropical and tropical latitudes. The large-scale meridional (north–south) baroclinic pressure gradient will result in zonal (east–west) geostrophic currents, not in mainly meridional currents, required for a meridionally overturning THC. A considerable change in the orientation of the large-scale density gradient by geostrophic adjustment is required to allow a (near-)geostrophic baroclinic meridional overturning circulation. Since the pressure gradient and the Coriolis force are both in a direction perpendicular to the velocity vector, no energy is transferred to the geostrophic flow; there is no real "driving" of the flow in the sense of transfer of potential to kinetic energy.

In fast currents through narrow gaps in the bottom topography and in fast, narrow boundary currents the geostrophic balance is not completely fulfilled, since there friction may play an additional role. For such situations a near-geostrophic balance is required between the pressure gradient force, the Coriolis force, and the friction force, in order to get a stationary current (Fig. 3.9). The actual friction in the ocean is not well understood. It may be caused by lateral as well as vertical turbulent exchange of momentum, to be parameterized with turbulent viscosity coefficients that are highly artificial in the large-scale global ocean circulation models. For the relation between the friction term at the bottom and the near-bottom velocity linear and quadratic friction parameterizations are used. In general the friction vector will more or less counter the direction of the flow. This will lead, as sketched in Fig. 3.9, to a cross-isobaric flow with a small component from high to low pressure. Hereby potential energy is converted to kinetic energy. In this situation one can state that the pressure gradient, caused by horizontal gradients in temperature and salinity, drives the circulation. The friction in its turn dissipates the kinetic energy of the flow. The magnitudes of the Coriolis force and the pressure gradient force are in general much larger than the magnitude of the friction. Then the use of the geostrophic balance in the dynamic method still will lead to a reasonable, although not perfect estimate of the current speed.

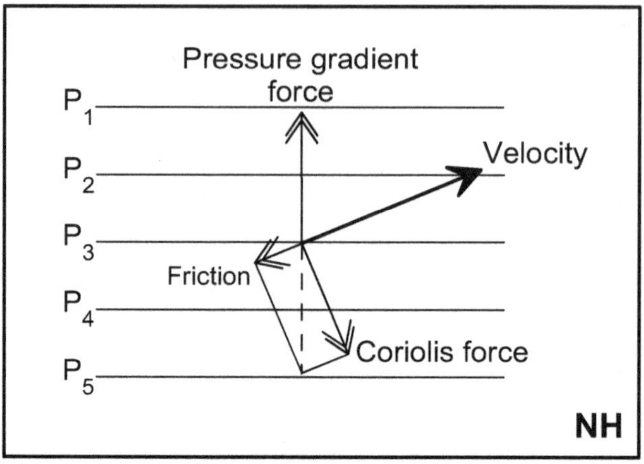

Fig. 3.9. Example of the near-geostrophic balance of forces (double arrows) in the northern hemisphere when in a stationary current friction no longer can be ignored. For simplicity it is assumed that the direction of the friction directly counters the direction of the velocity vector (filled arrow). The isobars are shown as thin lines, with pressure increasing from P_1 to P_5.

The Coriolis parameter f equals zero at the equator. There the geostrophic balance no longer can be applied. The dynamic modeling of cross-equatorial flow requires knowledge of frictional forces and nonlinear acceleration terms.

3.9. Friction and transport

As described above, friction in the ocean allows velocities and mass transport in a direction other than parallel to the isobars, contrary to the geostrophic balance. In the ocean friction plays a role in the near-surface layer due to wind forcing and in the bottom layer due to the flow over a nonmoving rough bottom. In these cases the frictional effects are related to turbulence generated in a vertical velocity shear. In fast currents, like, e.g., the Gulf Stream and Antarctic Circumpolar Current, mesoscale eddies also can play a frictional role, however, that is even more difficult to model than turbulent friction.

In the surface layer or bottom layer of the ocean, where turbulence cannot be ignored, generally a near-balance is assumed between the turbulent forces and the Coriolis forces related to the frictionally determined part of the velocity field. The layer where this balance approximately holds is named the Ekman layer after the Swedish mathematician and oceanographer Walfrid Ekman. The thickness of an Ekman layer in the ocean depends on the wind stress at the sea surface or the turbulent stress at the bottom, as well as on the Coriolis parameter f. The turbulent stress at the sea surface or sea bottom, $\vec{\tau}$, represents the vertical transport of horizontal momentum per surface area or a horizontal force per horizontal surface area. Similar to pressure the physical unit for such a stress in the SI system is Pascal. Sometimes the magnitude of the stress is expressed as a friction velocity, u_*, according to

$$\left|\vec{\tau}\right| = \rho u_*^2 \quad , \tag{3.18}$$

where ρ is the water density, while the friction velocity is assumed to be of the magnitude of the turbulent velocity components. Dimensional analysis easily shows that the thickness D_E of an Ekman layer will depend on the ratio of u_* and f, according to

$$D_E \approx C_E \frac{u_*}{f} = C_E \frac{\sqrt{\left|\vec{\tau}\right|/\rho}}{f} \quad , \tag{3.19}$$

where the constant C_E is probably of the order 1.

The magnitude of wind stress at the sea surface equals

$$\left|\vec{\tau}\right| = C_w \rho_a W^2 \quad , \tag{3.20}$$

where ρ_a is the air density ($\approx$1.2 kg/m^3) and W the wind speed, while C_w, the wind stress coefficient, is of the order of 10^{-3}. It now is easy to derive that at midlatitudes, where $f \approx 10^{-4}$s^{-1} a wind speed of 10 m/s will generate an Ekman layer with a thickness of the order of ~100 m. The magnitude of the bottom stress in the ocean relates to the water speed over the bottom, u, in a way similar to Eq. (3.20) as

$$\left|\vec{\tau}\right| = C\rho u^2 \quad , \tag{3.21}$$

where the friction coefficient C is also of the order of 10^{-3}. At midlatitudes a near-bottom flow will generate a bottom Ekman layer thickness of about 60 m for a water speed of 20 cm/s and about 30 m for a bottom speed of 10 cm/s. So the part of the water column where turbulent friction is important, either at the sea surface or near the bottom, is only a minor fraction of the total depth of the ocean which is a few kilometers.

As stated above, in the Ekman layer a balance is assumed between the friction and the Coriolis force due to the frictionally determined part of the water velocity. In the wind-driven Ekman layer the magnitude of the net wind stress is given by Eq. (3.20) and in the bottom Ekman layer by Eq.

(3.21), while the total Coriolis force equals the Coriolis parameter times the wind-driven mass or volume transport $\vec{V}_E$ in the Ekman layer, or

$$|\vec{\tau}| = f\rho|\vec{V}_E| \quad , \tag{3.22}$$

where $\vec{V}_E$ is the Ekman volume transport. The direction of volume transport will be perpendicular to the stress, to the right in the northern hemisphere, to the left in the southern. The direction of the wind stress is generally in the direction to which the wind blows. The direction of the stress, exercised by the bottom on the overlying moving water, is more or less against the direction of the flow, while $\vec{V}_E$ is the cross-isobaric volume transport. The SI unit to express a volume flux $\vec{V}_E$ in the ocean is m^3/s. In oceanography often an unofficial unit, the Sverdrup, is used (symbol Sv). One Sv equals 10^6 m^3/s.

Wind stress may influence the THC mainly in the near-surface return flow, where low-density water in the surface layers moves toward the regions where the deep water masses are formed in winter. The bottom stress is probably of minor importance for the THC in most parts of the oceans. Only where the water velocity over the bottom reaches high values, the bottom friction may induce a considerable cross-isobar flow in the bottom Ekman layer. For instance, over the East Greenland Slope where very cold water enters the North Atlantic between Greenland and Iceland, the near-geostrophic velocity over the bottom reaches values of 25 to 50 cm/s (Dickson and Brown 1994). There the cross-isobaric Ekman transport will induce a downslope transport into the deep Atlantic Ocean.

3.10. Vertical motion and mass conservation

The vertical momentum budget is approximated quite accurately by the hydrostatic equilibrium presented in Eq. (3.1), where the vertical velocity component of the THC, w, does not play any role. Therefore w is generally derived kinematically from the horizontal velocity distribution.

The ocean bottom is impermeable for ocean currents. This forces the velocity in the near-bottom layer to flow parallel to the bottom at a depth $z = -H(x,y)$. When the horizontal velocity near the bottom flows in the x direction, it is easy to derive from geometric arguments (Fig. 3.10) that the vertical velocity component w equals

$$w = -u\frac{\partial H}{\partial x} \quad . \tag{3.23}$$

Knowledge of the horizontal flow from geostrophy or current measurements, will supply an estimate for the vertical motion near the bottom via Eq. (3.23). In the outflow of cold arctic water from Denmark Strait over the East Greenland continental slope into the deep Irminger Basin, the typical along-flow bottom slope amounts to about 2500 m/1000 km with u of the order of 25 to 50 cm/s (Dickson and Brown 1994). This gives a typical near-bottom value of the downward velocity of $w \approx 0.6$ to 1.2 mm/s. Although some oceanographers consider that the sinking phase shows "the most spectacular features of the thermohaline circulation" (Toggweiler and Key 2001), in the ocean even spectacular processes in the THC go slowly in the vertical direction.

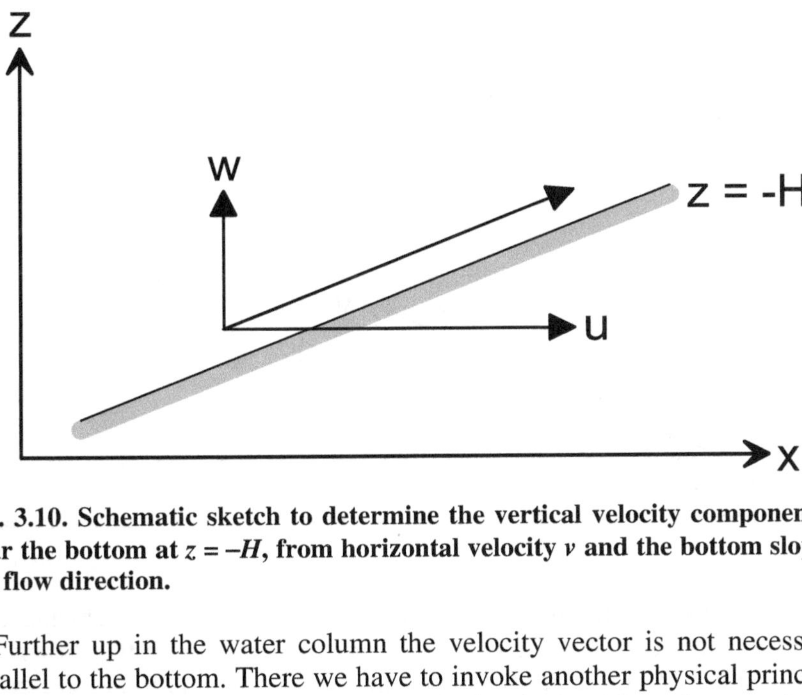

Fig. 3.10. Schematic sketch to determine the vertical velocity component, *w*, near the bottom at $z = -H$, from horizontal velocity *v* and the bottom slope in the flow direction.

Further up in the water column the velocity vector is not necessarily parallel to the bottom. There we have to invoke another physical principle, mass conservation, to estimate the vertical velocity. If we have an infinitesimal cubic volume in space with sides dx, dy, and dz (Fig. 3.11), then the change of mass in this volume is connected with the divergence of the advective mass transport. The rate of local density change in this volume is described by

$$\frac{\partial \rho}{\partial t} + \frac{\partial}{\partial x}\rho u + \frac{\partial}{\partial y}\rho v + \frac{\partial}{\partial z}\rho w = 0 \quad . \tag{3.24}$$

Equation (3.19) applies, even if diffusive processes of heat and salt are active. If that occurs, a diffusive or compressive change of density in the volume, e.g., a density decrease, will lead to a net advective flow out of the volume to compensate for the expansion of the water. It is easy to rewrite Eq. (3.24) as

$$\frac{1}{\rho}\frac{d\rho}{dt}+\frac{\partial u}{\partial x}+\frac{\partial v}{\partial y}+\frac{\partial w}{\partial z}=0 \quad . \tag{3.25}$$

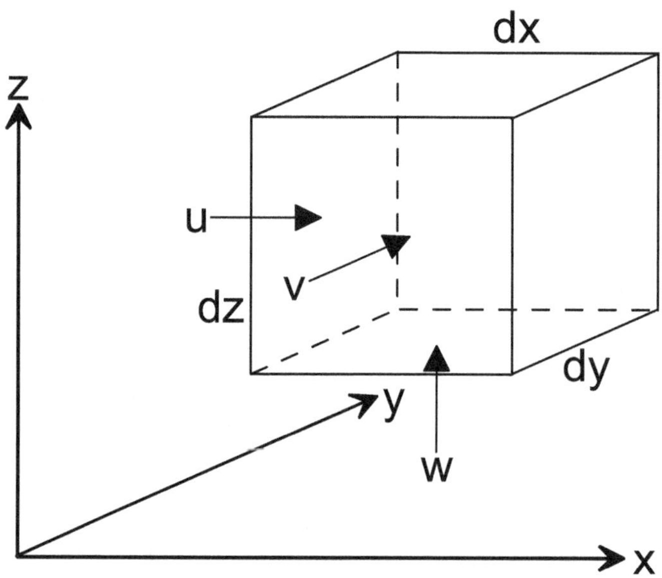

Fig. 3.11. Sketch to illustrate the derivation of the equation of mass conservation.

The complete differential $d\rho/dt$, representing the density change of a water particle moving in the ocean, is here connected with the divergence of the velocity field. That density change of a water particle can be ascribed to changes in potential temperature, salinity, and pressure of the particle according to

$$\frac{d\rho}{dt}=\left.\frac{\partial\rho}{\partial\theta}\right|_{S,P}\frac{d\theta}{dt}+\left.\frac{\partial\rho}{\partial S}\right|_{\theta,P}\frac{dS}{dt}+\left.\frac{\partial\rho}{\partial P}\right|_{\theta,S}\frac{dP}{dt} \quad . \tag{3.26}$$

From comparison of the density variation in the oceanographic range of temperature and salinity (~6 kg/m^3) shown by the isopycnals in Fig. 3.3, with the density variation due to realistic pressure changes in the THC from Fig. 3.4b (~25 kg/m^3) it becomes clear that in magnitude the last (compression) term in Eq. (3.26) is the most important in the right-hand side of this equation. Therefore we will assume here that the density change in Eq. (3.26) is in a reasonable approximation adiabatic ($d\theta/dt \approx 0$, $dS/dt \approx 0$). From Eq. (3.1) it can be derived that dP/dt can be written as

$$\frac{dP}{dt} = -\rho g \frac{dz}{dt} = -\rho g w \quad . \tag{3.27}$$

The adiabatic pressure derivative of the density equals the inverse of the sound velocity c squared, or

$$\left.\frac{\partial \rho}{\partial P}\right|_{\theta,S} = \frac{1}{c^2} \quad . \tag{3.28}$$

Substitution of Eqs. (3.27) and (3.28) into Eq. (3.25), ignoring non-adiabatic change, then leads to

$$\frac{\partial u}{\partial x} + \frac{\partial v}{\partial y} + \frac{\partial w}{\partial z} - \frac{gw}{c^2} = 0 \quad . \tag{3.29}$$

This equation shows that if the motion is horizontal ($w \equiv 0$) the horizontal flow is in good approximation nondivergent, and that any compressibility effect then can be ignored.

The magnitude of the third term of the left-hand side of Eq. (3.29) is O($w/H°$), where $H°$ is the typical vertical scale of the overturning oceanic motion, O($4 \cdot 10^3$ m). Since c in the ocean is about 1500 m/s, the resulting vertical scale for the fourth term, c^2/g, is O(200 km), indicating that the fourth term in Eq. (3.29) can be ignored. This leads to the continuity equation for an adiabatic incompressible flow,

$$\frac{\partial u}{\partial x} + \frac{\partial v}{\partial y} + \frac{\partial w}{\partial z} = 0 \tag{3.30}$$

as an accurate approximation for the divergence field in the real ocean, except for sound waves, for which the compressibility is essential.

By integration of Eq. (3.30) in the vertical, using Eq. (3.23) as boundary condition, the vertical velocity component w can be estimated reliably if the (quasi-) geostrophic horizontal flow field is known. A further estimate of the first two terms in Eq. (3.30) is $V°/L°$, with $L°$ the characteristic horizontal scale of the flow in the THC (10^7 m) and $V°$ the characteristic horizontal velocity. Then the typical magnitude of the vertical velocity, $W°$, will be of the order

$$W^o = \frac{H^o}{L^o} V^o = 4 \cdot 10^{-4} V^o \quad , \tag{3.31}$$

several orders of magnitude smaller than the typical horizontal velocity.

Note that in the open ocean convective mixing driven by surface heat loss of the ocean in winter, where convection cells have similar horizontal and vertical scales L^o and H^o (Schott et al. 1993; Marshall and Schott 1999), equal magnitudes of the horizontal and vertical velocities are expected. This was confirmed by several oceanographic experiments in different ocean areas, where vertical velocity components between 5 and 13 cm/s were observed in individual down-welling plumes (Gascard and Clarke 1983; Leaman and Schott 1991; Schott and Leaman 1991; Schott et al. 1993). This is two orders of magnitude larger than the estimate of the typical large-scale down-welling near-bottom velocity when arctic water from the Denmark Strait descends into the deep Atlantic Ocean over the East-Greenland slope, presented above.

4. Water mass and tracer analysis of the deep flow in the Atlantic Ocean

4.1. Meridional sections of temperature, salinity and density

The THC is essentially a meridional, vertically overturning circulation system with its sources of deep and bottom water at high latitudes, both in the northern and the southern hemisphere. A useful way to study the THC, following the example set by the German oceanographer Georg Wüst (1935), is the analysis of meridional vertical sections of the tracers potential temperature, θ, and salinity, *S*. Both parameters are conservative, which means that outside the direct influence of air–sea interaction in the upper ocean the salinity and potential temperature of a water particle only change due to mixing. The conservation of potential enthalpy is fundamentally a better approximation of the first law of thermodynamics (Fofonoff 1962; McDougall 2003), but for most oceanographic applications the conservation of potential temperature and of salinity function quite well. The "spreading path" of the cold deepwater, formed in high latitude source regions, often is derived from sections of potential temperature and salinity by a combination of intuition and physical reasoning. In this section we start, as an example, with the distribution of potential temperature and salinity in the Atlantic Ocean, since the main sources of deepwater, both arctic and antarctic, are found in that ocean.

In 1988 and 1989, American oceanographers surveyed a meridional section through the Atlantic Ocean between 63°N and 54°S (Tsuchiya et al. 1992), now known as the WOCE A16 line from the southern coast of Iceland to South Georgia Island in the southern Ocean (Fig. 4.1). Although this section did not extend into the main source regions of the deepwater formation, the hydrographic properties along this section illustrate some main characteristics of the cold branches of the THC in the Atlantic Ocean quite well. The temperature and salinity structure, derived from the A16 data set, agrees in its general aspects with the Meteor data from 1925 to

1927 (Wüst 1935), but with a better horizontal and vertical resolution and a strongly improved accuracy. The use of a two-dimensional hydrographic section like depicted in Fig. 4.2 does not imply that all meridional spreading of the deepwater masses takes place in this vertical surface. As will be shown in following chapters the main meridional transport of deepwater takes place in narrow deep western boundary currents. Lateral recirculation of water from that boundary current brings the hydrographic properties of the boundary current to the ocean interior.

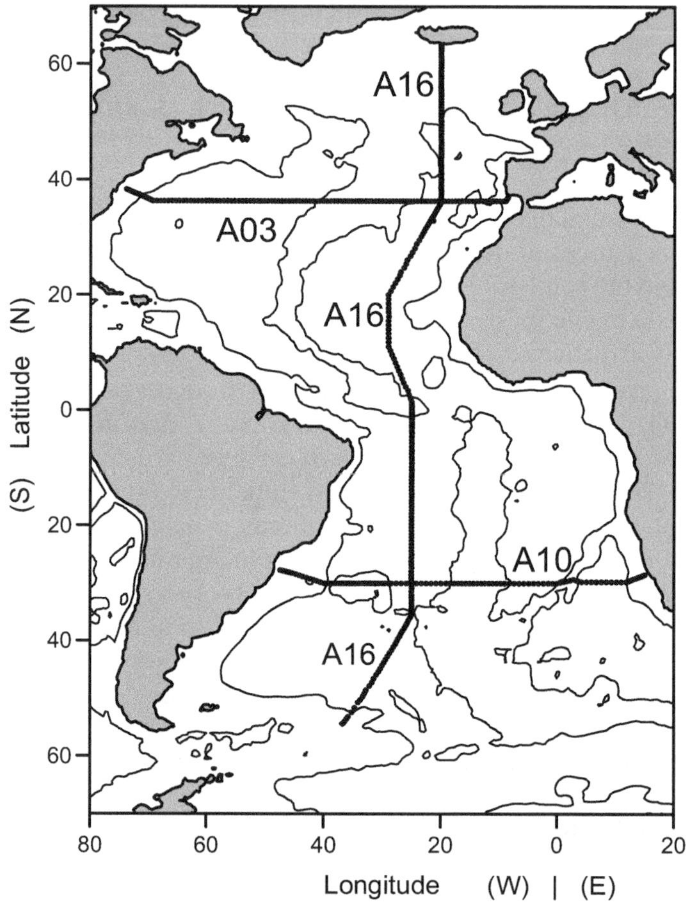

Fig. 4.1. The sections of the WOCE Hydrographic Program used in this chapter. The depth contour shows the 4000 m isobath.

A striking feature of the Atlantic Ocean, and of all oceans in general, is that below ~2500 dbar the potential temperature is low ($\theta < 3.0$°C), even near the equator (Fig. 4.2, upper panel). These low temperatures at low latitudes cannot be derived by mixing from the overlying warm (sub)tropical surface waters, since heat diffuses from high to low temperatures. The cold water has been brought to the tropics from high latitudes, either arctic or antarctic, by a deep flow. The lowest bottom temperatures are observed in the South Atlantic Ocean, an indication of the antarctic origin of these bottom waters. The salinity section shows that in the Atlantic Ocean this cold Antarctic Bottom Water (AABW) has a relatively low salinity ($S < 34.8$) compared to the overlying deepwater between 2000 and 4000 m (Fig. 4.2, lower panel). Less clearly visible in Fig. 4.2 is the presence of another cold bottom layer ($\theta < 2.5$°C) over the Icelandic slope at the northern end of the A16 section. Here the cold bottom water is relatively saline ($S > 34.96$), in contrast with the AABW. This water originates from overflow of cold water from the Norwegian Sea across the Iceland–Scotland Ridge. In Chapters 7 and 9 more attention will be given to this northern boundary of the deep Atlantic Ocean.

From a northern source near Iceland a cold deep water mass with a relatively high salinity, the North Atlantic Deep Water (NADW) spreads southward through the Atlantic between about 1500 and 4000 m. South of 40°S the salinity maximum, connected with the NADW, ascends from 3000 dbar into the Antarctic Circumpolar Current, reaching a depth of 1300 dbar near 54°S. In this region it often is referred to as Circumpolar Deep Water (CDW). North of 40°N a low-salinity water mass ($S < 34.92$), originating from the Labrador Sea, is observed above the deep high-salinity core (Labrador Sea Water or LSW). South of this latitude the LSW is replaced by water with a high-salinity core, Mediterranean Sea Outflow Water (MSOW), which enters the Atlantic Ocean via the Gibraltar Strait. This latter water mass obtains its high salinity from the evaporation excess in the Mediterranean Sea. The intrusion of MSOW into the Atlantic raises the salinity of the upper layers of the NADW core. The isolated high-salinity core near ~20°S at 2000 dbar suggests a sideward inflow of less diluted NADW; in this case an eastward flow from the South American continental slope in the direction of southern Africa.

At ~800 dbar a low-salinity core is observed to spread northward from ~50°S to north of the equator. This is Antarctic Intermediate Water (AAIW), formed in the Southern Ocean. In the upper 600 to 1000 dbar, where we find the permanent thermocline, the so-called Central Water is found. This water is characterized by a large vertical gradient in both θ and S. It is formed by subduction of winter mixed-layer water from the subtropical gyres into the thermocline. The flow of the Central Water is confined to the wind-driven layers.

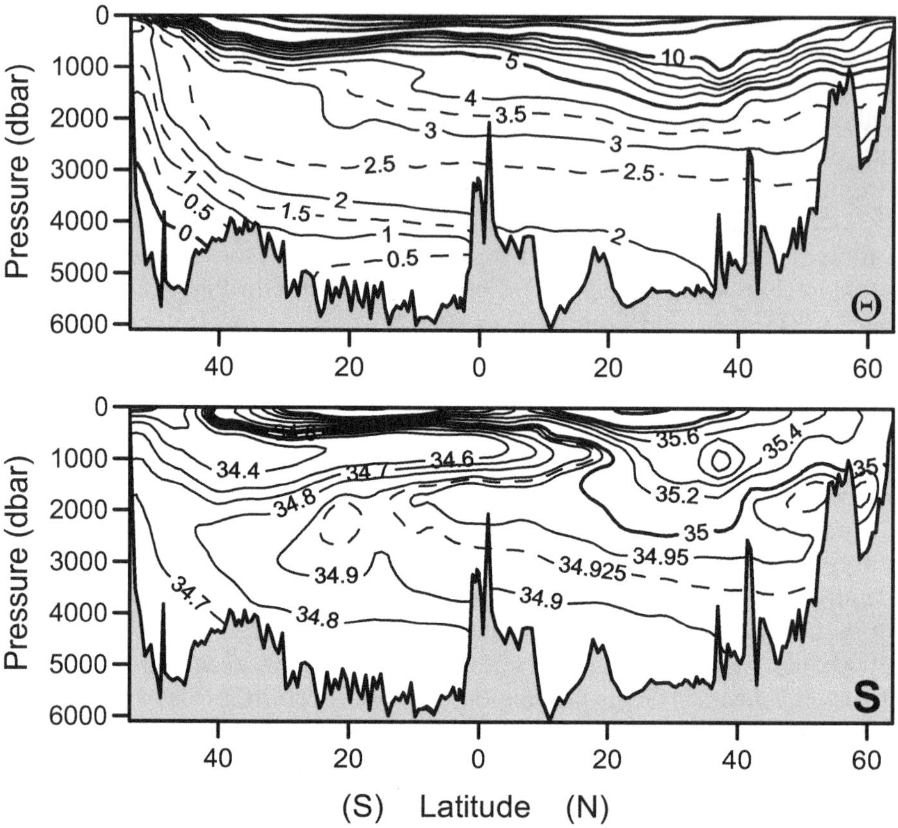

Fig. 4.2. Meridional hydrographic section of the Atlantic Ocean between approximately 20° and 30°W, showing the distribution of potential temperature θ (upper panel) and salinity *S* (lower panel). Pressure has been used as a vertical coordinate. The data, used to draw these sections, were obtained from the WOCE A16 section. The northern boundary of the section is formed by Iceland, the southern boundary by South Georgia Island. The shallow plateau near the equator is the Mid-Atlantic Ridge. South of this ridge the section runs through the western Atlantic basins, north of the ridge through the eastern basins.

The motion of the water in a stationary ocean approximately follows isopycnal surfaces (surfaces of constant potential density) except in regions where strong turbulent mixing may cause a nonnegligible diapycnal velocity component (velocity that crosses isopycnal surfaces). The course of the isopycnals along WOCE section A16 (Fig. 4.3) suggests a descending southward flow of NADW between 1000 and 3500 dbar north of 50°N and a descending northward flow of AABW in the lowest layers south of 40°S. In the South Atlantic Ocean the salinity maximum, connected with the NADW core, follows the ascent of the isopycnals to the south from ~3000 dbar at 40°S to ~1300 dbar at 54°S. Between 40°S and 40°N no large-scale meridional density gradient can be discerned which, according to popular definitions of the THC, should drive the deep flow.

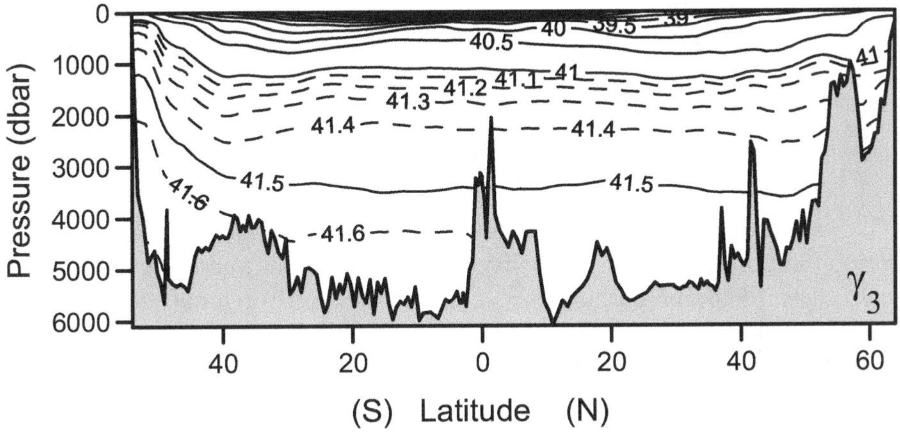

Fig. 4.3. The meridional distribution of the potential density anomaly γ_3 (kg/m^3) relative to a pressure of 3000 dbar along the meridional WOCE A16 section through the Atlantic Ocean. The full lines are 0.5 kg/m^3 apart, the dashed lines have been drawn each 0.1 kg/m^3 below the 41.0 kg/m^3 isopycnal.

Figure 4.3 shows also that the large-scale meridional density gradients in the wind-driven upper 1000 dbar are at least one to two orders of magnitude larger than the meridional density gradients in the deep ocean. These meridional density contrasts can be used to infer the zonal upper (and abyssal) ocean circulation assuming the geostrophic equilibrium, as argued in the previous chapter. Apparently these large but shallow meridional density gradients are not directly responsible for the meridional overturning circulation of the THC.

4.2. Deriving the deep circulation from tracer distributions

The interpretation of the spatial distribution of potential temperature and salinity in terms of the meridional advection and diffusion, as presented above, implicitly assumes that the equations that describe the spatial development of tracer fields can be inverted, at least in part. These equations describe the distribution of, e.g., θ and S, given the velocity field, the diffusivity, and the boundary conditions. The deep circulation and diffusion are assumed to leave their traces in the distribution of scalar properties or "tracers" of the ocean water which allow reconstruction of the advective flow and the diffusion; the inversion of the conservation equations. A complete inversion of the equations will derive velocity components and diffusion coefficients from the tracer distributions. An incomplete inversion at least should derive the direction of the deep flow.

The general equation for the development of a scalar variable or tracer C (e.g., salinity) is

$$\frac{\partial C}{\partial t}+u\frac{\partial C}{\partial x}+v\frac{\partial C}{\partial y}+w\frac{\partial C}{\partial z}=\nabla_h\cdot\left(K_h\nabla_h C\right)+\frac{\partial}{\partial z}\left(K_v\frac{\partial C}{\partial z}\right)+Q \quad . \tag{4.1}$$

The variable C is the tracer value and t is time, while u, v, and w are the velocity components in, respectively, the horizontal x and y and the vertical z directions. The parameter K_h is the lateral (or isopycnal) turbulent diffusion coefficient, and ∇_h is the horizontal component of the gradient operator. The K_v represents the vertical (or diapycnal) turbulent diffusion coefficient. Both diffusion coefficients are positive because of thermodynamic constraints that only allow a diffusive transport from high to low concentrations of C. Due to the small aspect ratio of the oceanic motions and the generally stable stratification, K_h is assumed to be at least several orders of magnitude larger than K_v. The turbulent diffusion coefficients are bulk parameters representing that part of the system that is not modeled explicitly, the "turbulent" motion. The term Q represents the source or sink term for non-conservative processes like, for example, the formation of the ^3He isotope due to the radioactive decay of ^3H in seawater or the consumption of dissolved oxygen when organic material in the ocean is remineralized by microbial activity. Surface heat and freshwater fluxes that influence the temperature and salinity of the surface water are not source terms but boundary conditions for Eq. (4.1).

Appropriate boundary conditions and known fields for u, v, w, K_v, and K_h allow in principle the forward solution of Eq. (4.1) either with analytical or numerical methods. When the boundary conditions for certain tracers vary strongly over time (e.g., tracers of anthropogenic origin like CFCs and bomb Tritium), we speak of transient tracers. For transient tracers the time derivative in Eq. (4.1) may be maintained, while for the more common natural tracers Eq. (4.1) often is approximated as stationary by removing the time derivative.

There is no general rule how to solve the inversion of Eq. (4.1). For different situations different inverse solution methods have been proposed to derive the deep ocean circulation from the distribution of one or several tracers, sometimes in combination with the geostrophic balance applied to the density distribution (Tarantola 1987; Wunsch 1985, 1996). By approximating Eq. (4.1) the problem to be solved inversely can be simplified. While sometimes the flow velocity is derived explicitly, more often constraints on the deep circulation, e.g., flow directions, are obtained. An example of a strong simplification of Eq. (4.1) has been given above with the description of the meridional sections of θ and S in the Atlantic Ocean shown in Fig. 4.2. The ocean was considered to be approximately two-dimensional (vertical–meridional plane) and stationary. The fact that even in equatorial regions of the oceans the deep water masses have temperatures below 3°C led to the conclusion that deepwater flows from high latitudes to low latitudes. With a positivity constraint for the diffusion coefficients the sign of the diffusive terms in Eq. (4.1) in principle can be derived from the change of gradients of C. This in turn may set constraints on the possible directions of the velocity vectors. From the salinity maximum in the NADW core, decreasing to the south due to mixing, a southward advective transport to supply salt to the S-maximum core can be derived. The northward increasing near-bottom temperature minimum implies a northward flow of AABW which is warmed by mixing with the overlying warmer NADW. In this reasoning the southward decrease of the NADW salinity maximum and the northward increase of the near-bottom temperature minimum are implicitly ascribed to diapycnal mixing, not to advection or diffusion perpendicular to the section's plane.

For the stationary distribution of a conservative tracer both the time derivative and the source term in Eq. (4.1) can be ignored. The simplest inversion of the remaining advective–diffusive balance that is the solution of (u, v, w, K_h, K_v) with its norm minimized in some way is the trivial solution with zero advection and diffusion, a physically unlikely stagnant ocean. In order to interpret stationary tracer distributions in terms of non-zero advection and diffusion generally additional *a priori* knowledge is required.

4.3. Wüst's core method

A simplification for the interpretation of tracer distributions is the "Kernschichtmethode" or core layer method. This method was developed by the German oceanographer Georg Wüst (1935) when he analyzed the results of the German Meteor Expedition. Wüst noted that several of the main water masses involved in the deep circulation of the Atlantic Ocean were characterized by extremes in the vertical distribution of hydrographic parameters, e.g., a salinity maximum or an oxygen minimum. The position of the extremes was assumed to coincide with the (ill-defined) core of the water mass. Following the so-defined core layer of the water mass emerging from the assumed source region would show the main spreading paths of the water mass. We use here the term spreading, because the method does not discriminate between advection and diffusion. This can be illustrated with a two-dimensional reduction of Eq. (4.1). First we assume that a stationary two-dimensional tracer pattern is caused by an advection–diffusion balance:

$$u\frac{\partial C}{\partial x} = K_v \frac{\partial^2 C}{\partial z^2} \quad . \tag{4.2}$$

As boundary conditions for C in this equation we assume further that

$$C = A_0 \cos\left(z/B\right) \text{ at } x = 0 \quad , \tag{4.3a}$$

and that

$$C = 0 \text{ at } z = \pm\frac{B\pi}{2} \quad . \tag{4.3b}$$

The solution of Eq. (4.2) with boundary conditions (4.3a, b) is

$$C = A_0 \cos\left(z/B\right) e^{-\left(K_z/_{uB^2}\right)\cdot x} \quad . \tag{4.4}$$

The form of the distribution of C according to Eq. (4.4) is illustrated in thc upper panel of Fig. 4.4.

When the horizontal spreading of the tracer is mainly caused by strong lateral diffusion instead of advection Eq. (4.2) is replaced by

$$K_h \frac{\partial^2 C}{\partial x^2} + K_v \frac{\partial^2 C}{\partial z^2} = 0 \quad , \tag{4.5}$$

with the boundary conditions of Eq. (4.3a, b) and the additional condition that C stays finite for increasing positive x, the solution of Eq. (4.5) is

$$C = A_0 \cos\left(z/B\right) e^{-\left(\sqrt{K_v/K_h}\cdot/_B\right)\cdot x} \quad , \tag{4.6}$$

a tracer distribution with a mathematical form similar to Eq. (4.4). This shows that from a stationary tracer distribution one cannot discriminate between horizontal advection and horizontal diffusion. Additional *a priori* information is required to interpret the spreading as solely due to diffusion or due to an advection–diffusion balance.

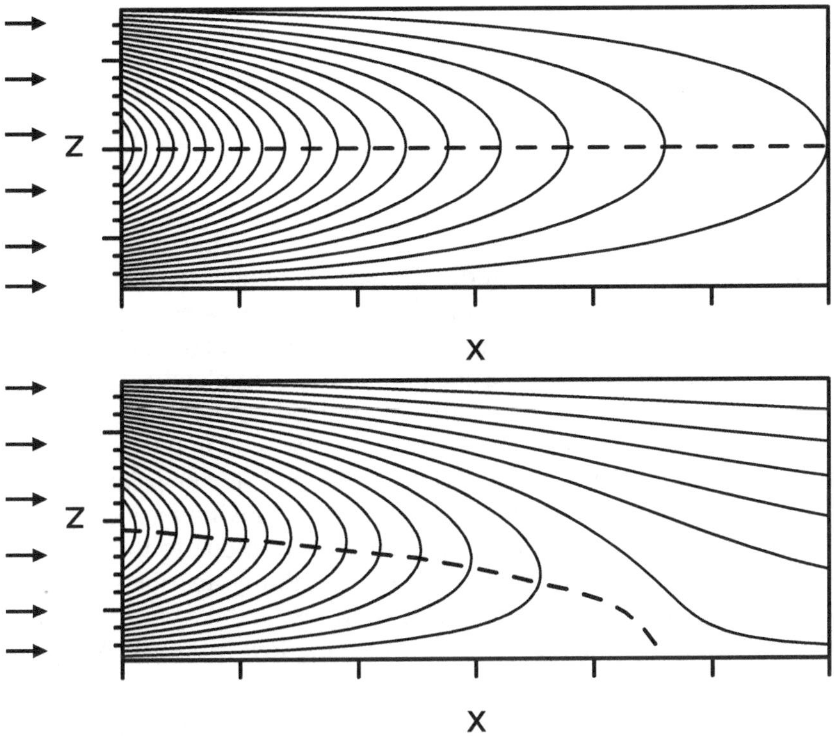

Fig. 4.4. The tracer distribution derived from Eq. (4.1) with boundary conditions (4.3a, b) (upper panel) and the tracer distribution derived from Eq. (4.1) and boundary conditions (4.7a, b)(lower panel). In both cases a cosine tracer core was advected by a vertically constant horizontal velocity (indicated by the arrows), in the first case without background stratification, and in the latter case with background stratification of 10% of the amplitude of the initial tracer distribution at $x = 0$. The broken line follows the vertical tracer maxima.

In oceanographic literature it often is assumed implicitly that advection is the main cause of the horizontal spreading of tracers This assumption also was made for the interpretation of the distributions in Fig. 4.2. The basic assumption for the core layer method that the vertical extreme of a hydrographic parameter, e.g. the salinity maximum, coincides with the core of the advected water mass is not always fulfilled. It is evident that the presence of a vertical extreme in the core of a water mass also depends on the tracer concentrations in the overlying and underlying water. To simulate this we adapt boundary conditions (4.3a, b) by adding a small (10% of the initial core amplitude) vertical background gradient. Then the boundary conditions become

$$C = A_0 \cos\left(z / B\right) + \frac{z}{B\pi}\frac{A_0}{10} \quad \text{at } x = 0 \quad , \tag{4.7a}$$

and

$$C = \pm\frac{A_0}{20} \quad \text{at } z = \pm\frac{B\pi}{2} \quad . \tag{4.7b}$$

The resulting tracer distribution, a linear combination of the rightmost term of (4.7a) with the solution from Eq. (4.4), shows a proceeding descent of the tracer maximum in absence of any vertical velocity. At some downstream distance from $x = 0$ the maximum will completely disappear and there the tracer distribution is mainly determined by the background stratification. An example of this phenomenon can be seen in the meridional salinity distribution in the Atlantic Ocean (Fig. 4.2, lower panel). North of 40°N the NADW is characterized by a salinity maximum at ~2600 dbar. Between 15 and 40°N the background stratification due to the presence of the overlying MSOW determines the vertical salinity distribution. South of 15°N the salinity maximum, connected with the NADW core, reappears at ~1600 dbar due to the presence of the overlying low-salinity AAIW.

4.4. Water mass, water type, and the temperature–salinity diagram

In the description of the hydrography of the deep Atlantic Ocean terms like water type and water mass were used and several were named without much emphasis on what was precisely meant with these terms. Here we will explicitly spend attention to these concepts and their relation with the analysis of θ–S diagrams.

Contrary to the flushing of the thermocline waters, where water from a range of latitudes in the subtropical gyre subducts from the winter mixed layer into the thermocline to form the so-called Central Water of the subtropical gyres, the intermediate and deepwater masses in the oceans are formed in only a few restricted locations. There, surface water is cooled and modified in the winter season by convective mixing, forced by the surface heat loss. This results in new winter water with regionally nearly homogeneous hydrographic properties. At the end of the cooling season the newly emerging seasonal stratification in the upper layer of the ocean isolates this water from further atmospheric influences. The newly formed homogeneous water subducts and spreads from its formation region at intermediate and deep levels, depending on its density.

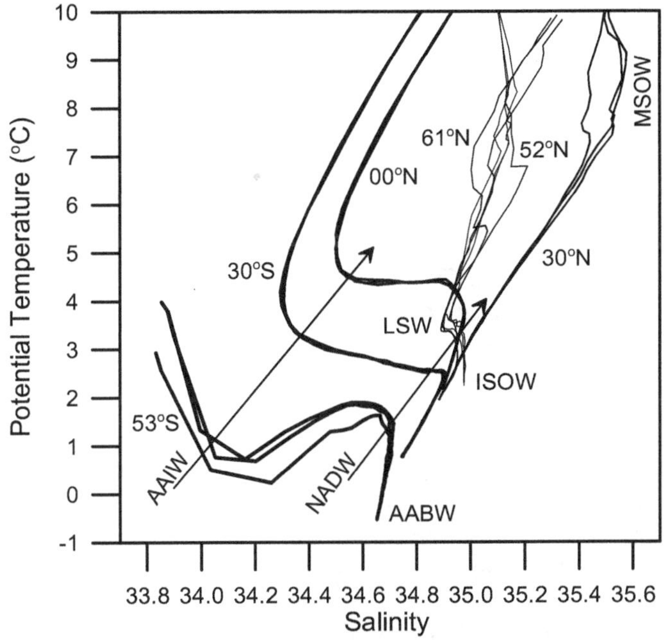

Fig. 4.5. Example of θ–*S* relations from the deep Atlantic Ocean. For six different latitude belts three neighboring stations are shown. The position of the main water types and water masses are indicated: Antarctic Bottom Water (AABW), North Atlantic Deep Water (NADW), Antarctic Intermediate Water (AAIW), Labrador Sea Water (LSW) Mediterranean Sea Outflow water (MSOW) and Iceland–Scotland Overflow Water (ISOW). The data from these curves were taken from the WOCE A16 section.

So the main sources of deepwater in the ocean produce water with a θ–*S* combination characteristic for its source region. The water that we encounter in the deep ocean is generally a mixture of water from only a few source regions. This leads to a predominant presence of certain θ–*S* combinations in the deep ocean or water mass modes, as can be seen in two-dimensional frequency distributions in θ–*S* space, derived from a thermohaline census (Fig. 4.6). The most common class of deepwater along the meridional WOCE section A16 in the Atlantic Ocean is cold and relatively saline ($\theta = 2.0$ °C, $S = 34.900$), reflecting the saline North Atlantic origin of NADW. Note, however, that this class of water is found below the salinity maximum of the NADW core layer. Apparently it also contains some cold and less saline AABW, introduced into the NADW core by upwelling and mixing. The thermohaline frequency distribution along section A16 shows two more peaks, $\theta = -0.2$ °C, $S = 34.675$, the cold AABW in the Argentine Basin, and θ = 0.3 °C, *S* = 34.700, the slightly warmed AABW in the Brazil Basin.

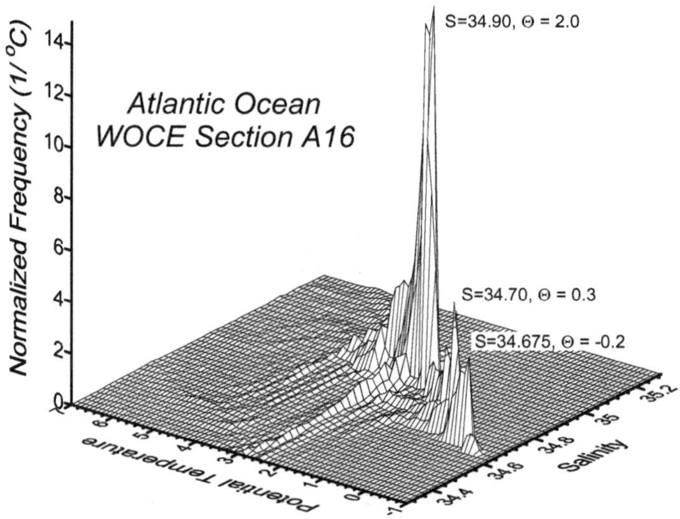

Fig. 4.6. Two-dimensional frequency distribution in θ–*S* space of the deepwater colder than 8°C along the meridional WOCE A16 section in the Atlantic Ocean, determined with a $\Delta S \times \Delta\theta = 0.025 \times 0.1$°C resolution. The integral of this frequency distribution over all possible salinities and temperatures equals 1.

Via their slowly changing θ–*S* characteristics, the large volumes of deepwater in the ocean can be traced back to their source regions. The continuous thermohaline structure of deepwater masses in the ocean is formed by advection and mutual mixing of water from different source regions. In analogy with the concept of air masses in the atmosphere, developed in the early 19th century, this has led to the concepts of water types and water masses. Unfortunately, different authors use different definitions for these concepts or different names for similar concepts, while most do not bother at all to refer to such definitions. Also in the discussion of the deep circulation in the Atlantic Ocean, presented above, the terms water type and water mass were used quite intuitively. Here we want to introduce the different concepts of water type, source water type, and water mass in more detail. Note that in literature sometimes the term water mass also is used for the concept we will name water type. A classic monograph on water mass analysis has been written by the Russian oceanographer Mamayev (1975), while a more modern overview of different aspects of water types and water masses is given by the Australian oceanographer Tomczak (1999).

A *water type* is a large volume of nearly homogeneous water with a common formation history having its origin in a particular geographic formation region in the ocean. It obtains its (nearly homogeneous) θ–S properties (locus and range) in the formation region due to a specific climatic process like air–sea interaction or overflow of dense water across a sill with subsequent entrainment and subsurface mixing with overlying water. When formed by convective mixing, driven by a surface heat loss, the formation rate of a water type will show a seasonal variation. If a water type moves through the ocean without mixing, the position and range of the water type in a θ–S diagram will not change. Water types are generally, but not always, connected with core layers. Due to changing climatic circumstances that influence the formation of a water type by air–sea interaction, its properties may show variation of formation rate and hydrographic properties over time. Because of their close connection with regional formation phenomena, water types often are named after their region of origin or region of occurrence, e.g., Red Sea Outflow Water or Lower Circumpolar Deep Water. From their formation region the water types are advected by regional current systems into the interior of the deep ocean.

An idealization of the water type concept is the *source water type*. A source water type is assumed to be perfectly homogeneous (only locus, no range) and is characterized by a single θ–S combination, or a point in a θ–S diagram. A source water type is a mathematical construct and does not occupy any volume in physical space. When using this source water type concept one has to be aware of the high degree of idealization.

A *water mass* can be imagined as a continuum of two or more water types transformed due to mutual mixing, with a finite horizontal and vertical extent. Near the regions of origin of the constituting water types the water mass converges to the regionally formed water type. From there advection and mixing bring the water mass through phases of evolution and decay until the water mass loses its identity through a final transformation. An example is the NADW in the deep Atlantic Ocean, characterized by a salinity maximum of ~34.95 to 34.80 between ~1500 and 3500 m. This water mass originates from the input of three northern water types, as will be discussed in Chapter 7, and mixing with the surrounding fresher water types of southern origin. Water masses are physical entities with a measurable extent that occupy a finite volume in the ocean. They are reservoirs of heat, salt, nutrients (nitrate, phosphate, and silicate), and dissolved gases (oxygen and carbon dioxide). Their formation, spreading, evolution, and decay form important processes in the oceanic branch of the climate machine. Mathematically a water mass can be considered to be a linear combination of source water types due to mutual mixing in varying degrees.

When using these concepts one always has to keep in mind that they were introduced as a shorthand to cluster large numbers of observations. They are concepts, not items with fixed physical properties. Water mass analysis is a tool from the oceanographic workshop used to study ocean circulation problems. A real proliferation of water types can be found for well-studied oceans to address details of the hydrography of regional ocean areas. It can be stated that the nomenclature of the water masses in the North Atlantic Ocean has become bewilderingly confused in the past half century, although it is apparent that the water masses themselves have remained essentially the same (Wright and Worthington 1970). In a recent publication on the Nordic seas (Greenland Sea, Iceland Sea, and Norwegian Sea) 35 different water types were discriminated for this very small part of the world ocean (Hopkins 1991). A real danger exists that water mass analysis develops into a kind of stamp collecting instead of a useful tool for oceanographic investigations. In an effort to generalize the understanding of water masses in the world ocean Emery and Meincke (1986) listed only 19 water types to describe the global intermediate and deep water mass structure.

4.5. Quantitative water mass analysis

The idealized evolution of a water mass in an advective–diffusive balance is illustrated in Fig. 4.7. The original stratification of source water types has been set up with three homogeneous layers (the dots in the θ–S diagram). The lowest salinity is located in the intermediate layer, similar to, e.g., the AAIW in the Atlantic or Pacific oceans. The water mass, described by a continuous line in the θ–S diagram, evolves downstream from the source of the intermediate water as mixing progressively erodes the salinity minimum. At a certain stage in the development of the water mass the salinity minimum disappears from the salinity profile and the θ–S diagram (thick line in Fig. 4.7). Only the curvature of the θ–S line at intermediate temperatures reminds one of the original salinity minimum. A similar evolution of a water mass is spatially illustrated in the lower panel of Fig. 4.4.

The points on the θ–S line, characteristic for the water mass, result from mixing, in different ratios, of water from the three original source water types. To quantify this we assume that a water particle with a unit mass located at a certain position (θ, *S*) on the θ–*S* line is formed by mixing of a mass m_1 of source water type 1 (θ_1, S_1), m_2 of type 2 (θ_2, S_2), and m_3 of type 3 (θ_3, S_3). Then mass conservation can be described as

$$m_1 + m_2 + m_3 = 1 \quad , \tag{4.8a}$$

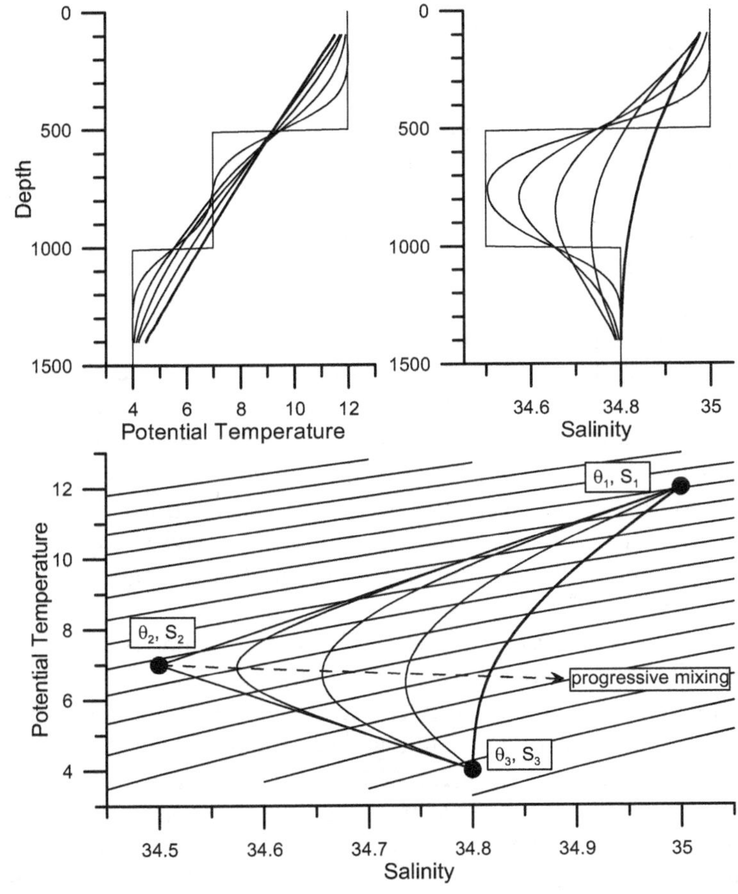

Fig. 4.7. Schematic evolution of a water mass, set up by three source water types (dots in the lower panel), due to progressive mixing in an advective–diffusive balance. The vertical stratification of potential temperature and salinity is shown in the upper panel, the θ–*S* diagram in the lower panel. In the θ–*S* diagram the set of thin lines present isopycnals, with the density increasing from the upper left corner to the lower right corner.

and the conservation of heat and salt, respectively, as

$$m_1\theta_1 + m_2\theta_2 + m_3\theta_3 = \theta \quad , \tag{4.8b}$$

and

$$m_1 S_1 + m_2 S_2 + m_3 S_3 = S \quad . \tag{4.8c}$$

For the set of linear equations (4.8) the mass contributions m_1, m_2, and m_3 can be solved as long as the three known source water types form a triangle in θ–S space and are not aligned along a straight line. A physical constraint that $m_i \geq 0$.for i=1 to 3 should be added.

It easily can be derived that this implies that all possible points θ, S should be located within the triangle set up by the three source water types, the mixing triangle. When only two source water types are involved (e.g., types 1 and 2, $m_3 \equiv 0$) all mixing products will be found on a straight mixing line between the end members (θ_1, S_1) and (θ_2, S_2). When all three source water types are involved, the mixing products will set up a water mass characterized by a continuous curved line in θ–S space. This type of water mass analysis formed a major tool for the study of central aspects of the oceanic circulation before 1940 and was well summarized in the handbooks by Sverdrup et al. (1942) and Mamayev (1975). The results suggested a view of the circulation in the deep ocean dominated by slow advection and diffusion. The large-scale θ–S distribution is a long-term integrated result of oceanic motions which themselves may show strong variability in space and time at relatively small scales and high frequencies.

Because of the idealized nature of the source water type concept with constant and homogeneous properties and also because of the limited accuracy of tracer measurements, it may occur that some data points are found outside of the mixing triangle. In that case at least one of the contributions m_i (i=1, 2, or 3) derived from Eq. (4.8) will be negative. This violates the nonnegativity constraints for m_i. In those cases an optimal (but not exact) solution has to be found according to some predefined optimality measure. That makes weighting of the conservation Eqs. (4.8) necessary. In many cases the water mass is determined by more than three water types, and then the quantitative θ–S analysis cannot produce any reliable results, unless additional assumptions are made, since the set of equations for θ, S, and mass is underdetermined. For example, the NADW core in Fig. 4.5 appears to interact with AABW, AAIW, LSW, and MSOW. With the increased capacity to measure other hydrographic parameters but θ and S, multiparameter water mass analysis has become a possibility, including conservation of oxygen, nutrients, and several transient tracers next to θ and S. Equations (4.8) can be extended with additional tracers and additional water types. Nonconservative behavior can be modeled by incorporating an integrated source or sink term as additional unknown.

When only three source water types are involved in the generation of the tracer fields, addition to Eqs. (4.8) of extra conservation equations for the other tracer fields may generate an overdetermined linear system for the solution of the unknowns m_1, m_2, and m_3. Such a system can be solved with a least-squares approximation to obtain an optimal solution. In that case each conservation equation should be given a certain weight to account for differences in range, accuracy, variability, and physical units for the different tracers. Extension of the number of conservation equations also allows the possibility to describe the water mass as a linear combination of more than three source water types. The maximum number of source water types equals the number of independent tracers plus one (mass conservation). This is the research field of the optimal multiparameter analysis (OMPA) (Tomczak 1981). The classic θ–S diagram is the projection of the hydrographic observations in a multitracer parameter space onto the two-dimensional θ–S plane.

We have to note here that the calculation of the water type contributions m_i from the observed hydrographic tracers is a mathematical transformation of the observations based on *a priori* assumptions about the properties of source water types involved in the water mass evolution. As such the quantitative water mass analysis does not add any new and independent information. The required definition of source water types replaces the boundary conditions for the inversion of Eq. (4.1). However, the water mass analysis changes the perspective from continuous tracer fields with prescribed boundary conditions to advection and dilution of independent water types that originate from different restricted ocean regions and are formed there by specific processes.

4.6. The use of biogeochemical tracers

The minor constituents of dissolved sea salt account for less than 0.1% of the dissolved material in seawater. Most major constituents exhibit a conservative behavior so that their spatial distribution in the ocean is mirrored in the salinity distribution. An exception is calcium carbonate ($CaCO_4$) which forms hard skeletons for some plankton species. When after cell death these skeletons rain down to deep pressure levels, the $CaCO_4$ may dissolve again below the so-called lysocline, forming a source term for the dissolved calcium budget. This leads to slightly enhanced deep calcium concentrations in the deep and bottom waters of the ocean.

Many of the minor constituents of sea salt are nonconservative, being in some way or another involved in biological processes. Biological activity may change their concentrations significantly, characteristic for the aging of the water type after it left its source region. With knowledge of the source or sink terms for these elements, they can be used as tracers with a clock function. Even if the rate of the clock is not known, at least the sign of the aging is.

For the primary production of organic material in the near-surface photic zone, dissolved inorganic nutrients are required, next to photosynthetically active radiation and carbon dioxide. Oxygen is produced in this process. When organic material is remineralized by microbial activity the inorganic nutrients and carbon dioxide become available again. This will lead to the increase of their concentration in seawater while the oxygen concentration decreases. In the photic zone, where primary production by phytoplankton consumes large amounts of nutrients, the nutrient concentrations often decrease to near-zero in the summer season. The excess oxygen, produced in the photosynthesis, may be released to the atmosphere, while CO_2 from the atmosphere is, with some delay, dissolved into the surface ocean.

The main nutrients for the production of soft organic matter are nitrate (NO_3^-) and phosphate (PO_4^{3-}). Dissolved inorganic carbon in the oceans is available as CO_2, HCO_3^-, and CO_3^{2-}, lumped together as total inorganic carbon (*TIC*), also referred to as total CO_2 or ΣCO_2. The synthesis of organic material and the subsequent remineralization or respiration often is represented by the generic equation

$$1H_3PO_4 + \alpha HNO_3 + \beta CO_2 + \lambda H_2O \leftrightarrow C_\beta H_\zeta O_\varphi N_\alpha P + \gamma O_2 \quad , \qquad 1:\alpha:\beta:\gamma = 1:16:106:170 \quad . \tag{4.9}$$

The positive proportionality constants α, β, and γ represent the molar ratios between the fluxes of nitrate, *TIC*, and oxygen relative to phosphate involved in the production or remineralization of organic material. These stochiometric ratios often are assumed to be constant. They are named Redfield ratios, after A.C. Redfield, who studied these ratios in plankton and ocean water in the early 20th century (Redfield 1934).

In the deepwater of the oceans, well below the photic zone, remineralization of exported organic material occurs, which releases nutrients and carbon dioxide and consumes oxygen, according to Eq. (4.9). Exceptions are the small deep ecosystems near hot springs or methane seeps, where organic matter is produced by chemosynthesis, anoxic remineralization in oxygen-depleted environments like the deep Black Sea, and probably some deep production by archaea. Such specific processes generally are ignored in the study of the deep circulation.

The remineralization may occur on particulate organic material (POM) or on dissolved organic material (DOM). Examples of POM are dead plankton cells, fecal pellets or so-called marine snow. DOM is produced as exudates of phytoplankton during photosynthesis (excess organic material released by the cells) or by the transformation of POM by microbial activity. The remineralization processes may take place in the water column as well as in the sediment. In the latter case sediment–water interaction can transport the remineralization products back into the overlying water, while it also transports oxygen from the water into the sediment.

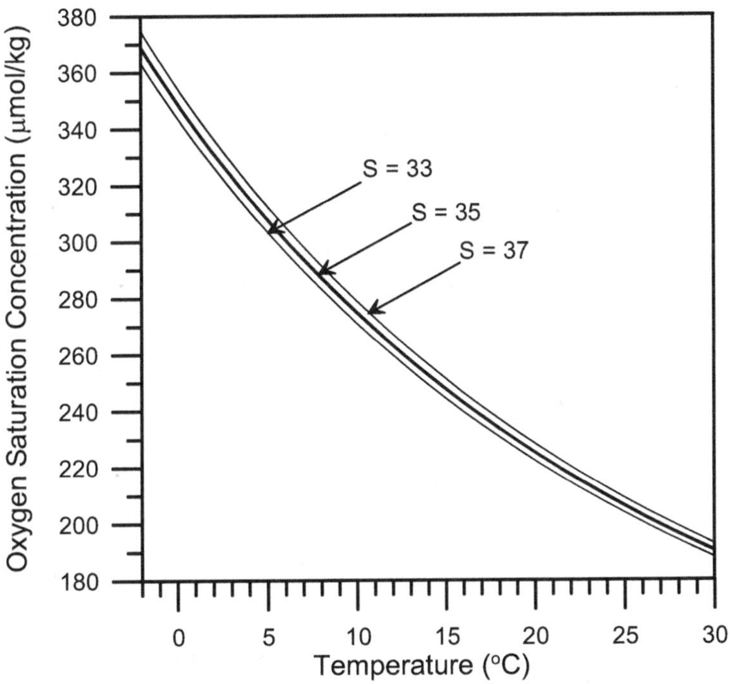

Fig. 4.8. The value of the oxygen saturation concentration in seawater in equilibrium with the atmosphere as a function of temperature for the oceanic temperature range. The thick line represents the function for a salinity $S = 35$, while the lower and upper thin lines show the function for salinities of respectively $S = 33$ and $S = 37$.

The spatial distributions of dissolved oxygen, nitrate, phosphate, and *TIC* can be used as tracers for the deep circulation, next to the classic θ–S analysis. Aging of water since it left the near-surface layer, in direct contact with the atmosphere, is reflected by the increase of the concentrations of dissolved inorganic phosphate, nitrate, and *TIC* and by the decrease of the concentration of dissolved oxygen. When deepwater is formed in winter by deep convective mixing, the oxygen concentration generally is close to its saturation value in balance with the atmosphere, O_{2sat}. That value strongly depends on temperature and to a lesser extent on salinity (Fig. 4.8). The high oxygen saturation concentrations make the formation of cold deep water masses at high latitudes and their subsequent spreading in the deep branch of the THC the oxygen valve of the ocean. The temperature dependence of the oxygen concentration in the source regions of the deepwater has a significant influence on the spatial distribution of the dissolved oxygen concentration, even after oxygen consumption due to aging has taken place.

A useful parameter to visualize the effects of aging on the oxygen distribution separated from that temperature effect is the apparent oxygen utilization (*AOU*), defined as

$$AOU = O_{2sat} - O_2 \quad , \tag{4.10}$$

where O_2 is the in situ concentration of dissolved oxygen. With regard to the effects of aging by remineralization of organic material the distributions of *AOU*, phosphate, nitrate, and *TIC* are mutually related via the Redfield ratios.

Plankton not only contains "soft" organic material. Several phytoplankton species have hard parts consisting of either calcium carbonate ($CaCO_3$) or biogenic opal or silicate (H_2SiO_4). In the deep ocean and the sediment these hard parts may be dissolved, which leads to an increase in the concentrations of dissolved silicate, calcium, and *TIC*. Since these physical processes occur independently from the microbial remineralization of soft organic material, the source term Q in Eq. (4.1) for the dissolution of hard parts is not coupled to the mineralization effects via a constant Redfield ratio. The spatial distribution of dissolved silicate therefore contains information not found in the distribution of the other nutrients. The effect of the dissolution of $CaCO_3$ on the Ca distribution is small, because of the large background concentration of the major salt constituent Ca which mainly reflects the variability of the salinity. In the deep ocean the Ca concentration is only slightly enhanced. The effect of the dissolution of $CaCO_3$ on the *TIC* distribution, however, may be significant.

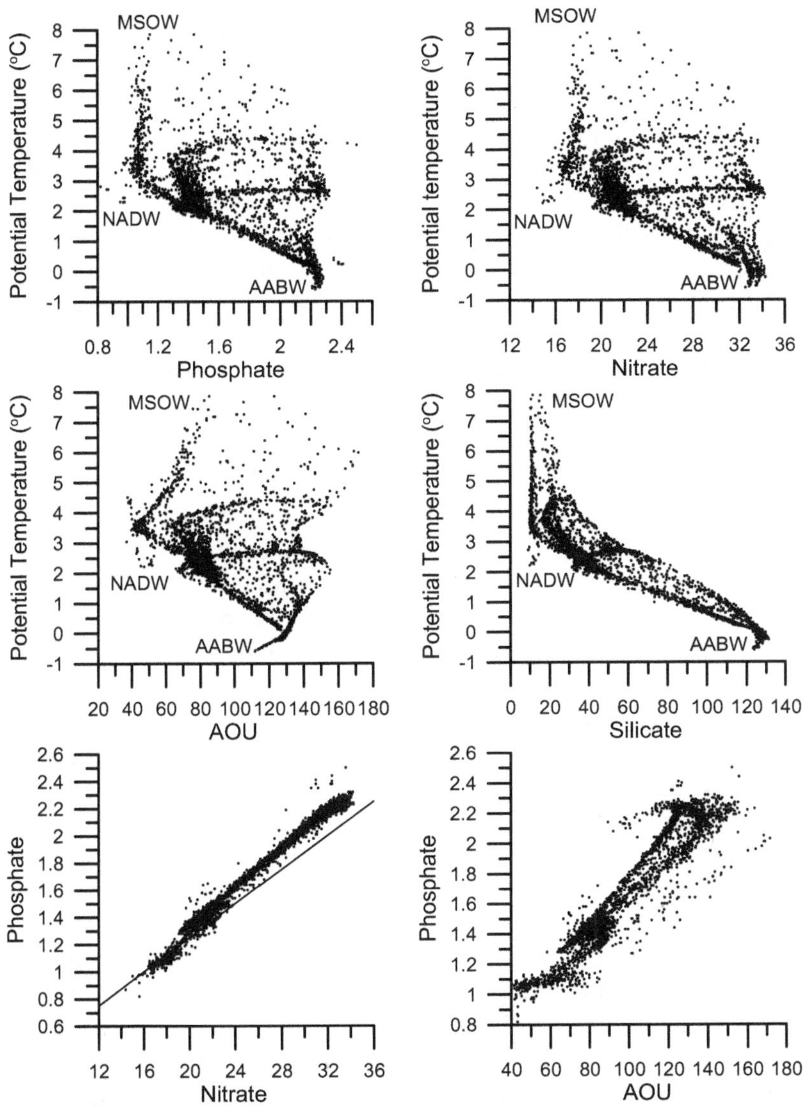

Fig. 4.9. Property–property plots for the concentrations of the biogeochemical tracers, based on data obtained along the meridional WOCE A16 line in the Atlantic Ocean. Phosphate, nitrate, *AOU* and silicate (in μmol/kg) have been plotted versus the potential temperature (upper four panels) with an indication of the main deepwater types. Property–property plots of phosphate versus nitrate and *AOU* (lowest two panels) show the mutual relations between these nonconservative tracers. In the phosphate–nitrate plot the straight *N/P* = 16 line has been added.

4.7. Biogeochemical tracers in the Atlantic Ocean

When we plot the concentrations of the biogeochemical tracers *AOU*, phosphate, nitrate, and silicate versus the potential temperature for samples from the deep Atlantic Ocean below 1000 m depth (Fig. 4.9, upper four panels) a water mass structure is shown that is determined by the varying concentrations of the original deep water types as well as by the aging effects due to remineralization of organic material and dissolution of biogenic silicate. The water types NADW, AABW, and MSOW can be recognized from extremes in these plots. Due to the coupling via Redfield ratios, the structure of the deep water mass in θ–PO_4 space is quite similar to the structure in θ–NO_3 space and has large agreements with the θ–AOU structure. The θ–SiO_4 structure is quite different, mainly because of different SiO_4 conditions for the deep water types. The highest silicate concentrations (> 120 µmol/kg) are found in the low-temperature range of the South Atlantic AABW (–1 to 0°C). In the temperature range of the North Atlantic water types NADW (2 to 3°C) and MSOW (> 7°C) the silicate concentrations are definitely lower. The lowest phosphate, nitrate, and *AOU* values are found in the NADW in the Iceland Basin.

Dissolved inorganic phosphate and nitrate are highly correlated (lower left panel of Fig. 4.9), which causes the similar θ–parameter structure for both nutrients. This is in part caused by the correlation between PO_4 and NO_3 in the original water types and in part by the similar effects of aging on these nutrients which are coupled by a Redfield ratio of 1:16. It appears that the N/P ratio for dissolved phosphate and nitrate in the ocean is quite close to this ratio (the straight line in the lower left panel).

The phosphate concentrations in the deep Atlantic Ocean have a high correlation with *AOU* ($r = 0.94$), confirming the role of mineralization of organic material during aging of the deep water masses for the deep structure of *AOU* and nutrients (Fig. 4.9, lower right panel). The high correlation between *AOU* and dissolved nitrate and phosphate and the relatively narrow (although nonlinear) relation between the potential temperature and dissolved silicate show that the extra information, to be obtained from the biogeochemical tracer fields, is limited. That information mainly deals with common aging effects.

The constant Redfield ratios describing the coupling between the different biogeochemical tracers can be used to define quasiconservative tracers. Preformed phosphate (PO_4°) and nitrate (NO_3°) are defined as

$$PO_4^\circ = PO_4 - AOU/170, \quad \text{and} \quad NO_3^\circ = NO_3 - AOU \times 16/170 \ . \qquad (4.11)$$

Another type of quasiconservative tracers are the parameters "*NO*" and "*PO*" (Broecker et al. 1985), defined as

$$\text{"}PO\text{"} = 170 \times PO_4 + O_2, \quad \text{and "}NO\text{"} = (170/16) \times NO_3 + O_2 \ . \qquad (4.12)$$

In these definitions it is implicitly assumed that during the formation of the water types by air–sea interaction the surface water was oxygen saturated (for the preformed tracers), or had at least a constant saturation rate ("*NO*" and "*PO*"). According to Eq. (4.1) for conservative tracers ($Q = 0$) the hydrographic structure of these tracers is influenced only by advection and mixing.

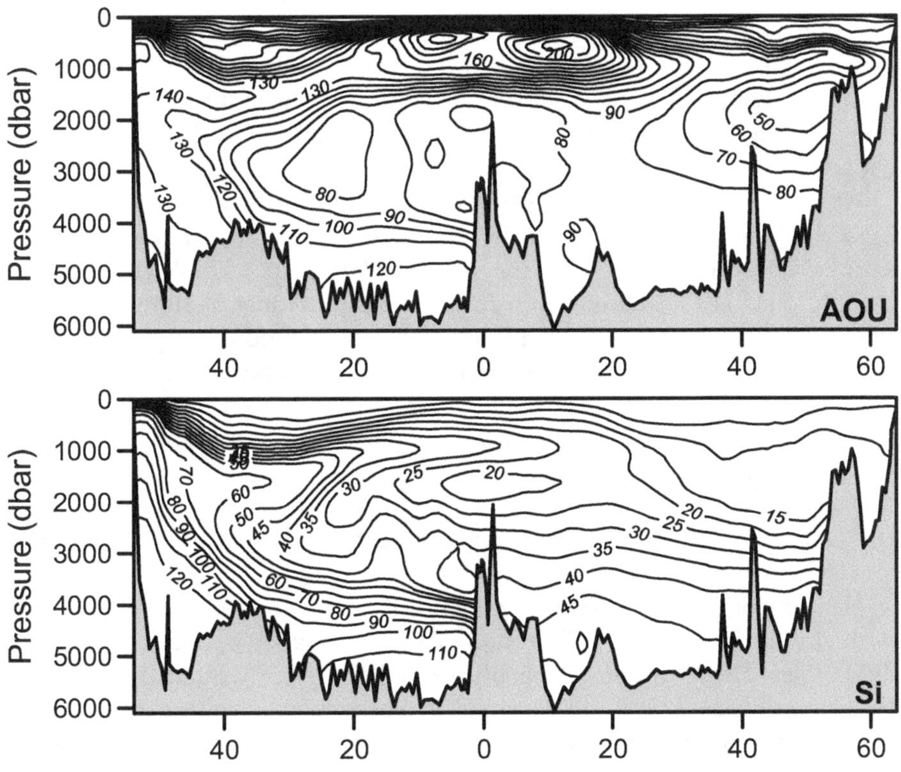

Fig. 4.10. Meridional section of *AOU* (upper panel) and dissolved silicate (lower panel), both in μmol/kg, through the Atlantic Ocean. These sections are based on the observations along WOCE section A16.

The vertical profiles of *AOU* and dissolved silicate along the meridional WOCE A16 section in the Atlantic Ocean (Fig. 4.10) show a structure below 1000 dbar depth with some similarities relative to the salinity distribution along this section (Fig. 4.2). In the high-salinity core of NADW between ~1500 and 4000 m relatively low values of *AOU* and SiO_3 concentrations are found. The lowest values ($AOU < 50$ μmol/kg, $SiO_4 < 15$ μmol/kg) are observed in the upper part of the NADW core north of 40°N, mainly determined by the properties of the LSW water type. Very high silicate concentrations (> 120 μmol/kg) are found in the near-bottom AABW core south of the equator, characteristic for its southern origin. In the Argentine Basin south of 35°S an *AOU* maximum ($AOU > 130$ μmol/kg) is observed in the upper parts of the AABW core, while in the near-bottom layer south of 50°S *AOU* values below 120 μmol/kg are found. Apparently the near-bottom layer with $SiO_4 > 120$ μmol/kg is less aged, in terms of remineralization of organic material, than the overlying layer with silicate concentrations between 90 and 120 μmol/kg, agreeing with a northward and upward flow in the deepest layers.

At intermediate depths (1000 to 1500 dbar) south of ~15°S, a core of high *AOU* and silicate values is present, seemingly similar to the low-salinity core connected with the AAIW water type. However, the low-salinity core is located at depth levels about 500 dbar shallower than the maxima in *AOU* and dissolved silicate that are found at the boundary between the low-salinity AAIW core and the high-salinity NADW core. This aged core extends from the Antarctic Circumpolar Current where it is positioned above the CDW salinity maximum in the circumpolar Southern Ocean connected to the NADW core in the south Atlantic. The latter, saline water type therefore is often named Lower Circumpolar Deep Water (LCDW), the former high-*AOU* water type Upper Circumpolar Deep Water (UCDW).

Near the equator a zone of relatively low *AOU* and SiO_4 values is found with minima between 1500 and 2500 dbar and between 3000 and 4000 dbar. These minima probably extend from the western boundary of the Atlantic ocean and are assumed to be connected with separate cores within the NADW, originating from the Greenland Sea and the Labrador Sea. Between 20 and 30°S another nearly separate core with relatively low *AOU* and SiO_4 concentrations is observed at the NADW levels, coinciding with a similar isolated salinity core ($S > 34.925$) in this latitude band. This core also extends from the western boundary of the ocean and reaches the continental margin off southwest Africa. Apparently here a branch of NADW crosses the Atlantic Ocean to the east. This core has been observed to follow the African Continental Slope around southern Africa into the

Mozambique Channel in the Indian Ocean (van Aken et al. 2004), as will be presented in Chapter 5.

4.8. A natural radioactive tracer: radiocarbon

A naturally occurring clock tracer with a known characteristic decay time-scale can be used to estimate age differences in the deep ocean. An example of such a clock tracer is ^{14}C or radiocarbon. This radioactive isotope of carbon is used in archaeology and earth sciences for the determination of the age of the carbon-containing compounds. The atomic nucleus of ^{14}C consists of six protons (p) and eight neutrons (n). The ^{14}C isotope is produced in the atmosphere as a result of incoming cosmic radiation. High-energy protons collide with atmospheric atomic nuclei and produce in this process low-energy neutrons. These neutrons are caught by nitrogen nuclei to form ^{14}C according to

$$^{14}N(7p+7n)+n \rightarrow {}^{14}C(6p+8n)+p \quad . \tag{4.13}$$

The newly formed carbon nucleus is unstable and ultimately will decay with the emission of a β particle (electron) to ^{14}N

$$^{14}C(6p+8n) \rightarrow {}^{14}N(7p+7n)+e^{-} \quad . \tag{4.14}$$

The number of ^{14}C nuclei in a sample, decaying in a short interval Δt, is proportional to the number of ^{14}C present, $[^{14}C]$. Radioactive decay therefore is a first-order process. Such a process can be described as

$$\left[^{14}C(t)\right] = \left[^{14}C_0\right]e^{-t/\tau} = \left[^{14}C_0\right]2^{-t/\tau_{0.5}} \quad , \tag{4.15}$$

where τ is the e-folding time (8266 year) and $\tau_{0.5}$ is the half-life of ^{14}C (5730 year). With the onset of the massive use of fossil fuels (coal, oil, and natural gas) ^{14}C-depleted CO_2 was brought into the atmosphere. The nuclear test explosions from the 1950s and early 1960s brought huge amounts of ^{14}C into the environment (bomb radiocarbon). Only before about 1890 the amount of ^{14}C in the atmosphere relative to the stable isotope ^{12}C was nearly constant due to the balance between the production (Eq. 4.13) and decay (Eq. 4.14). The amount of ^{14}C relative to the stable isotope ^{12}C is expressed as a deficit $\delta^{14}C$ (negative) referenced to the atmospheric equilibrium value $[^{14}C_0]$ in 1950, in per mille (‰), extrapolated from the preindustrial era before 1890.

When carbon has become isolated from its atmospheric source by incorporation into organic material (e.g., wood or plankton) or dissolution of

CO_2 into the ocean, $\delta^{14}C$ will decrease due to radioactive decay in a way described by Eq. (4.15). The Nobel laureate Willard F. Libby (chemistry, 1960) of the University of Chicago in 1947 was the first to use this property to estimate the age of carbon-containing samples. For such age determinations an earlier estimate of τ, 8033 year, often is used, which results in the so-called conventional radiocarbon age.

Until recently $\delta^{14}C$ was determined by the measurement of the electrons released by the ^{14}C β decay in counting chambers. Because of the low number of disintegrations per hour, large carbon samples were needed as well as extended counting periods, in the order of days. For the determination of ^{14}C in seawater CO_2 samples obtained from ~100 liter of seawater were used. Currently accelerator mass spectrometers (AMS) are used to determine $\delta^{14}C$ from the $^{14}C/^{12}C$ ratio. This requires only CO_2 from a 1 liter sample. At the Woods Hole Oceanographic Institution an AMS has been installed to handle the large amount of samples from the WOCE Hydrographic Program. Several other institutes also purchased an AMS in the 1990s. However, the accuracy of counting chambers in general is superior to the accuracy of AMS measurements.

In the surface layers of the ocean $\delta^{14}C$ is lower than in the atmosphere since CO_2 from the atmosphere is introduced into a relatively old pool of carbon, due to the large overturning time of the oceans (~2000 year). This is called the reservoir effect. Surface waters in tropical and moderate latitudes had a radiocarbon age of about 400 year ($\delta^{14}C = -47$ ‰) before the introduction of bomb radiocarbon into the atmosphere. When the air–sea exchange of CO_2 between the atmosphere and the ocean is reduced due to the presence of sea ice, the reservoir effect may be larger. Broecker et al. (1995) mentioned a reservoir effect of –70 ‰ (~800 year) for the NADW source regions in the northern North Atlantic, while Schlosser et al. (1994) found a reservoir effect of –140 ‰ (~1250 year) for the source regions of AABW in the Southern Ocean. The $\delta^{14}C$ in dissolved inorganic carbon in the deep ocean may get "contaminated" by dissolution of old carbonate, by the oxidation of old DOM as well as by the mineralization of relatively new organic particles that rain down from the surface layers.

Examples of the ^{14}C structure in the western Atlantic Ocean are given in Fig. 4.11. The potential density anomaly relative to 3000 dbar (γ_3) is used as vertical coordinate. In the thermocline ($\gamma_3 < 41.1$ kg/m^3) the presence of bomb radiocarbon causes positive values of δ^{14}C. In the upper parts of the NADW, from the LSW salinity minimum to the NADW salinity maximum, ($41.25 < \gamma_3 < 41.36$ kg/m^3) $\delta^{14}C$ was quite similar for 60°N and 42°N but still higher than the value of –47 ‰ expected because of the reservoir effect for prebomb ^{14}C. In the early 1990s bomb radiocarbon apparently had penetrated to these levels at 42°N.

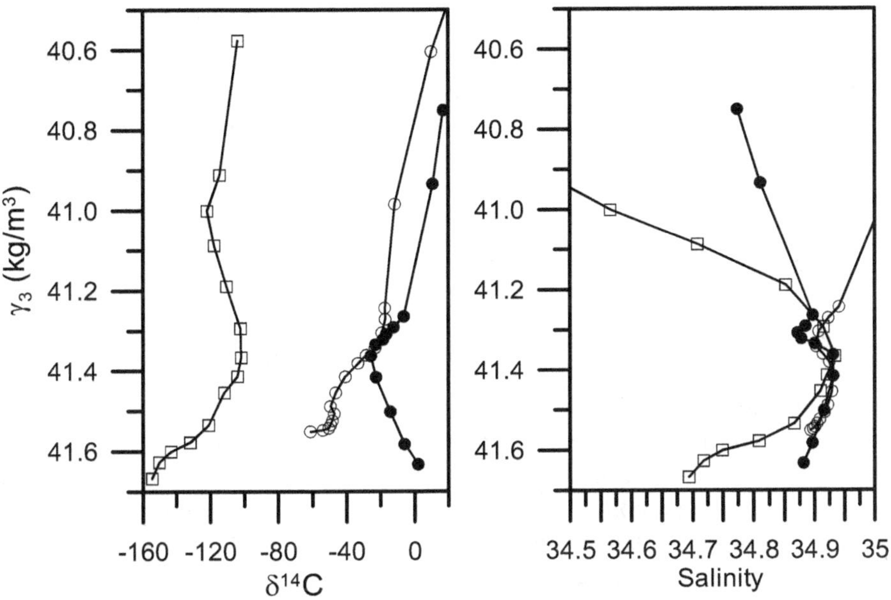

Fig. 4.11. A plot of $\delta^{14}C$ (left panel) and S (right panel) versus the potential density anomaly relative to a pressure of 3000 dbar in the western Atlantic Ocean. The black symbols represent the smoothed mean profiles over the East Greenland continental slope at 60°N (WOCE section A1E), the open circles represent the smoothed mean profiles over the Canadian continental slope at 42°N (WOCE section A2), while the open squares shows the smoothed profiles for GEOSECS station 54 at 15°S.

Below the NADW salinity maximum at ~2400 m the $\delta^{14}C$–γ_3 curves for 60°N and 42°N diverged. Where at 60°N the coldest NADW near the bottom at ~3000 m had positive $\delta^{14}C$ values because of the bomb radiocarbon in the overflow water from the Greenland Sea, at 42°N the values of $\delta^{14}C$ varied from ~–40 ‰ at ~3000 m to ~–60 ‰ near the bottom at ~4800 m. This is caused by the presence of older water at these depths, where less bomb radiocarbon had penetrated into the bottom water at that latitude. In the South Atlantic Ocean at 15°S the salinity maximum near $\gamma_3 = 41.4$ kg/m^3, characteristic for the NADW core, also is characterized by a $\delta^{14}C$ maximum. NADW is clearly the best ventilated deep water mass in the South Atlantic Ocean. Both the underlying AABW and the overlying UCDW have lower $\delta^{14}C$ values. The $\delta^{14}C$ values in the AABW core increase northward, similar to the NADW core, showing that in that water mass mixing with the overlying well-ventilated NADW is more important than radioactive decay. A consequence is that the meridional gradient of ^{14}C ages in the NADW core is stronger than can be expected from the transport of this water mass, since mixing with the older AABW also contributes to the southward decrease of the $\delta^{14}C$ values in the NADW core. Ignoring these mixing effect and using a reservoir effect for NADW of −70 ‰, the $\delta^{14}C$ value of −102 ‰ in the NADW core at 15 °S gives a ventilation age of nearly 275 year.

4.9. Halocarbons as tracers

In the 20th century a number of industrially produced chemical substances of the halocarbon group (halogenated organic compounds) in increasing amounts have been introduced into the atmosphere where they spread globally from their sources, which are mainly in the industrialized nations. Some of these substances that have entered the surface water of the ocean appear to be quite stable in seawater. This makes them suitable as transient tracers of the THC since they enter the deep ocean with the deep water types formed in the surface layer by air–sea interaction. Halocarbons that are used as tracer are the chlorofluorocarbons (CFCs) CFC-11, CFC-12 and CFC-113 and carbon tetrachloride (CCl_4). The CFCs, produced since the late 1930s are common coolants that were used in refrigerators, transformers, and spray cans and in a large number of other technical applications all around the world. CCl_4 was introduced in the early 20th century as a dry cleaning fluid and as a solvent for organic substances. It also has been used in large quantities as a feedstock for the production of CFC-11 and CFC-12. A problem with the halocarbons is that they also have entered the stratosphere where they cause depletion of the ozone layer which protects us from harmful ultraviolet radiation. Therefore an international agreement was signed in 1987 to reduce worldwide halocarbon emissions, the Montreal Protocol, which has led to a considerable decline of the emissions in the 1990s.

The history of the annual mean atmospheric concentrations for CFC-11, CFC-12, CFC-113, and CCl_4 has been reconstructed by Walker et al. (2000). Since these substances are mainly produced in the industrialized nations, a slight difference between the northern and southern hemisphere could be established. Figure 4.12 shows their global mean atmospheric concentrations in mole fractions in part per trillion (10^{-12}). The effect of the Montreal Protocol is visible in the slight decline after 1990 of the atmospheric concentrations of CFC-11, CFC-113, and CCl_4 and the strongly reduced increase of CFC-12. Because of their increase in the reduced increase of CFC-12. Because of their increase in the atmosphere until 1990 it can be expected that recently formed deepwater will have higher CFC and CCl_4 concentrations than water that was in contact with the atmosphere at some earlier time.

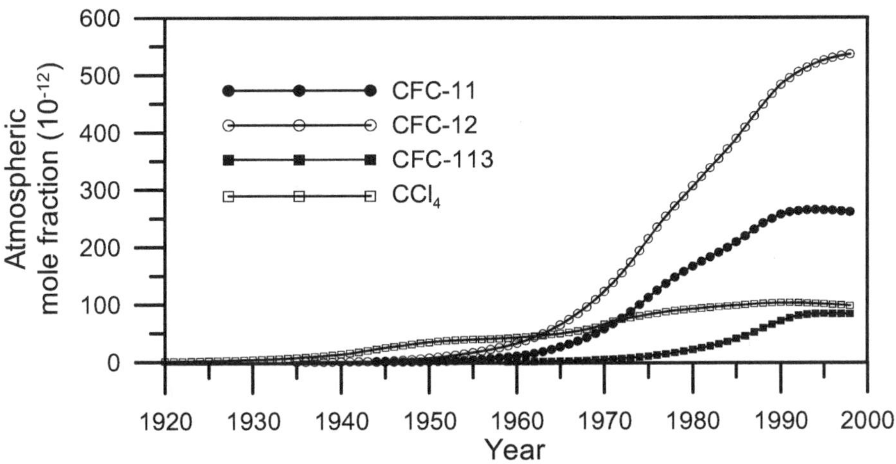

Fig. 4.12. Reconstructed history of the globally mean atmospheric concentration of the tracer halocarbons as a mole fraction in part per trillion (10^{-12}), derived from Walker et al. (2000).

Not only the atmospheric concentrations of the CFCs but also their atmospheric mutual mixing ratios MR_a changed over time. This enables us to define a halocarbon age based on a reconstructed atmospheric mixing ratio *MR* which is defined as

$$MR = \frac{C_1 / F_1(\theta, S)}{C_2 / F_2(\theta, S)} , \tag{4.16}$$

where C_1 and C_2 are the in situ measured seawater concentrations of compounds 1 and 2, and F_1 and F_2 are their solubility functions. Warner and Weiss (1985), Bu and Warner (1995), and Bullister and Wisegarver (1998) have determined the dependency of the solubility *F* for, respectively, CFC-11 and CFC-12, CFC-113, and CCl_4 from temperature and salinity. By comparing the reconstructed MR with the atmospheric time-dependent mixing ratio $MR_a(t)$, derived from Fig. 4.12, the year of equilibration can be determined. Note that the resulting age is the age of the dissolved halocarbons, since dilution by mixing with water free of halocarbon compounds 1 and 2 will not alter the result. This age will be the true halocarbon age, only if mixing occurs with halocarbon free water and if, during the formation of the deep water types, the ocean mixed layer was in equilibrium with the atmosphere or had the same percentage of undersaturation for both compounds. Because of the differing time histories of the different halocarbon compounds, different combinations can be used for specific age ranges. The CFC-11/CFC-12 ratio can be used for water formed from the 1950s to 1970, the CFC-11/CFC-113 ratio for the period 1970 to 1990, and the CFC-11/CCl_4 ratio for the 1950s to 1990. The decrease of the atmospheric halocarbons after the Montreal Protocol became operational has caused an age ambiguity for deepwater formed after 1990.

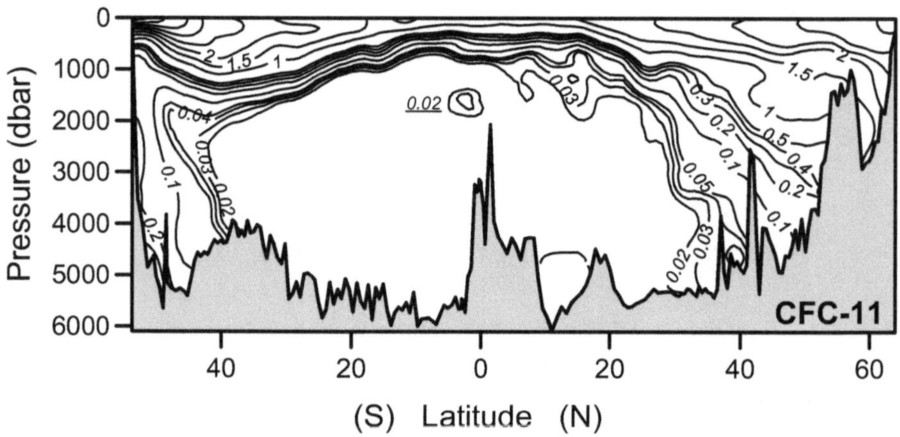

Fig. 4.13. Distribution of the concentration of dissolved CFC-11 (pmol/kg) along the meridional WOCE A16 section in the Atlantic Ocean, surveyed in 1988/89.

The distribution of the different CFCs also is used to follow qualitatively the spreading of recently ventilated deep and bottom water in the ocean (e.g., Fine and Molinari 1988; Smethie et al. 2000). The distribution of CFC-11 in the Atlantic Ocean along the meridional A16 WOCE section (Fig. 4.13) shows that this halocarbon compound has penetrated into the deep Atlantic from northern as well as southern sources. At the levels of the NADW (~1500 to 4000 m) spreading from the northern source has proceeded to at least 30°N (in 1988). Due to diapycnal mixing from recirculating NADW CFC-11 also has entered the near-bottom low-salinity layer at northern latitudes originating from the AABW spreading from the south. At the southern end of the A16 section the highest CFC-11 concentrations in the deepwater (> 0.3 pmol/kg) are found (in 1989) in the bottom layer of the Argentine Basin, south of 50°S. This maximum coincides with the near-bottom *AOU* minimum (see Fig. 4.10, upper panel). This confirms that the youngest AABW is found near the bottom and agrees with a northward and upward spreading of AABW in the Atlantic Ocean.

A remarkable feature in Fig. 4.13 is the CFC-11 maximum in the upper parts of the NADW water mass (> 0.03 pmol/kg), centered at 1650 dbar between the equator and 5°S, where also relative minima in *AOU* and dissolved silicate were observed (Fig. 4.10). The distribution of CFC-12 (not shown) displays the same feature. This supports the concept of an eastward south-equatorial jet that transports relatively young NADW from the western boundary into the ocean interior. From the CFC-11/CFC-12 ratio in this halocarbon maximum (1.6±0.2) the formation year is estimated to be 1965 (±3). Since this part of the section was surveyed in 1989 this suggests an age of 24 year since the last contact with the atmosphere. The upper layers of the NADW are assumed to be formed in the Labrador Sea at a distance of about 11,000 km. This implies an advection velocity of this water of about 1.5 cm/s. The maximum concentration in the CFC maximum is a factor 10 lower than the concentration in equilibrium with the 1965 atmosphere, indicative of a 10 times dilution with CFC free water on its way to the southern hemisphere. However, the actual dilution may be less, since CFC saturation down to about 60% regularly has been observed in the surface mixed layers of deepwater formation regions.

4.10. Zonal hydrographic sections in the Atlantic Ocean

As can be seen from Fig. 4.2, the salinity along section A16 does not show a clear maximum salinity core of the NADW at a latitude of about 35°N because of the presence of the overlying saline MSOW. The zonal A03 section near that latitude (Fig. 4.1) has been surveyed in 1993 by the Russian ship Prof. Multanovskiy. This sections shows a zonal asymmetry in the intermediate and deepwater properties.

The salinity along the A03 section (upper panel in Fig. 4.14) shows west of the Mid-Atlantic Ridge (MAR) only minor subsurface salinity minima ($S < 34.95$) originating from the LSW core, especially along the American continental slope at ~1500 dbar. These minima are traces of south flowing LSW in the upper parts of the NADW water mass. The slight salinity maximum below the LSW core represents the NADW core. The *AOU* distribution shows a coinciding high-oxygen core (low *AOU*, light gray in Fig. 4.14) west of the Mid-Atlantic Ridge, representative of the relatively young NADW core. At intermediate levels (~1000 dbar) a high-salinity core of MSOW is clearly visible in the eastern half of the section. Below the Gulf Stream along the American continental slope an *AOU* maximum is observed (*AOU* > 100 μmol/kg), connected with a north flowing core of AAIW. In the deep basins below 4000 dbar low-salinity AABW (S < 34.9) is found. The *AOU* distribution at these levels show a zonal asymmetry. East of the MAR bottom water with an *AOU* slightly above 80 μmol/kg is found, about 10 μmol/kg higher than in the western basin.

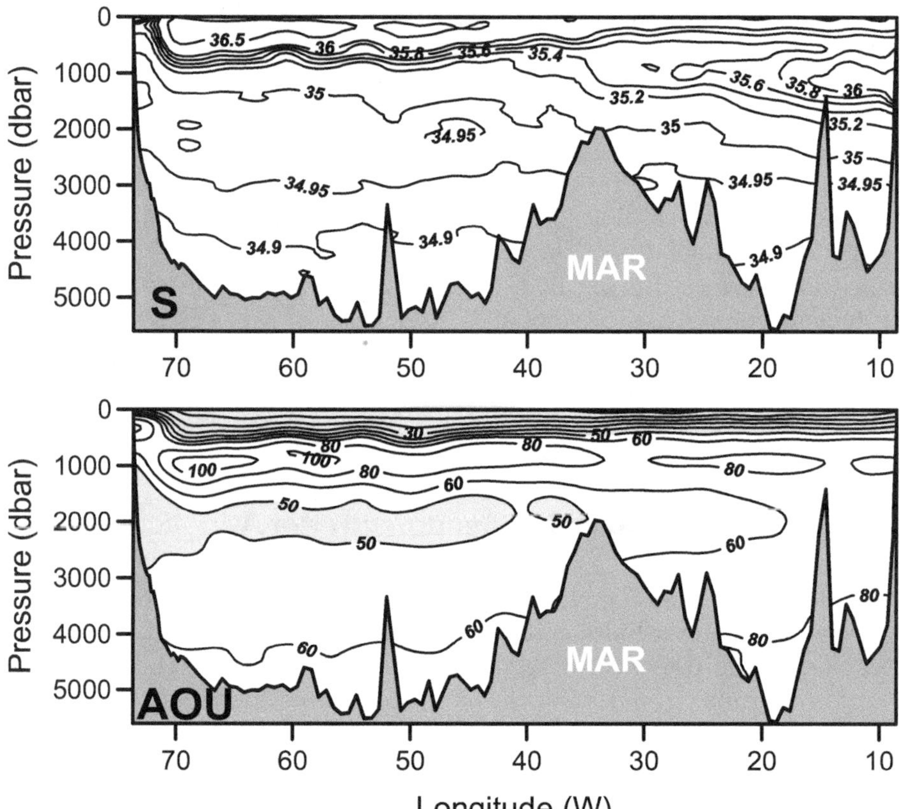

Fig. 4.14. Distributions of salinity, *S*, and apparent oxygen utilization, *AOU*, (μmol/kg) along the zonal A03 section near 36°30'N in the North Atlantic Ocean. The Mid-Atlantic Ridge (MAR) divides the deep Atlantic zonally. *AOU* values below 50 μmol/kg are shown in light gray.

Along the zonal A10 section near 30°S in the south Atlantic (Fig. 4.15) a salinity maximum (*S* > 34.9) was observed connected with the south flowing NADW west of the MAR. This NADW core also was characterized by low dissolved silicate values (*Si* < 45 μmol/kg). East of the Walvis Ridge (WR) a similar subsurface *Si* minimum was found off Africa, with slightly higher *Si* values at salinity values above 34.85. Apparently part of the low-*Si* NADW is transported southward in a branch separate from the NADW core near the South American continent, emerging from the zonally spreading high-salinity NADW core at ~20°S (see Fig. 4.2).

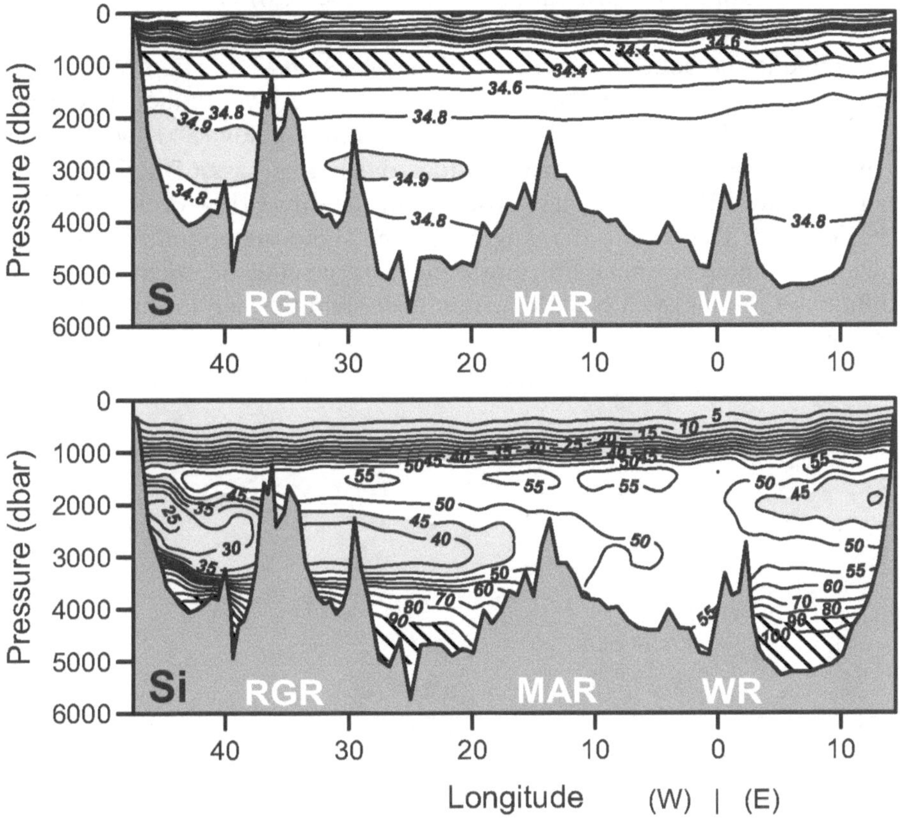

Fig. 4.15. Distributions of salinity, *S*, and dissolved silicate, *Si* (μmol/kg) along the zonal A10 section near 30°S in the South Atlantic Ocean. The subsurface salinity maximum (*S* > 34.9) is shown in gray, the overlying salinity minimum (*S* < 34.4) is hatched. *Si* values below 45 μmol/kg are shown in light gray, while the deep *Si* maximum in the bottom layer (> 90 μmol/kg) is hatched. The Angola Basin is located between the Mid-Atlantic Ridge (MAR) and the Walvis Ridge (WR). The Rio Grande Rise (RGR) forms a shallow obstacle for the meridional flow in the West Atlantic Basin.

Overlying the NADW core at 30°S is a basinwide salinity minimum at the bottom of the thermocline near approximately 800 to 1000 dbar. At these levels AAIW is the dominant water type, with low salinities. This AAIW core is located above the basinwide *Si* maximum with cores of over

55 μmol/kg, characteristic of the UCDW. Both water type cores are found over the whole width of the South Atlantic Ocean, not only in a narrow boundary current, suggesting that they are incorporated in the circulation of the wind-driven subtropical gyre.

In the bottom layer low-salinity AABW ($S < 34.8$) is found west of the MAR and east of the WR. Similarly high *Si* concentrations (> 90 μmol/kg) are found in the bottom layers west of the MAR and east of the WR. These are characteristic for the antarctic origin of the AABW. In the Angola Basin between the MAR and the WR no salinities below 34.8 were found while the near-bottom *Si* concentrations did not surpass 65 μmol/kg. Apparently the WR is an efficient obstacle for the direct flow of AABW from the Southern Ocean into the Angola Basin. There the bottom water originates from the Romanche Fracture Zone near the equator, where AABW is diluted by mixing with NADW with higher salinities and lower *Si* concentrations.

5. The deep flow in the Southern, Indian, and Pacific oceans

5.1. Hydrography of the Southern Ocean

As was described in the previous chapter, the cold branch of the THC in the Atlantic Ocean brings the saline core of NADW to the latitude of the southern tip of Africa (~35°S) from where it can enter the southern Ocean (Fig. 5.1). The NADW core overlies fresher AABW which appears to spread northward from the Southern Ocean into the Atlantic.

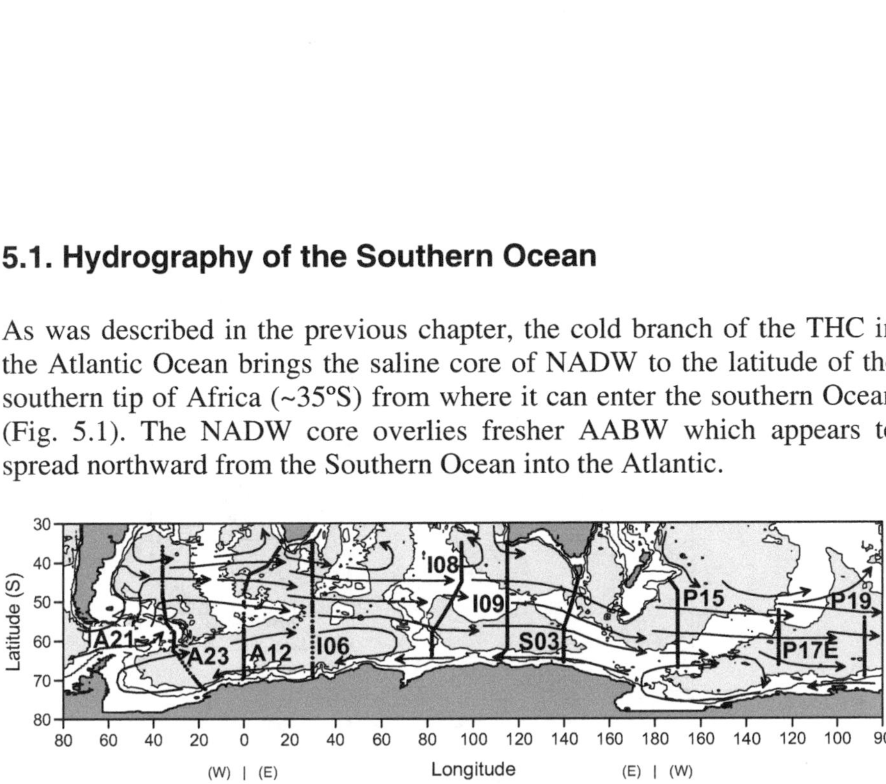

Fig. 5.1. Topography of the Southern Ocean with the prevailing currents (arrows). The dots show the hydrographic stations of the WOCE sections that are used in this section. The light gray areas are deeper than 4000 m.

The geography of the Southern Ocean allows a continuous circumpolar flow around Antarctica, driven by the prevailing winds (Fig. 5.1). Most of the southern ocean is covered by the zone of westerly winds in the southern hemisphere (the roaring forties and screaming fifties from the ages of sail). Polar easterlies blow along the Antarctic coast, driven by the permanent thermal high air pressure area over the continent. Since the Southern Ocean is not interrupted by a meridional continental barrier, the wind stress exerted by this atmospheric circulation forces a circumpolar current system with a strong eastward wind drift. This is the Antarctic Circumpolar Current (ACC) between 35 to 38°S and 60 to 65°S, depending on the sector of the Southern Ocean. The reversal of the wind direction to the polar easterlies generates an opposing narrow westward Peri-Antarctic Coastal Current. Where the latter encounters partial meridional barriers, e.g., the Antarctic Peninsula, a cyclonic circulation is formed south of the ACC: the Weddell Sea gyre and Ross Sea gyre. Since the mean vertical and horizontal density gradients in the ACC are relatively small, the geostrophic adjustment of the density field to the wind forcing is generally not able to reverse the direction of the ACC at greater depths. Therefore, the ACC is a mixed barotropic/baroclinic current and the NADW brought south into the ACC is transported eastward.

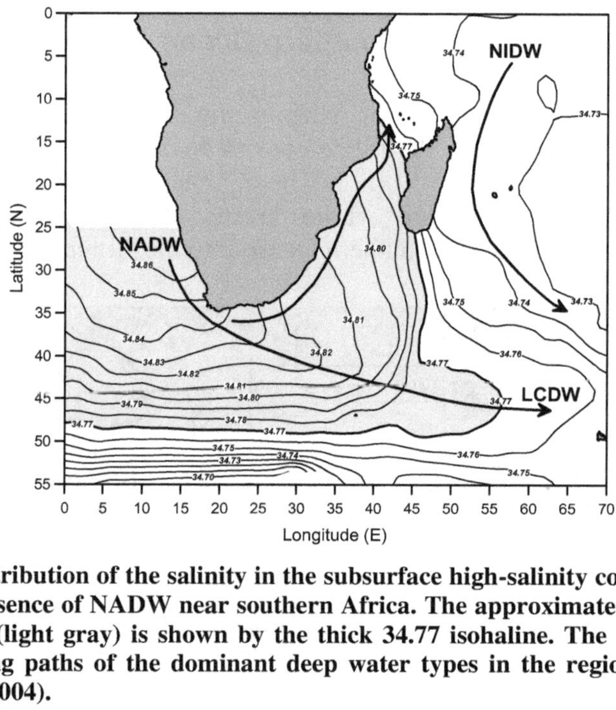

Fig. 5.2. Distribution of the salinity in the subsurface high-salinity core connected with the presence of NADW near southern Africa. The approximate boundary of the NADW (light gray) is shown by the thick 34.77 isohaline. The arrows show the spreading paths of the dominant deep water types in the region (after van Aken et al. 2004).

Directly south of South Africa more or less undiluted NADW from the South Atlantic flows anticlockwise along the continental slope around Africa toward the Mozambique Channel between Africa and Madagascar (Fig. 5.2). This flow is fed by the eastern branch of the NADW transport, discernable in the dissolved silicate minimum shown in Fig. 4.15. On its way around southern Africa this NADW core loses salt due to diapycnal mixing with the overlying and underlying fresher water types. Current measurement have shown that about 2×10^6 m^3/s (= 2 Sv) NADW directly enters the Indian Ocean following this route (van Aken et al. 2004). Most of the NADW that enters the Southern Ocean is incorporated into the ACC, where it is quickly transformed by mixing with the surrounding water mass. The resulting saline water type in the ACC is the LCDW, introduced in Section 4.7.

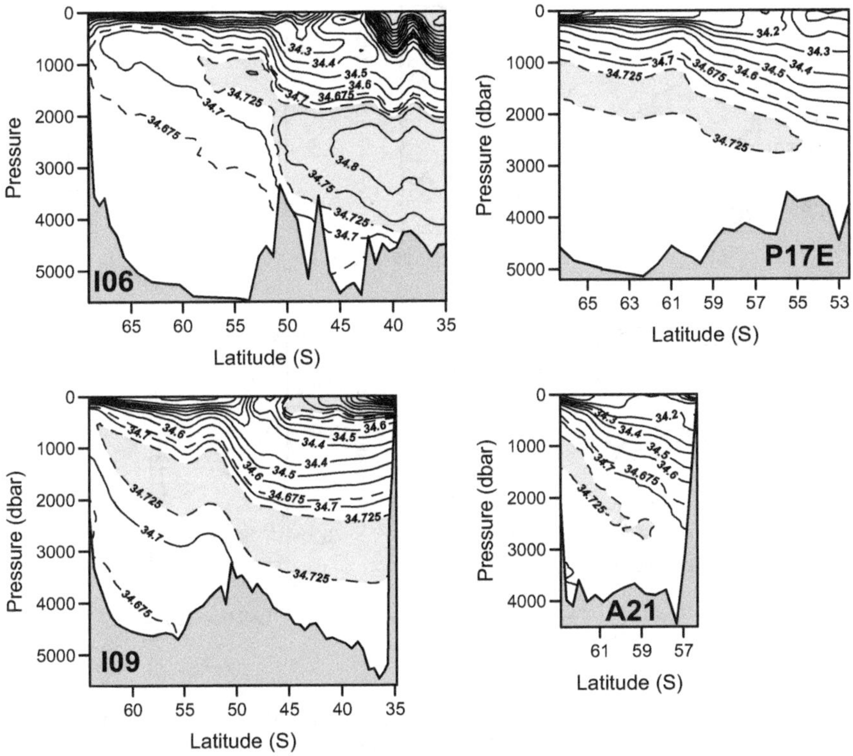

Fig. 5.3. Meridional sections of salinity (PSS-78) in the Southern Ocean. The location of these WOCE sections is shown in Fig. 5.1. The high-salinity NADW/LCDW core ($S > 34.725$) is highlighted in light gray.

The LCDW is one of the four main core layers that can be recognized from hydrographic sections across the ACC. As examples for the hydrographic structure of the Southern Ocean the WOCE sections I06, I09, P17E, and A21 have been chosen (Fig. 5.1). The first three sections cross, respectively, the Atlantic–Indian Basin, the Australian–Antarctic Basin, and the Pacific–Antarctic Basin, while the latter section transects the Drake Passage. The salinity distribution along these sections (Fig. 5.3) shows that everywhere in the ACC a subsurface high-salinity core ($S > 34.725$) is found, representing the LCDW. Following the sloping isopycnals in the ACC (see, e.g., Fig. 4.2) the high-salinity LCDW core ascends to the south where it reaches depths of less than 300 m in the center of the Weddell Sea gyre.

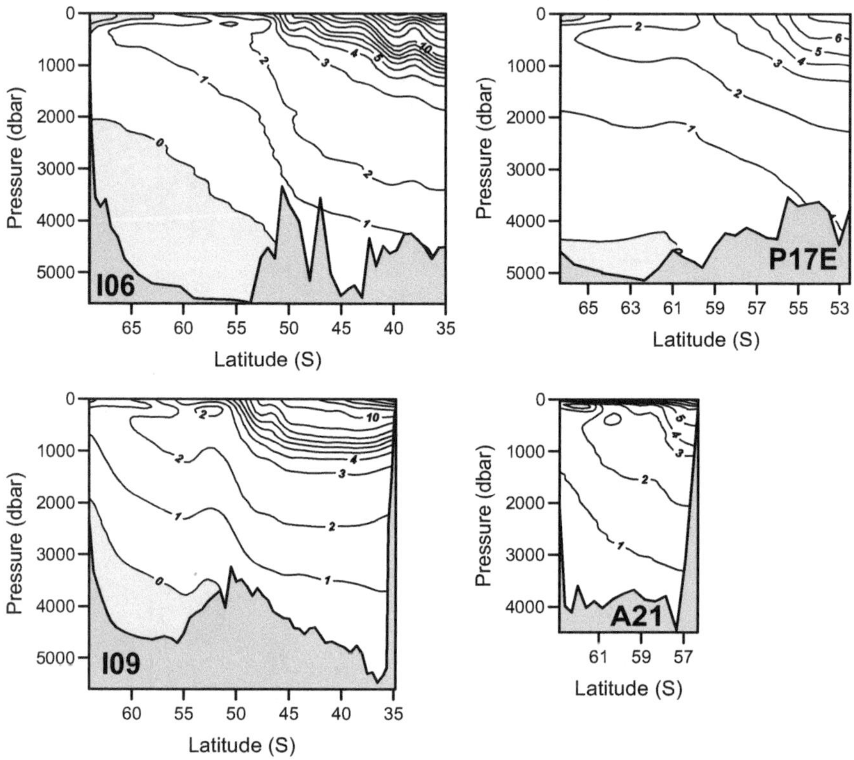

Fig. 5.4. Meridional sections of potential temperature (°C) in the Southern Ocean. Temperatures below 0°C are highlighted in light gray.

Over the Atlantic–Indian Basin (section I06) the transition to the deep-lying more saline NADW that flows around Africa coincides with the Sub-Antarctic Front near 52°S. In the southern half of the ACC the LCDW core is generally overlain by a fresher surface layer. From there a salinity minimum connected with the AAIW core descends to the north. The thickness and salinity of the LCDW core varies from basin to basin. A low-salinity core is encountered in the bottom layer of the Southern Ocean basins, increasing its salinity from the Weddell Sea to the east. This core is identified as the AABW which was shown in the previous chapter to flow along the bottom northward into the Atlantic Ocean.

The high-salinity core of LCDW in the ACC is found in the potential temperature range of 1 to 2°C (Fig. 5.4), definitely colder than the potential temperature of over 2°C where the NADW core is observed in the Atlantic

Ocean (Fig. 4.2) and directly south of Africa (section I06). Erosion of the LCDW layer by convective mixing, driven by surface heat loss in those parts of the ACC where the LCDW is very shallow, is assumed to play an important role. Such a process will bring the upper parts of the NADW/LCDW back to the ocean surface layer where they are modified by air–sea interaction. Only the colder parts of NADW survive as LCDW. In the southern parts of the ACC, where a cold surface layer is encountered, the surface layer is generally less saline than the underlying more saline LCDW, representing the effects of a precipitation excess and sea ice melting. Here the salinity stratification prevents deep convective mixing. The potential temperature in the low-salinity AABW, found in the deep Southern Ocean basins, is below zero (Fig. 5.4). However, this cold AABW does not pass through the shallower Drake Passage, where temperatures below zero are not encountered in the bottom layers.

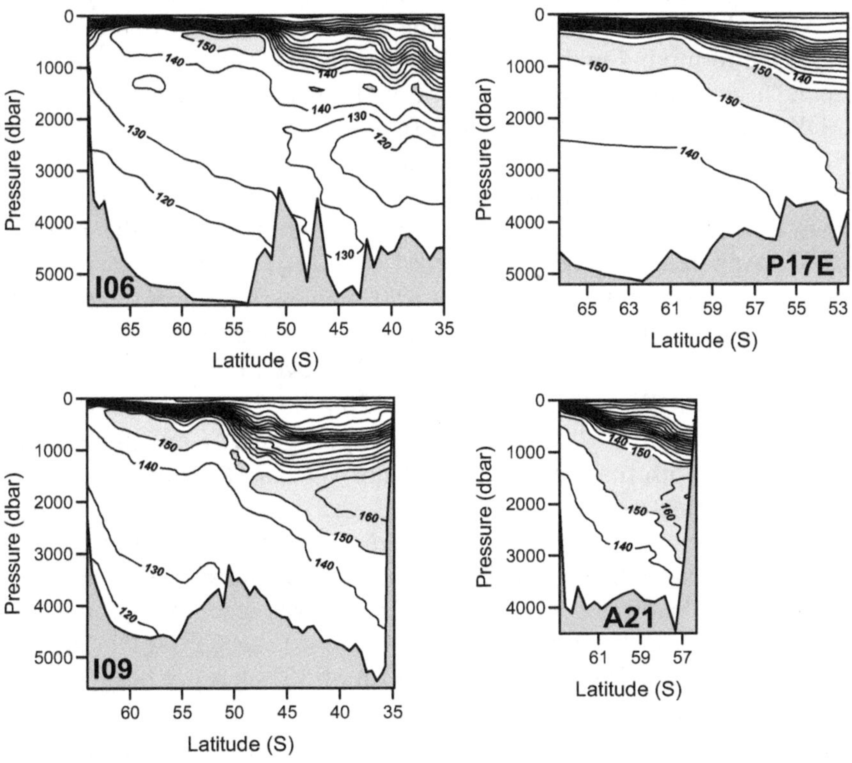

Fig. 5.5. Meridional sections of apparent oxygen utilization (μmol/kg) in the Southern Ocean. High *AOU* values (*AOU* > 150 μmol/kg) are highlighted in light gray.

The *AOU* distributions (Fig. 5.5) show another shallow core layer, characterized by high *AOU* values (> 150 µmol/kg). This layer is found between the high-salinity LCDW core and the overlying fresher surface water or subsurface AAIW core, the UCDW, introduced for the southern Atlantic Ocean in section 4.7. UCDW owes its low oxygen content (high *AOU*) to southward recirculation of aged NADW/LCDW which originally entered the Indian and Pacific oceans from the south (Reid 1981). While the water stayed in these oceans it aged and the *AOU* increased. Due to deep upwelling and diapycnal mixing potential density and salinity decreased slightly during the aging, which brought the recirculated UCDW above the high-salinity LCDW core. An example of such low-oxygen, high-*AOU* recirculating water is the North Indian Deep Water (NIDW) sketched in Fig. 5.2. The highest *AOU* values in the UCDW core in the ACC are found at the northern ends of sections I09 and A21, where recirculating deepwater from the Indian and Pacific Ocean flows eastward in the ACC along, respectively, the Australian and South American continents. Due to mixing with the low-oxygen UCDW the *AOU* in the LCDW core is high compared with the relatively well-oxygenated (low *AOU*) NADW core observed along WOCE sections A16 and I06 (Figs. 4.10 and 5.5). The cold low-salinity AABW in the deep basins of the Southern Ocean appears to be characterized by relatively low *AOU* values. This shows that the AABW is relatively young, compared to the overlying LCDW. Apparently AABW is formed regularly around Antarctica. The eastward increase of the near-bottom *AOU* values from the Atlantic–Indian Basin to the Pacific–Antarctic Basin suggests that most newly formed AABW originates from the Weddell Sea. But even in the Weddell Sea the *AOU* in the AABW has values (110 to 120 µmol/kg) well above those of NADW in the North Atlantic (< 50 µmol/kg, Fig. 4.10), indicating that the gas exchange with the atmosphere did not yet lead to equilibrium during the AABW formation.

The concentration of the anthropogenic transient tracer CFC-11 (Fig. 5.6) along the WOCE sections in the Southern Ocean shows traces of ventilation of the deepwater with cold water that was recently in contact with the atmosphere. In the upper 500 to 1000 dbar the CFC-11 is generally above 0.5 pmol/kg. The NADW core north of 45°S has not been ventilated recently, since its CFC-11 concentration is well below 0.05 pmol/kg, while in the whole water column of the Atlantic–Indian Basin the CFC-11 concentration is above 0.2 pmol/kg and indication of more "recent" ventilation in the Southern Ocean. Along the continental slope of Antarctica high CFC-11 concentrations, above 1.0 pmol/kg, are indicative of the formation of new AABW. Further east along the I09 section in the Australian–Antarctic Basin high CFC-11 values also are found along the Antarctic continental slope and in the deep layers of the basin. In the northern half of this section, where also the high *AOU* of the UCDW was observed, low CFC-11 values reflect the presence of old deep and bottom water leaving the Indian Ocean, flowing east in the ACC along the Australian continent. The UCDW in the Australian–Antarctic Basin further south has higher CFC-11 values (0.5 to 1.0 pmol/kg), indicative for the mixing of this water type with the recently ventilated overlying water. In the Pacific–Antarctic Basin the low CFC layer stretches further south to at least 66°S on section P17E.

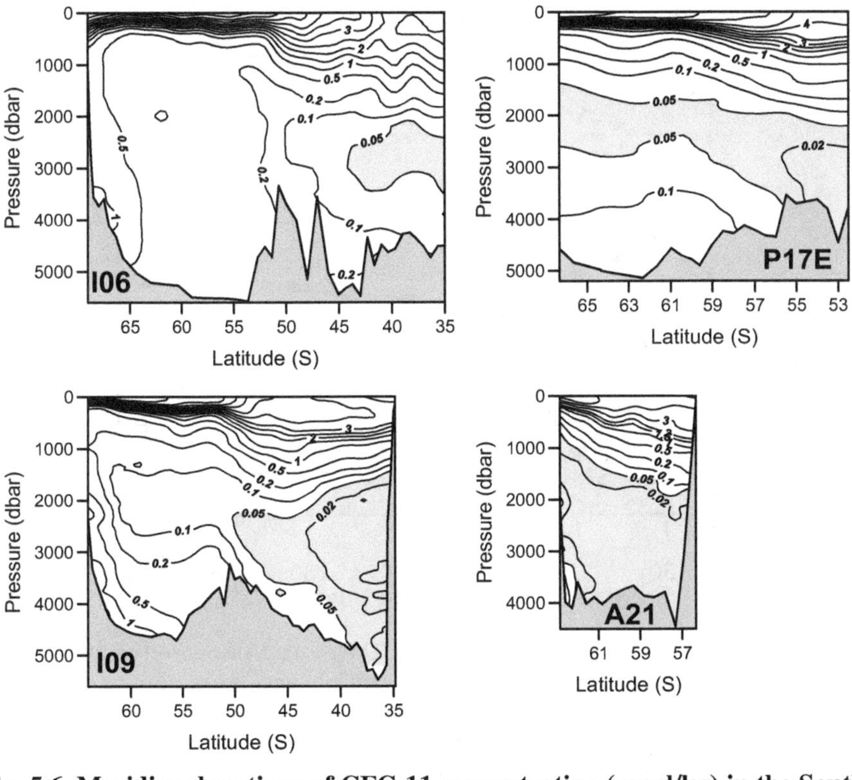

Fig. 5.6. Meridional sections of CFC-11 concentration (pmol/kg) in the Southern Ocean. Low CFC-11 values (CFC < 0.05 pmol/kg) are highlighted in light gray.

Section P17E did not reach close enough to the Antarctic continent to observe traces of the recent formation of AABW in the Ross Sea. In Drake Passage (section A21) nearly all the deep and bottom water below 1500 dbar had low CFC-11 concentrations (< 0.05 pmol/kg). Apparently most deep and bottom water here was not recently ventilated, probably originating from the recirculation of deep and bottom water from the Pacific Ocean. Only along the continental slope of the Antarctic Peninsula slightly elevated CFC-11 concentrations (> 0.1 pmol/kg) were observed along section A21.

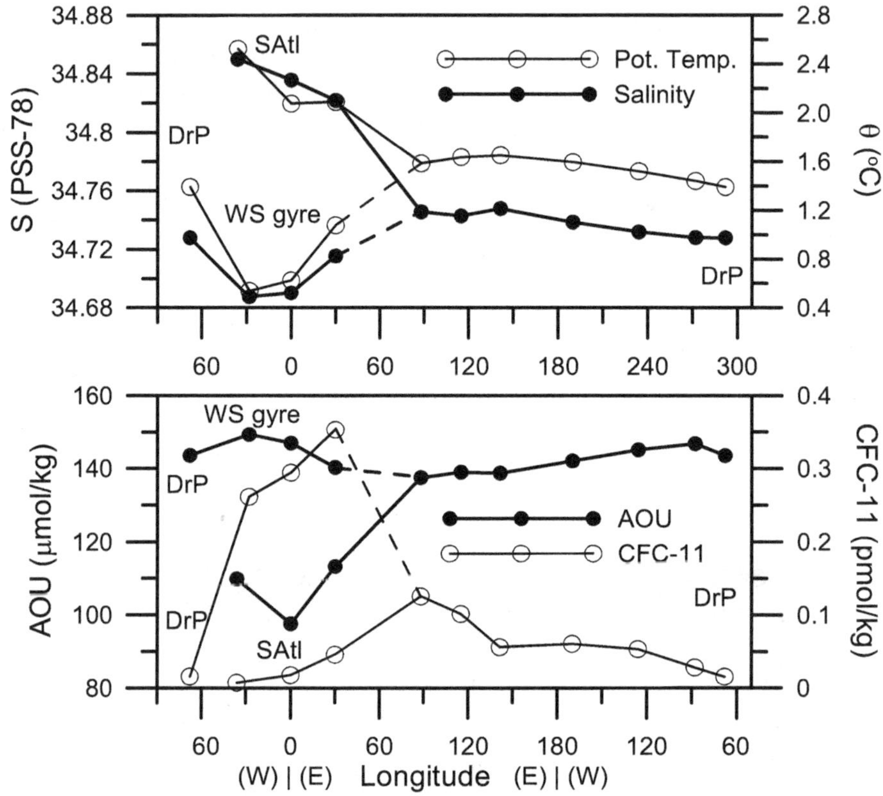

Fig. 5.7. Characteristics of the NADW/LCDW core in the Southern Ocean as function of the longitude, derived from the observations along the WOCE sections shown in Fig. 5.1. For the analysis the NADW/LCDW core has been defined as the subsurface salinity maximum. The data points for Drake Passage (DrP) have been repeated after 360°. In the sector of the Atlantic–Indian Basin data points have been given for the Weddell Sea gyre (WS gyre) as well as for the southern South Atlantic (SAtl).

The hydrography of the NADW/LCDW cores in the Southern Ocean can be summarized in the zonal distribution of its characteristic parameters (Fig. 5.7). The NADW in the South Atlantic Ocean is characterized by a relatively high salinity and potential temperature. In the Weddell Sea gyre, at the same longitudes, the coldest and least saline LCDW is encountered, even colder and fresher than the water entering from Drake Passage. This can be ascribed to either diapycnal mixing with the overlying UCDW from Drake Passage or lateral mixing with newly formed AABW, descending from the Antarctic continental slope. The strong increase of the CFC-11 concentration, a sign of recent ventilation, points to the latter process, while the relatively high *AOU* values in the Weddell Sea gyre show the importance of the former process.

Downstream from the Atlantic sector both water types merge in the ACC and a more homogeneous LCDW develops. Temperature and salinity slowly decrease toward Drake Passage, while *AOU* rises and the CFC-11 levels decrease. That can be ascribed to an increasing influence of cold and relatively fresh aged deepwater from the Indian and Pacific Ocean recirculating to the south, where they merge with the waters in the ACC.

With all the attention given here to the southern water types, characterized by typical cores in the distribution of hydrographic parameters, it is good to consider that in the deep Antarctic Basins huge amounts of water are present between the LCDW core and the AABW core at the bottom of the basins which build up the complete Southern Ocean water mass. The distribution of *AOU* (Fig. 5.5) shows that this water is less aged than most recirculating deepwater from the Indian and Pacific Ocean, while the slightly elevated CFC-11 concentrations (0.1 pmol/kg and higher) point to the influence of "recent" ventilation of this water mass. When discussing the formation rate of deep and bottom water in the Southern Ocean this large part of the Southern Ocean water mass should not be ignored!

5.2. The deep Indian Ocean

The hydrography of the deep and bottom water in the Indian Ocean is illustrated with observations along the meridional WOCE section I08 and the zonal section I05 (Fig. 5.8). The I08 section is a composite of three different surveys. It runs from an eastward extension of the Atlantic–Indian Basin, via the Australian–Antarctic Basin, across the Ninety East Ridge near 30°S through the Central Indian Basin toward the continental slope off Sri Lanka. The zonal I05 section between Africa and Australia was surveyed in 1987, before the proper WOCE period, by the British ship Charles Darwin.

The meridional distribution of salinity in the central Indian Ocean shows a subsurface maximum near 2500 dbar, descending from the Southern Ocean (Fig. 5.9, top panel). This maximum, observed at potential temperatures between 1.5 and 2.0°C, is connected with the LCDW encircling Antarctica in the ACC. This high-salinity core, extending from the LCDW core in the Southern Ocean, is occasionally referred to as Indian Deep Water (IDW). In the northern part of the section the subsurface salinity maximum meets the overlying high-salinity water originating from the Red Sea. That Red Sea Outflow Water (RSOW) is formed by the strong evaporation excess in the Red Sea, where it leads locally to salinities above 40. The IDW salinity maximum is possibly maintained in the northern Indian Ocean by some mixing with the overlying RSOW. Further south in the Indian Ocean a salinity minimum ($S < 34.7$), connected with the AAIW, is found above the IDW core. In the South Australian Basin (north of ~45°S) AABW with a salinity below 34.725 is found in the bottom layers of the Indian Ocean.

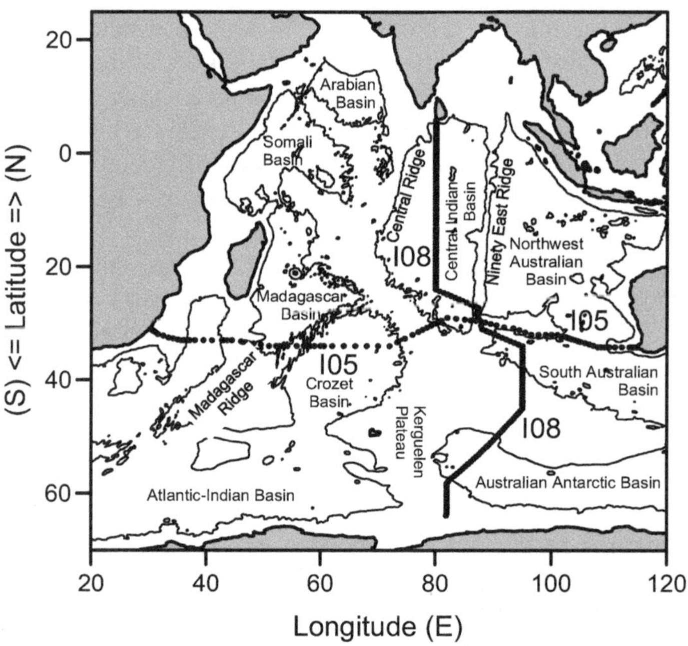

Fig. 5.8. The hydrographic stations along the sections of the WOCE Hydrographic Program in the Indian Ocean, used in this chapter. The depth contour shows the 4000 m isobath.

The potential temperature in the Indian Ocean (Fig. 5.9, second panel) shows cores of bottom water slightly below 1°C in the Central Indian Basin and South Australian Basin, warmer than the bottom water in the Australian–Antarctic Basin where subzero temperatures are observed. During its northward spreading the AABW has been warmed by entrainment of overlying water.

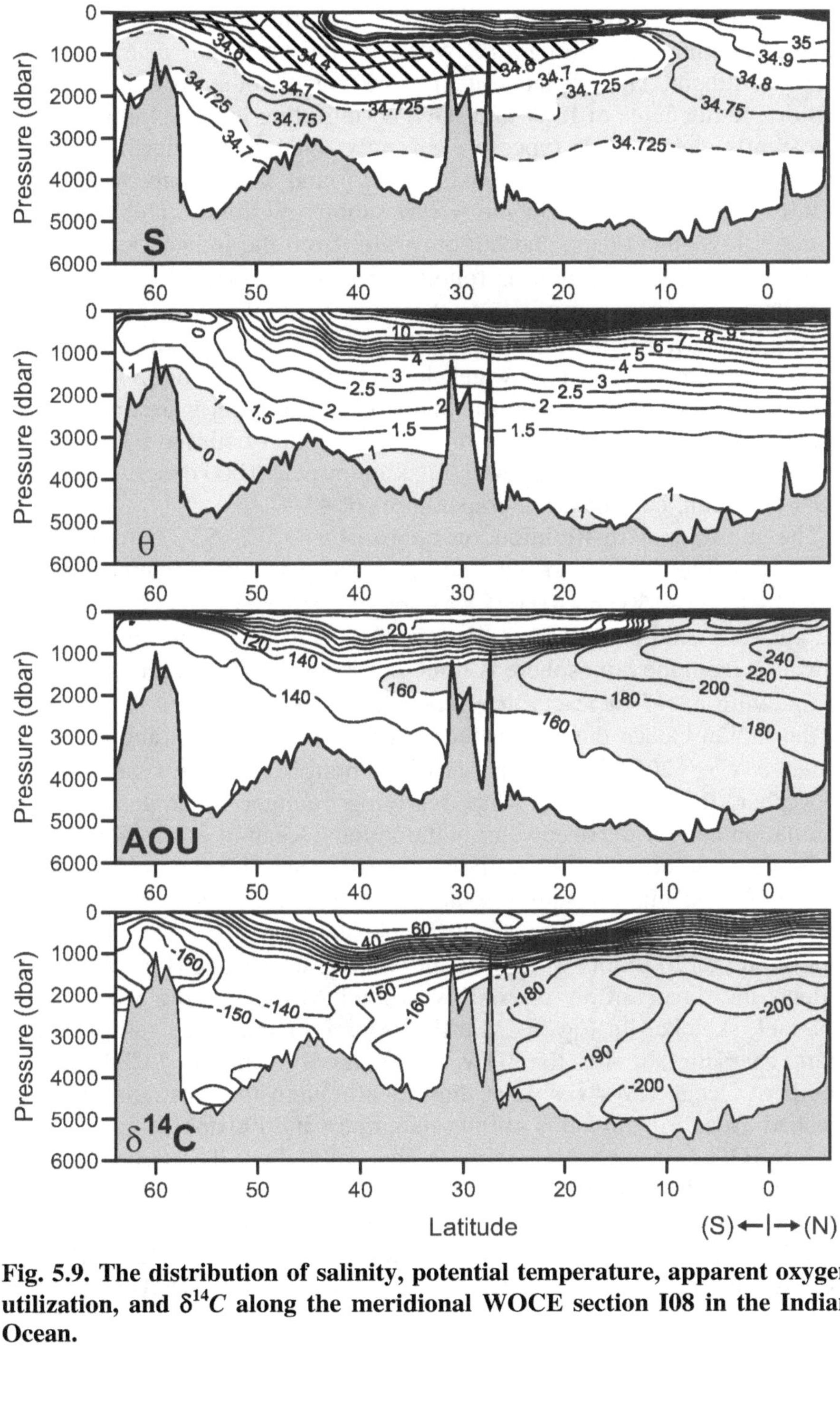

Fig. 5.9. The distribution of salinity, potential temperature, apparent oxygen utilization, and $\delta^{14}C$ along the meridional WOCE section I08 in the Indian Ocean.

The meridional *AOU* distribution (Fig. 5.9, third panel) shows a very strong maximum north of ~30°S with the highest values near the South Asian continent. The isopleths of *AOU* cross the isotherms and isohalines that define the cores of IDW and AABW, intruding into the Indian Ocean. Apparently these water types are strongly aged in the northern Indian Ocean. The core of the *AOU* maximum is found at levels above the IDW salinity maximum and below the AAIW salinity minimum. This is the core of the NIDW, aged deep and bottom water from the Indian Ocean. North of 20°S the *AOU* maximum is found at potential temperatures between 5 and 7°C, definitely warmer than the underlying IDW core. Apparently NIDW in the northern Indian Ocean is formed by aging of AABW and IDW as well as intermediate water, by mineralization of organic material. This oxygen-depleted water mass mixes with overlying warmer and more saline RSOW and the lower salinity AAIW which both have lower *AOU* levels, thereby generating an *AOU* maximum near 1000 dbar. At 30°S the *AOU* minimum has potential temperatures of ~3 °C.

The meridional distribution of radiocarbon (Fig. 5.9, bottom panel) shows that the AABW in the Southern Ocean has a characteristic $\delta^{14}C$ value between –150 and –160 ‰. Although that water mass is formed near the antarctic continent, the ventilation of the *TIC* in this water mass with new CO_2 from the atmosphere is limited, similar to the oxygen ventilation, so that with a –47 ‰ reservoir effect it has a ^{14}C age of nearly 950 years. In the Indian Ocean the deepwater ages further northward, and the oldest water ($\delta^{14}C \approx -205$ ‰) is found above the bottom layer between 3500 and 2500 dbar. Ignoring mixing with overlying younger water this implies a ventilation age for the deepwater in the Indian Ocean of at least 450 years.

On the southern side of the Indian Ocean, near 34°S along WOCE section I05P, the highest salinities at the levels of the deepwater (around 2800 dbar) are found in a subsurface *S* maximum (>34.80) and *AOU* minimum (< 120 µmol/kg) west of the Madagascar Ridge (Fig. 5.10). This reflects the direct inflow of NADW around Africa into the Mozambique Channel, sketched in Fig. 5.2. Further east a weaker deep salinity maximum characteristic for the IDW that extends from the LCDW in the Southern Ocean stays visible at these depths unto the Australian slope is reached. Near Australia this salinity maximum at 2800 dbar is found below a high-*AOU* core with its maximum near 1500 dbar. The two sources of high-salinity deepwater, NADW near Africa and RSOW in the northwestern Indian Ocean, do not show a strong influence further east, suggesting that the main flushing of the deepwaters occurs through the inflow of LCDW and AABW. The near-bottom salinity of the AABW increases from Africa (≈ 34.690) toward Australia (≈34.705), following the near-bottom salinity gradient in the Southern Ocean.

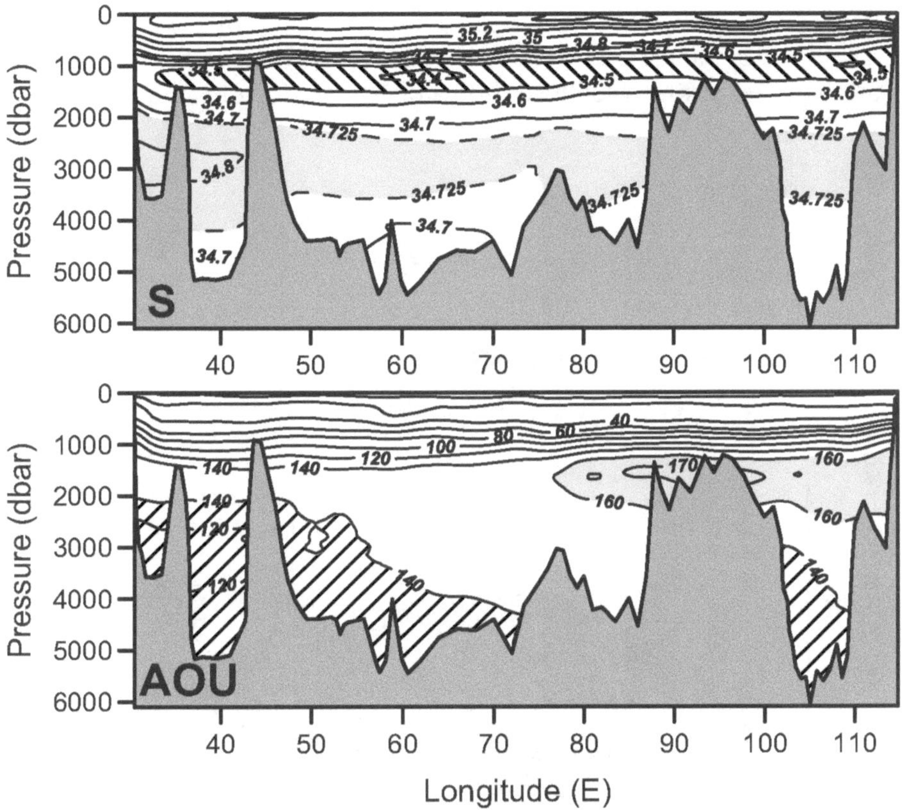

Fig. 5.10. Distribution of salinity and apparent oxygen utilization, along the zonal WOCE section I05P near 34°S between south Africa and Australia. Characteristic water mass cores are highlighted by shading and hatching.

At the levels of the deep and bottom waters an eastward increase of *AOU* is discerned both in the Crozet Basin and the South Australian Basin. This agrees with a northward inflow of LCDW on the western side of these deep basins and a clockwise recirculation of aged deep and bottom water toward the Southern Ocean in the eastern parts. In the eastern half of the I05P section high-*AOU* water ($AOU > 150$ μmol/kg) is present at ~1500 dbar, in between the low-salinity AAIW and high-salinity IDW. This agrees with a clockwise recirculation of the aged NIDW of the southern Indian Ocean. The large difference of the *AOU* near Australia compared with the northern Indian Ocean ($\Delta AOU > 80$ μmol/kg) shows that during this recirculation the NIDW from the northern Indian Ocean does not recirculate undiluted to the Southern Ocean. Probably most of the LCDW that entered the Indian Ocean from the south did not reach the northern hemisphere, but recirculated as IDW south of the equator. The highest *AOU* values are observed near the Ninety East Ridge (> 170 μmol/kg), suggesting a southward flow of NIDW along this ridge.

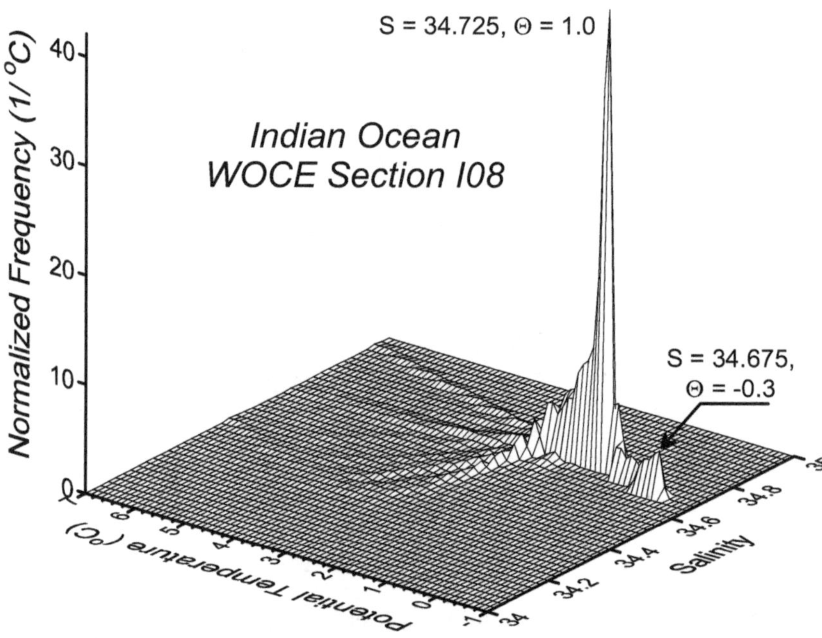

Fig. 5.11. Two-dimensional frequency distribution in θ-*S* space of the deepwater colder than 8°C along the meridional WOCE I08 section in the Indian Ocean, determined with a Δ*S*×Δθ = 0.025×0.1°C resolution. The integral of this frequency distribution over all possible salinities and temperatures equals 1.

The two-dimensional frequency distribution in θ–*S* space along the I08 section shows that the dominant mode of deepwater along this section in the Indian Ocean is found near θ = 1.0°C, *S* = 34.725 (Fig. 5.11). That water forms the transition between the IDW/LCDW and the AABW cores in the Indian Ocean. The frequency distribution shows a ridge along *S* = 34.725 toward the warmer θ–*S* characteristics of IDW. At lower temperatures a secondary mode of deepwater is found near $\theta = -0.3°C$, $S = 34.675$, the AABW in the Australian–Antarctic Basin.

5.3. The hydrography of the deep Pacific Ocean

The Pacific Ocean is the largest of the ocean basins. During the WOCE hydrographic program a series of meridional and zonal sections through this ocean was surveyed. Here we will use the meridional section P15 stretching from the Southern Ocean to the Aleutian island arc, the zonal section P03 in the subtropical North Pacific and the P06 section between Australia and Chile to illustrate the hydrography of the deep Pacific waters (Fig. 5.12). Additionally the P16N section in the Northeast Pacific Basin will show the structure of the radiocarbon distribution.

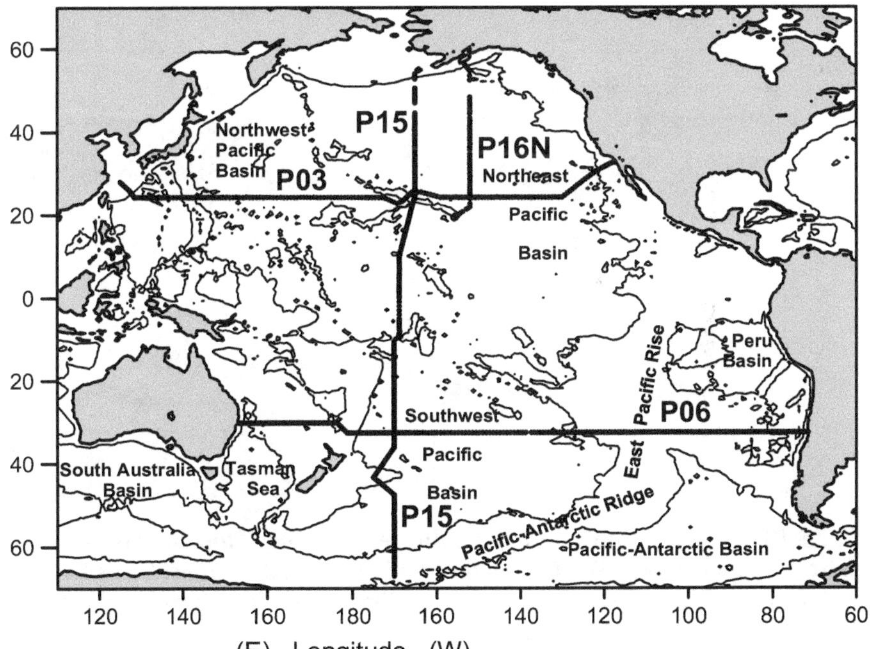

Fig. 5.12. The hydrographic stations along the sections of the WOCE Hydrographic Program from the Pacific Ocean, used in this chapter. The depth contour shows the 4000 m isobath.

WOCE hydrographic section P15 was surveyed in three legs by Canadian and American scientists on board the research vessels John P. Tully in 1994 and Discoverer in 1996. The distribution of the potential temperature along this line shows that cold water ($\theta < 1°C$) spreads in the bottom layer to at least 20°N north of the equator (Fig. 5.13, top panel). In this cold layer a salinity maximum ($S > 34.71$), emerging from the LCDW in the Southern Ocean, can be followed to ~10°S (Fig. 5.13, bottom panel). The underlying colder and fresher water is AABW. Due to diapycnal mixing with the fresher surrounding water on its way up north the LCDW salinity maximum core loses its identity.

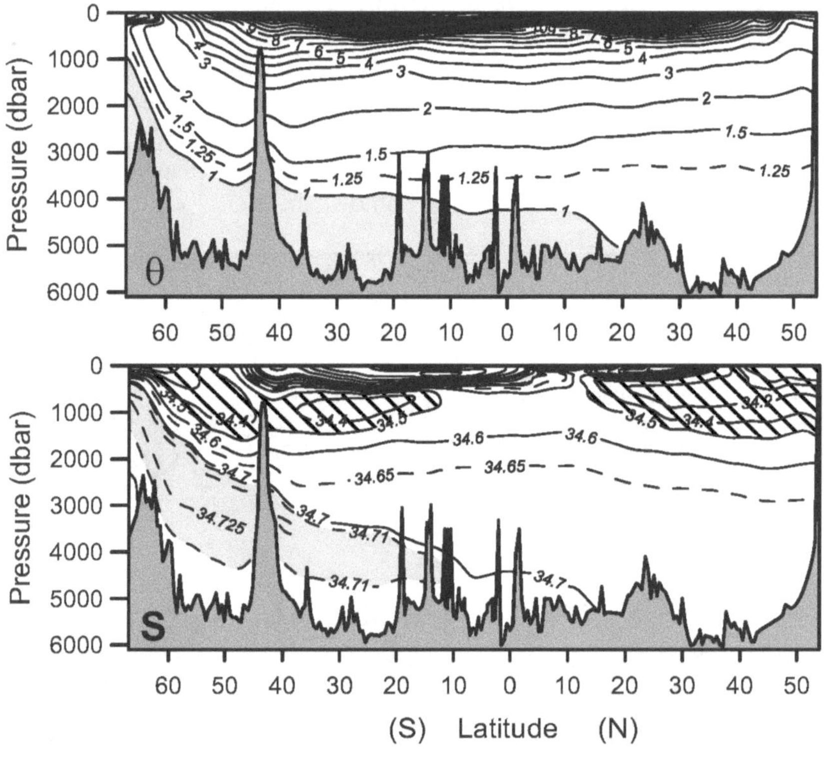

Fig. 5.13. The distribution of potential temperature (top panel) and salinity (bottom panel) along WOCE section P15. Potential temperatures below 1°C and salinities above 34.71 are shown in light gray. The low-salinity intermediate water with $S < 34.5$ is dashed.

In the northern hemisphere a nearly homogeneous water mass, the result of the mixing of AABW and LCDW, is formed [Pacific Deep Water (PDW)]. Deepwater formation does not occur in the northern Pacific Ocean. This is due to the very low surface salinity in the subarctic Pacific north of 45°N (less than 33.0). Even if that surface water is cooled to the freezing point its density will remain less than the density of the water at a depth of 200 m, preventing deep convection and water mass formation below that level. By subduction the relatively fresh ($S < 34.5$) Pacific Arctic Intermediate Water (PAIW), also known as North Pacific Intermediate Water, spreads equatorward below the main thermocline in the subtropical gyre circulation. From the Southern Ocean AAIW is transported northward.

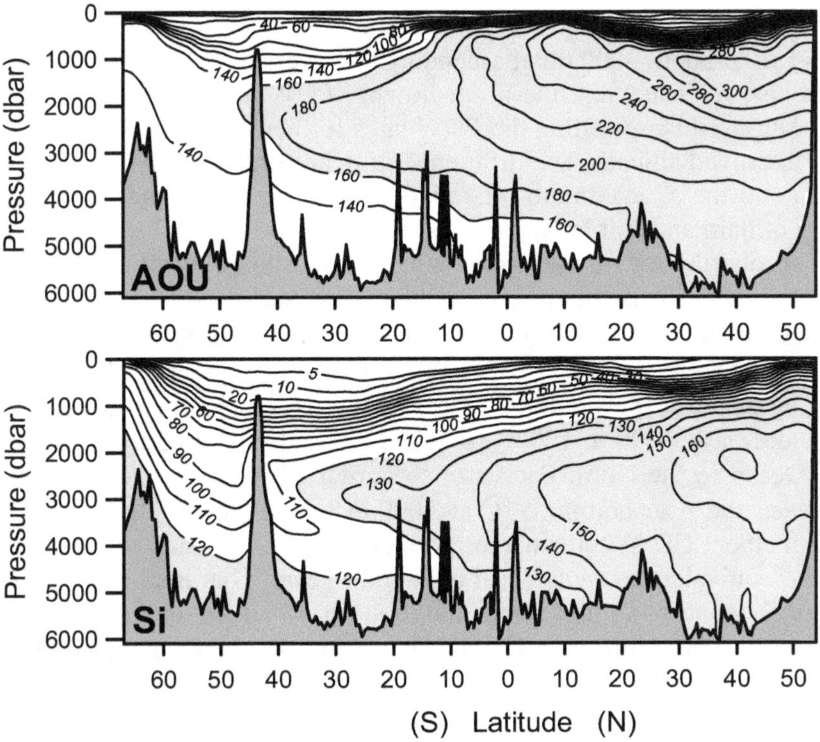

Fig. 5.14. The distribution of apparent oxygen utilization (top panel) and dissolved silicate (bottom panel) along WOCE section P15, both in μmol/kg. By the use of shading characteristic water mass cores are highlighted.

The inflow of relatively young AABW at ~40°S is shown by a near-bottom *AOU* minimum, while its antarctic origin is characterized by a near-bottom maximum in dissolved silicate (Fig. 5.14). The Atlantic origin of the deep high-salinity LCDW core is characterized by a coinciding deep *Si* minimum. At shallower levels an *AOU* maximum already is found south of 35°S at a pressure of about 2000 dbar. This core is connected with the *AOU* maximum of the UCDW core observed in section I09, south of Australia (Fig. 5.5). North of 10°S the meridional gradient in the high-*AOU* core increases suddenly, while the *AOU* increases to values above 300 μmol/kg at ~1000 to 1500 dbar, just below the salinity minimum of the PAIW core. That value corresponds to an oxygen concentration of only about 20 μmol/kg. The existence of such a relatively shallow *AOU* maximum below the PAIW in the northern Pacific Ocean suggests that this aged water mass is formed by the oldest AABW and LCDW, transformed by mixing to PDW, and then brought upward by deep upwelling. Between 35°S and 10°N the subsurface maximum of dissolved silicate is found at a pressure of about 3000 dbar, ascending to a pressure of 2200 dbar at 40°N. This also indicates that the north-flowing PDW ascends by deep upwelling. During this upwelling the PDW ages further, increasing both its *AOU* and dissolved silicate. The difference in depth of the northern *AOU* maximum and the *Si* maximum reflects the differences in biogeochemical cycling of hard and soft biogenic material.

The distribution of radiocarbon in the Northeast Pacific Ocean (Fig. 5.15) shows a maximum aging of the PDW near 45°N at a pressure of 2000 dbar ($\delta^{14}C \approx -250$ ‰). In the LCDW core in the Pacific sector of the Southern Ocean, south of 50°S, δ^{14}C had values of about –150 to –160 ‰. The aging from the Southern Ocean to the North Pacific, expressed in a $\delta^{14}C$ decrease of 90 to 100 ‰, requires a circulation time from the Southern Ocean to the North Pacific of the order of 950 years. The difference between the near-bottom $\delta^{14}C$ at 5000 m in the North Pacific (–220 ‰) and in the LCDW core in the Southern Ocean amounts to about 60 to 70 ‰, equivalent to an aging of about 650 years. That implies a mean meridional velocity in the deep and bottom water of the order of only 0.5 mm/s. The additional $\delta^{14}C$ decay during upwelling from the bottom layer to 2000 m requires an extra 300 years, equivalent to a deep upwelling velocity of the order of 10 m/year.

The two-dimensional frequency distribution in θ–S space of the deepwater along the P15 section shows two maxima, representative for dominant water mass modes (Fig. 5.16). They illustrate the transition of AABW and LCDW to PDW. The characteristic θ–S values for the highest peak in the frequency distribution refers to the main body of PDW in the North Pacific (1.1°C, 34.675). The AABW in the South Pacific and Pacific sector of the Southern Ocean is connected with the secondary peak (0.6°C, 34.70). The salinity maximum in the core of the LCDW represents only a small part of the volume, with a relatively large vertical temperature gradient, and therefore does not result in a separate peak in the frequency distribution.

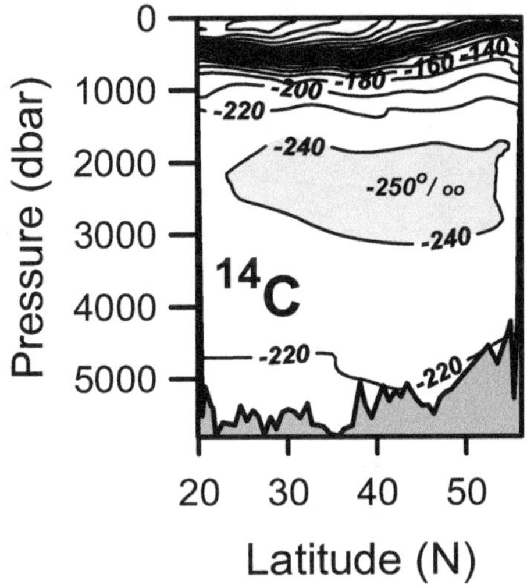

Fig. 5.15. The distribution of $\delta^{14}C$ (‰) along WOCE section P16N in the Northeast Pacific Basin.

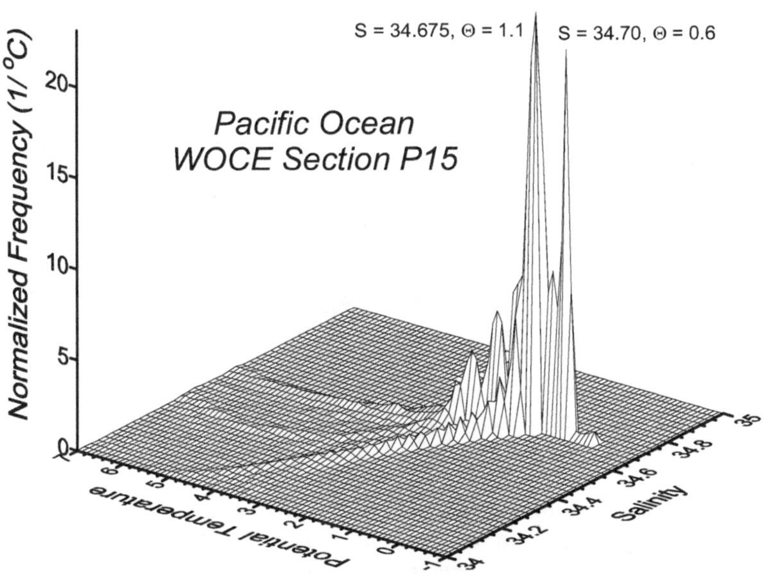

Fig. 5.16. Two-dimensional normalized frequency distribution in θ–S space of the deepwater colder than 8°C along the meridional WOCE P15 section in the Pacific Ocean, determined with a $\Delta S \times \Delta\theta = 0.025 \times 0.1$°C resolution.

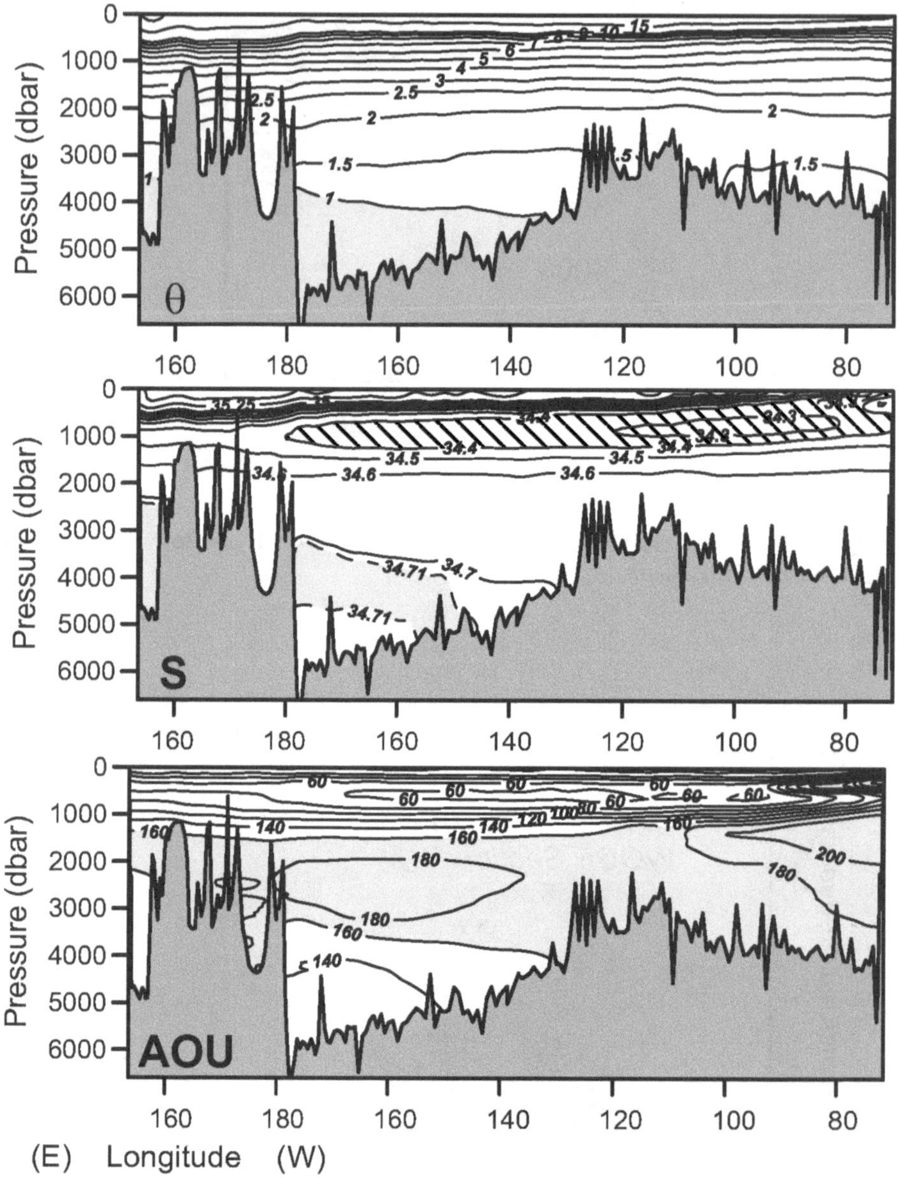

Fig. 5.17. Zonal distributions of the potential temperature (top panel), salinity (middle panel), and apparent oxygen utilization (bottom panel) along the South Pacific meridional WOCE section P06 near 35°S. Shading and dashing has been applied to highlight the main water mass cores.

The zonal distributions of hydrographic parameters along the WOCE sections P06 and P03 give additional information on the deep circulation in the Pacific Ocean. Along section P06 between Australia and Chile (Fig. 5.17) the relative high salinity and low *AOU* in the westernmost deep basin near Australia, the northern part of the Tasman Sea, reflects direct flushing of this deep basin from the Southern Ocean. However, this is not a main pathway for the ventilation of the deep Pacific, since this basin is isolated from the main deep Pacific basins by a ridge between Australia and New Zealand, with sill depths less than 3000 m. The main inflow path is through the Southwest Pacific Basin, where the temperature minimum below 5000 dbar, the salinity maximum near 4000 dbar and *AOU* maximum near 2500 dbar at lower salinities connected with, respectively, the cores of AABW, LCDW and UCDW, are found above each other. In the small Chile Basin east of the East Pacific Rise the AABW core is absent, since the ridge between this basin and the Southern Ocean is too shallow to allow a northward transport of this water type.

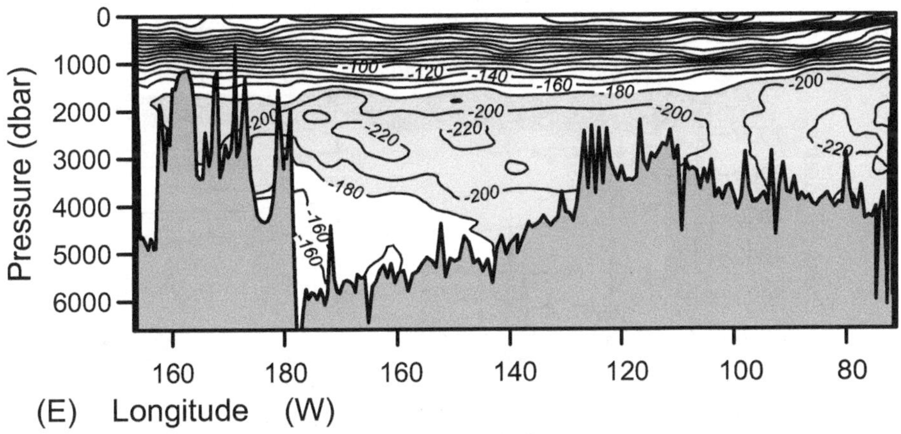

Fig. 5.18. Zonal distribution of $\delta^{14}C$ along WOCE section P06 near 35°S. Shading has been applied to highlight the most aged water mass with $\delta^{14}C < -180$ ‰.

Whereas the *AOU* maximum of the UCDW core (> 180 μmol/kg) in the Southwest Pacific Basin is found at a potential temperature of about 1.9°C, along the coast of Chile the *AOU* maximum (> 200 μmol/kg) is located at higher temperatures (~2.75°C) and lower densities. This suggests a southward outflow of South Pacific PDW in the eastern half of the ocean, after it has been transformed by upwelling, aging, and mixing with overlying warmer water. However, the UCDW which leaves the Pacific sector of the Southern Ocean through Drake Passage (section A23 in Fig. 5.5) has lower *AOU* values (~170 μmol/kg), showing that it is diluted in the South Pacific by mixing with the overlying AAIW. The salinity minimum, connected with the core of AAIW (dashed region in Fig. 5.17), is observed above the *AOU* maximum. The $S = 34.4$ isohaline outcrops in the eastern South Pacific, where surface salinities below 34.2 are observed near the Chilean coast. This suggests that the source of this water type is located near South America and that AAIW subducts from there equatorward below the main thermocline of the subtropical gyre.

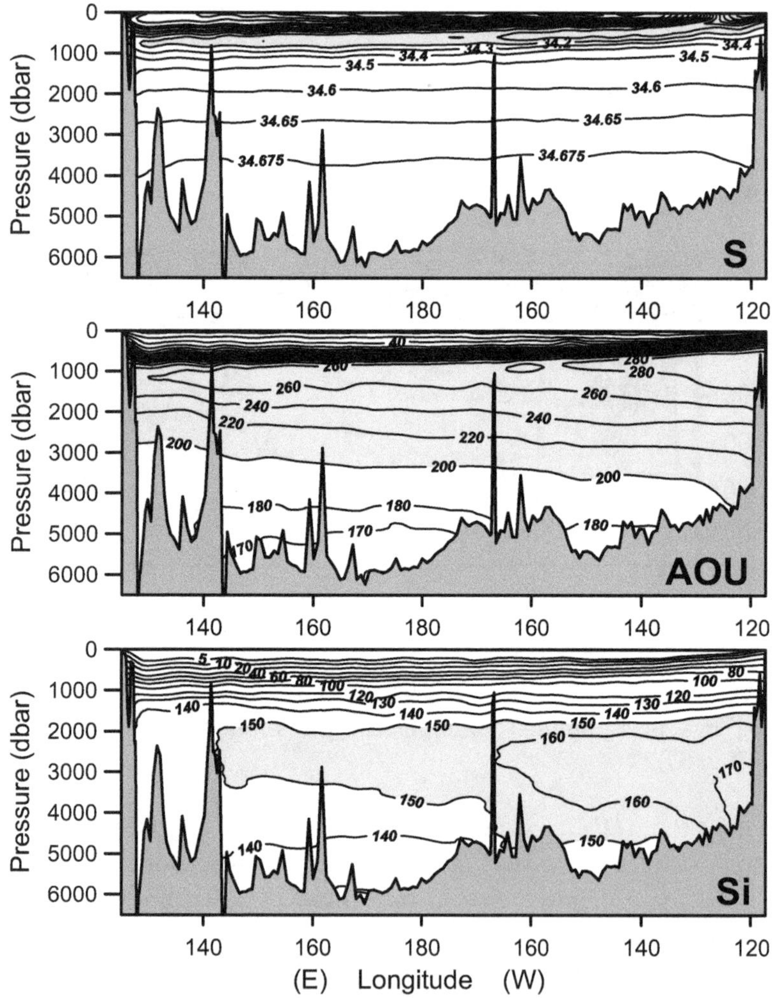

Fig. 5.19. Zonal distribution along WOCE section P03 near 25°N of the salinity (top panel), apparent oxygen utilization (middle panel), and dissolved silicate (bottom panel). Shading has been applied to highlight the main water mass cores.

The distribution of radiocarbon along the WOCE P06 section (Fig. 5.18) confirms that the layer with high *AOU* values indeed is aged during advection and deep upwelling, since the layer with $AOU > 180$ μmol/kg more or less coincides with the layer with $\delta^{14}C < -180$ ‰. The radiocarbon age difference between the inflowing AABW ($\delta^{14}C \approx -160$ ‰) and the core near 2500 dbar ($\delta^{14}C \sim -220$ ‰) is of the order of 600 years.

Along the P03 section in the North Pacific Ocean (Fig. 5.19) a deep salinity maximum is not observed at all, characteristic for the complete transformation of LCDW and AABW to PDW. Above the 1000 dbar isobar a subsurface salinity minimum is found, connected with the PAIW core. Similar to the South Pacific, the isohalines outcrop near the American continent. From there the PAIW core appears to subduct equatorward below the main thermocline of the North Pacific subtropical gyre. The highest *AOU* values (> 280 μmol/kg) are found near the North American continent directly below the PAIW core. Similar to the NIDW in the northern Indian Ocean, this aged PDW does not appear to recirculate undiluted to the southern hemisphere, but is probably transformed by mixing with the overlying relatively fresh high-oxygen PAIW. The zonal distribution of dissolved silicate also has its highest values near the American continent, probably due to the sediment water exchange with the aging PDW.

5.4. Deep upwelling

The tracer distributions, presented above, indicate that in the different ocean basins upwelling occurs. The Southern Ocean is a source of cold AABW and low-salinity AAIW which both spread northward into the Atlantic, Indian, and Pacific oceans. As will be discussed in a later chapter, the western winds over the Southern Ocean also drive a northward wind-driven near-surface flow. Mass conservation then requires a southward flow to feed this northward outflow from the Southern Ocean. The NADW from the North Atlantic supplies water to the shallower LCDW layer, while the overlying UCDW layer in the Southern Ocean receives aged water from the Indian and Pacific oceans. These are the only available sources to feed the northward outflow from the Southern Ocean. These water types appear to ascend poleward in the Southern Ocean by deep upwelling. In the near-surface layers apparently part of this upwelled water is transformed by air–sea–ice interaction. During the formation of AABW near the continental margin of Antarctica remnants of the freshened upwelling deepwater are cooled to temperatures near the freezing point before they descend again into the bottom waters of the world oceans. In this process most of the aged UCDW and the warmer and more saline part of the original NADW are transformed, leaving behind the cooler and fresher salinity maximum of the LCDW core. Apart from the direct inflow of NADW into the Indian Ocean around southern Africa the LCDW forms the deepwater which, with the underlying AABW enter the deep basins of the Indian and Pacific oceans.

In the Indian and Pacific oceans aging of the deepwater takes place. The distributions of *AOU*, dissolved silicate, and radiocarbon show that in these basins the deep and bottom waters age. The oldest water (highest *AOU* and silicate, lowest $\delta^{14}C$) are found at temperature and density levels well above those of the northward spreading deep and bottom water. This agrees with deep upwelling in these basins. By turbulent mixing with the overlying water the upwelling water is transformed, losing its characteristic properties. Ultimately the upwelled water may reach the surface water, where it will adapt to the atmospheric forcing during the return flow to the deepwater formation areas. However, not all the upwelling deep and bottom water reaches the surface layers in the Indian and Pacific oceans, since aged IDW and PDW appear to reenter the Southern Ocean at sub-surface levels above the LCDW core, where they contribute to the maintenance of the UCDW core.

6. The upper branch of the THC

6.1. Interocean exchange

The major features of the spreading of deep and bottom water from their source regions can be derived from the distributions of oceanographic tracers, as was illustrated in the previous two chapters. Mass conservation constraints require a return flow at shallower depths to bring the deep and bottom water back to their formation regions. Deep upwelling brings these cold water masses from abyssal depths in the ocean basins to shallower levels, either in the Southern Ocean or in the subtropical and tropical ocean basins and the subarctic, including the Atlantic Ocean. The temperature and salinity will adapt to the shallow boundary conditions by diapycnal mixing with the overlying waters, determined by atmosphere–ocean interaction, and the density of the upwelling waters will decrease. The upwelling in the Southern Ocean brings water directly into the ACC. The ACC enables zonal exchanges of this water between the different basins in the Southern Ocean where AABW and AAIW is formed. In Chapter 5 it was already shown that aged deepwater from the Indian and Pacific oceans is exchanged zonally between the different oceans by the ACC (UCDW). Deepwater that enters the thermocline in the Indian and Pacific oceans by upwelling has to follow another route to reach the source regions of NADW in the North Atlantic Ocean (Gordon 1986). Different pathways have been proposed for the interocean exchange of thermocline water (Gordon 2001). *First:* through the Bering Strait and the Indonesian seas North Pacific upper waters are transported to the Atlantic and Indian oceans. *Second:* the so-called Agulhas leakage brings Indian Ocean thermocline water and Subtropical Mode Water to the South Atlantic Ocean via large eddies or rings, shed from the Agulhas Current. From there it proceeds northward to the source regions of the NADW. *Third:* via subduction AAIW is brought into the south Atlantic gyre. From there this subantarctic water mass spreads into the Atlantic Ocean, crosses the equator northward; and ultimately reaches the northern North Atlantic. Nearly all of the return flow takes place in the wind-driven upper 1500 m of the ocean, linking this part of the THC directly to the climatic wind forcing. It can be expected that a considerable recirculation of the return flow will take place in the major wind-driven current systems like the subtropical gyres and the ACC, before the return flow reaches the formation areas of deep and bottom water.

6.2. The Bering Strait through-flow

Near 66°N, 169°W the Bering Strait forms an open connection between the northern part of the Pacific Ocean (Bering Sea) and the neighboring part of the Polar Ocean (Chukchi Sea). The strait is 80 km wide but very shallow, with a sill depth of about 45 m. Through this strait relatively freshwater from the upper layers of the Pacific Ocean with salinities between 30 and 33, including Alaskan coastal water, enters the Polar Ocean (Coachman and Aagaard 1966). This water can flow to the Atlantic Ocean via the Fram Strait between Greenland and Spitsbergen, or via the Canadian Archipelago.

Already in the 18th century the explorers Vitus Bering and James Cook observed from their ship's drift a northward flow in Bering Strait. Coachman and Aagaard (1966) analyzed existing and newly obtained current observations to determine the volume transport through the strait. The mean transport through the Bering Strait indeed was directed toward the Polar Ocean. The total volume transport for 5 to 7 August 1964 was estimated to be 1.4 Sv. Later transport observations with moored current meters, deployed for 10 months, showed a pronounced seasonal cycle around an annual mean of 0.8±0.2 Sv (Coachman and Aagaard 1981). The transport through the Bering Strait appeared to be highly correlated with the wind component in a direction of 12° true north. Interannual variability of the wind climate over the Bering Strait was assumed to be responsible for a considerable interannual variation in the volume exchange between the North Pacific and North Atlantic oceans. From meteorological observations Coachman and Aagaard (1988) reconstructed a 40-year time series of the annual mean Bering Sea transport from 1946 to 1985. In this period the volume transport varied between 0.55 and 1.0 Sv to the north and never became negative.

One may compare the effect of the Bering Strait on the freshwater budget of the polar and arctic seas with that of the main rivers that discharge freshwater to the northern ocean. The largest three of these rivers are the Yenissei, Lena, and Ob, which together have a total mean runoff of

46,600 m^3/s (Baumgartner and Reichel 1975). With reference to the typical salinity of the deep Greenland Sea (34.9) an inflow of 0.8 Sv seawater with a mean salinity of 32.5 represents an extra freshwater input of $(1-32.5/34.9) \times 800{,}000 = 55{,}000\ m^3/s$. So the Bering Strait through-flow is of similar importance as the river runoff for the surface salinity in the arctic seas where the source waters for the NADW are formed by deep convection. During glacials, when the sea level was about 100 m lower than at present, the freshwater source of the Bering Strait through-flow was completely blocked, inhibiting all possible feedbacks related to this return route.

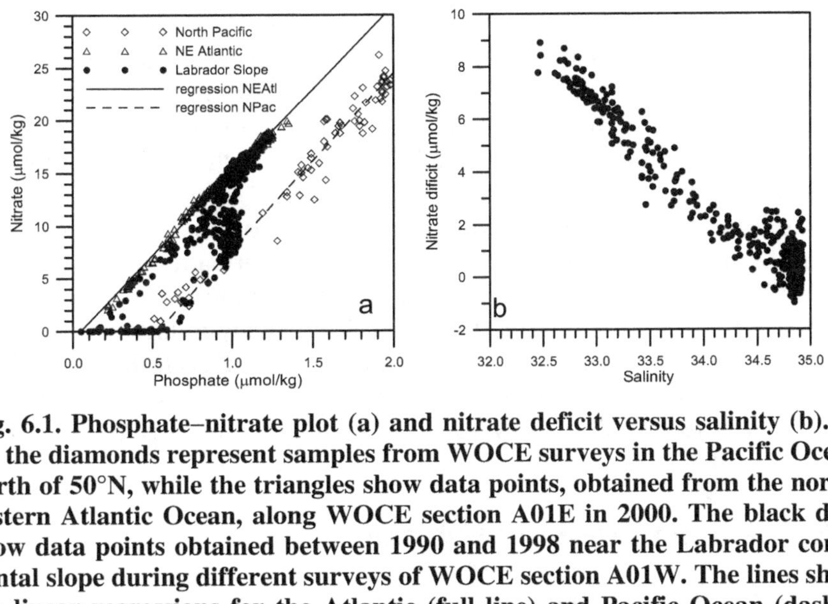

Fig. 6.1. Phosphate–nitrate plot (a) and nitrate deficit versus salinity (b). In (a) the diamonds represent samples from WOCE surveys in the Pacific Ocean north of 50°N, while the triangles show data points, obtained from the northeastern Atlantic Ocean, along WOCE section A01E in 2000. The black dots show data points obtained between 1990 and 1998 near the Labrador continental slope during different surveys of WOCE section A01W. The lines show the linear regressions for the Atlantic (full line) and Pacific Ocean (dashed line). In (b) water samples from the upper 30 m, where local primary production had exhausted the available nitrate [the near-zero dots in (a)] were removed.

It is difficult to distinguish Pacific water that entered the Atlantic via the Bering Strait from the local Atlantic low-salinity water, based on tracer properties. However, Jones et al. (2003) introduced a method using nutrient–nutrient relations that we will follow here. While the slopes for the phosphate–nitrate plots (Redfield ratios) are approximately identical for water samples from the North Pacific and North Atlantic oceans (triangles and diamonds in Fig. 6.1a) the Pacific waters have a much lower (~7 μmol/kg) nitrate content, compared with the waters from the northeastern Atlantic. This is probably due to denitrification by bacteria in the North Pacific, a process where bacteria reduce nitrate to molecular nitrogen. Since 1990 the WOCE hydrographic A01W section in the Labrador Sea was surveyed annually by Canadian and German ships. Over the continental slope of Labrador relatively fresh near-surface water was found during each cruise ($S < 34.0$). The phosphate–nitrate relation in this water (black dots in Fig. 6.1a) appeared to be a mixture of both the Atlantic and Pacific types, with many samples close to either the Atlantic or the Pacific regression line, showing that part of the fresher surface water had its origin in the Pacific Ocean.

We can take the nitrate deficit relative to the linear regression for the Atlantic Ocean as a measure for the amount of North Pacific water in the Labrador Sea. When data points from the upper 30 m that are clearly influenced by primary production are removed, this parameter shows a strong linear correlation with the salinity (Fig. 6.1b). The highest nitrate deficits correspond with the lowest salinities. This supports the hypothesis that the relatively fresh slope water in the Labrador Sea is transported from the Bering Strait through the Canadian Archipelago toward the Labrador Sea, where its presence potentially can influence the formation of LSW. Jones et al. (2003) carried out this analysis also for a number of hydrographic sections in the Fram Strait and Denmark Strait. They found that North Pacific water from the Bering Strait could be observed at all these locations. Apparently the Bering Strait through-flow contributes in measurable amounts to the return flow toward arctic source regions of NADW.

6.3. The Indonesian through-flow

Gordon (1986) assumed that the main return flow toward the formation areas of NADW is accomplished primarily in the main thermocline of the ocean. In this flow scheme, thermocline water from the tropical Pacific Ocean passes southward through the Indonesian seas and from there across the Indian Ocean (Fig. 6.2).

The main through-flow occurs via the Makassar Strait (MakS) between the islands of Kalimantan and Sulawesi (Borneo and Celebes). In the Indian Ocean the South Equatorial Current (SEC) brings this Indonesian Trough-Flow Water (ITFW) westward toward Africa, where it is fed into the East Madagascar Current (EMC) and Mozambique Current (MC) which are main sources for the Agulhas Current (AgC). Song et al. (2004) have suggested that a large part of the ITFW recirculates for about 50 years in the Indian Ocean before it reaches the Agulhas Current. Recirculation occurs in the northern Indian Ocean via the Somali Current (SC) and Equatorial Counter Current (ECC) while part of the recirculation even reaches the Arabian Sea. In the southern hemisphere recirculation occurs in the subtropical gyre via the Agulhas Retroflection (AgRf) and South Indian Ocean Current (SIOC). Occasionally a large eddy with Indian Ocean water is shed from the Agulhas Retroflection and drifts into the Atlantic Ocean (MARE). The WOCE hydrographic sections, depicted in Fig. 6.2, are used here to describe the transport and modification of ITFW.

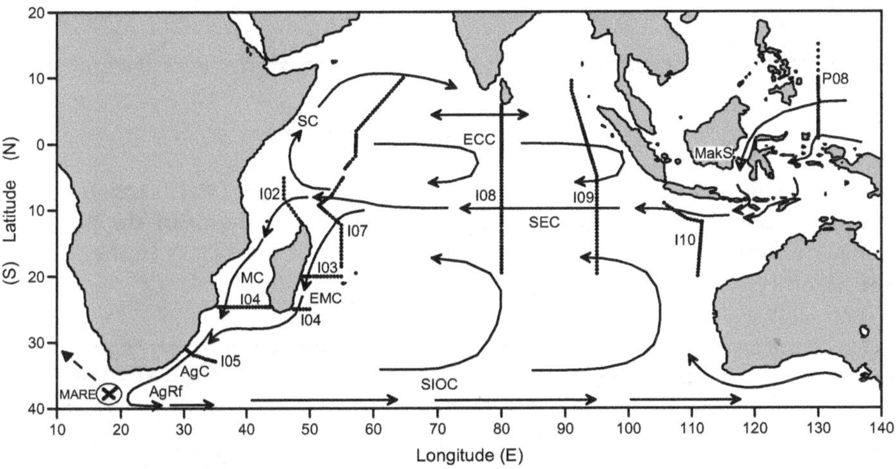

Fig. 6.2. Circulation scheme of tropical Pacific thermocline water via the Indonesian through-flow and the Indian Ocean toward the Atlantic Ocean. The dots show hydrographic stations along WOCE sections and additional programs used in this chapter. The acronyms are referred to in the text.

Current measurements have shown that about 9 Sv relatively freshwater ($S \approx 34.5$) passes through the Makassar Strait (Gordon et al. 1999). In the upper 100 m relatively fresh shelf water from the Java Sea, brought there by the northwest monsoon, blocks the through-flow of warm surface water in the upper 150 m for a part of the year, thereby lowering the mean temperature of the through-flow (Gordon et al. 2003). The origin of the ITFW that passes through Makassar Strait has to be sought in the relatively fresh core in the tropical Pacific near 7 to 12°N (Fig. 6.3). This water is drawn mainly from the North Pacific thermocline and intermediate water and is brought westward in the North Equatorial Current. The more saline water found further south, close to New Guinea, originates from the South Pacific and enters the Indonesian seas in small amounts through gaps east of Sulawesi. The transport weighted mean temperature of the southward flow through the Makassar Strait is slightly below 15°W (Vranes et al. 2002), a temperature found in the thermocline at a pressure of about 200 dbar.

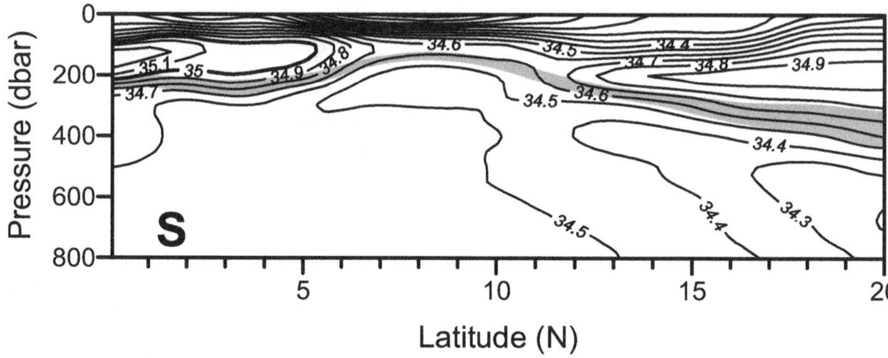

Fig. 6.3. The meridional distribution of salinity along WOCE section P08 in the western Pacific Ocean. The gray band shows the depth of the potential density anomaly interval coinciding with the core of the ITFW in the Makassar Strait (γ_θ = 25.35 to 25.85 kg/m^3).

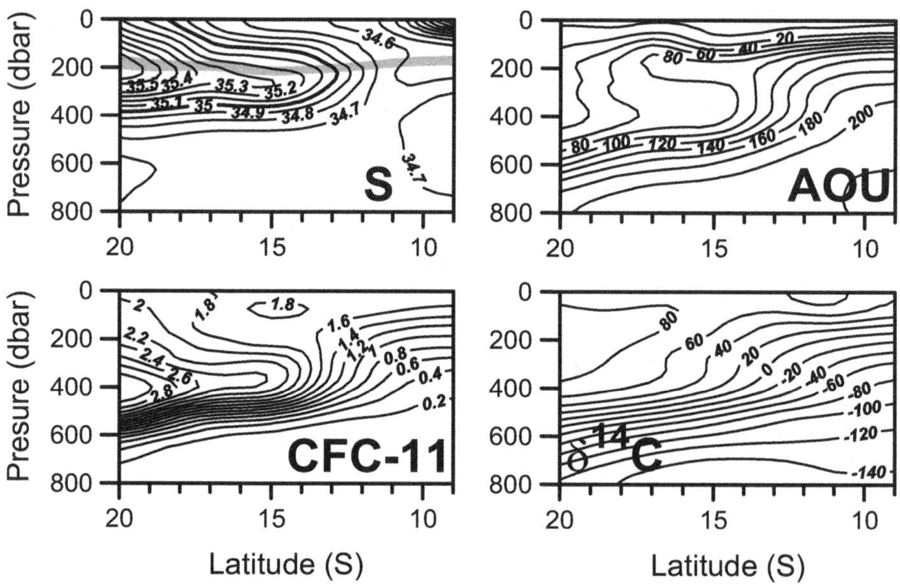

Fig. 6.4. The meridional distribution of salinity *(S)*, *AOU*, CFC-11, and radiocarbon ($\delta^{14}C$) along the WOCE section I10 south of Java. The gray band in the salinity section shows the depth of the ITF core density.

The potential density at this level is $\gamma(\theta,S)$ = 25.6 kg/m^3. Here we will consider a density interval of 0.5 kg/m^3 around that density (25.35 to

25.85 kg/m^3) as the core of the ITFW, referred to as the ITF core density (the gray band the upper left panel of Fig. 6.3). In the Indonesian seas the salinity of the ITFW is strongly homogenized by turbulent mixing.

In the Indian Ocean south of Java near 10°S the westward flowing ITFW that left the Indonesian seas can be followed as a relatively fresh core in the upper 800 m of the ocean. This water mass is well aged compared to the more saline water in the subtropical gyre further south, as is shown by the high *AOU* values, low CFC-11 concentrations, and low $\delta^{14}C$ values (Fig. 6.4). The ITFW is bound to the south by the presence of relatively young and saline subducted Subtropical Mode Water (STMW).

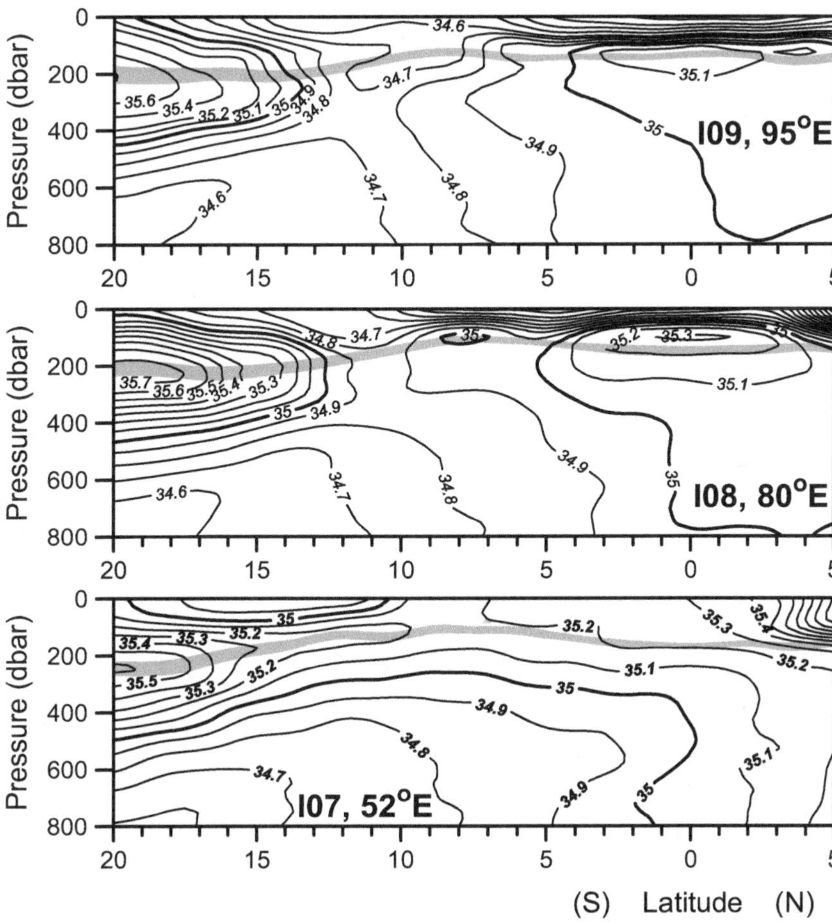

Fig. 6.5. The vertical distribution of salinity in the upper 800 dbar along three zonal WOCE sections in the Indian Ocean, I09 at ~95°E, I08 at 80°E, and I07 at ~55°E. The gray band represents the ITF core density.

Further west the ITFW can be followed as a zonal salinity minimum in the upper 400 dbar along the SEC near 10°S (Fig. 6.5). This ITFW core is located in between subducted saline STMW further south and the saline water of the Equatorial Undercurrent further north. The salinity in the ITFW appears to increase from Java westward in the direction of Madagascar. Along WOCE section I07, close to Madagascar the low-salinity ITFW core between 3 and 10°S is positioned north of the northern tip of Madagascar, and can be observed at section I02 in the Somali Basin. The part of the SEC feeding the EMC further south probably contains more subducted STMW and less ITFW, although the EMC is still slightly less saline than the water in the subtropical gyre further east.

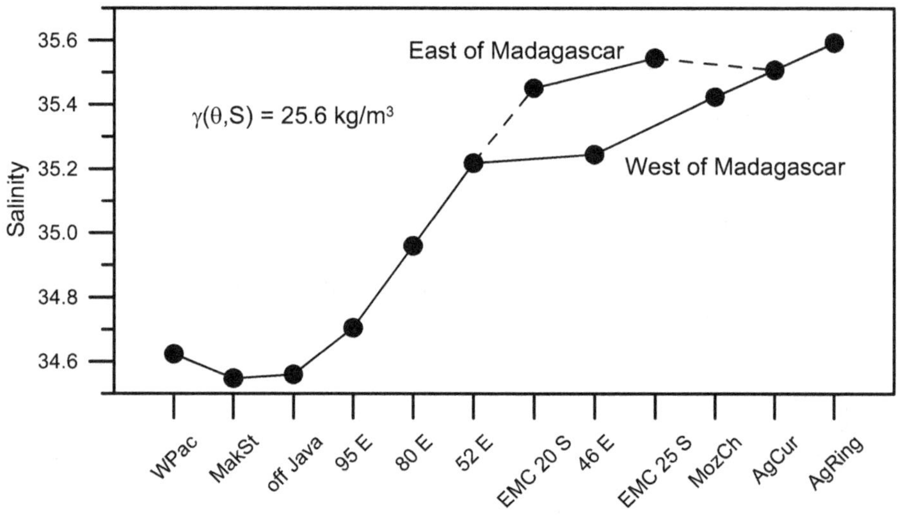

Fig. 6.6. The development of the salinity minimum in the isopycnal of the ITF core density layer from the western tropical Pacific to an Agulhas Ring in the Atlantic Ocean near South Africa. The dots represent ITFW properties measured at the hydrographic sections shown in Fig. 6.2.

The development of the salinity minimum in the ITFW core density layer from the western tropical Pacific toward the Atlantic shows a slight decrease toward the eastern Indian Ocean south of Java, due to mixing with overlying tropical surface water (Ffield and Gordon 1992). From the eastern Indian Ocean toward the South Atlantic Ocean, where an Agulhas Ring was observed during the MARE experiment in 2000 (van Aken et al. 2003), we see a trend of increasing salinity along the main ITFW pathway as a large-scale phenomenon, continuing until the ITFW reaches the Agulhas Ring (Fig. 6.6). This large salinification of ITFW is probably due to lateral mixing with neighboring more saline water masses, since over most of the SEC precipitation is larger than evaporation. The latter effect is responsible for the low-salinity surface layers shown in Fig. 6.5. Near Madagascar the transport via the East Madagascar Current is more saline than the western route via the Mozambique Channel, as can be expected when most of the low-salinity ITFW core passes north of Madagascar.

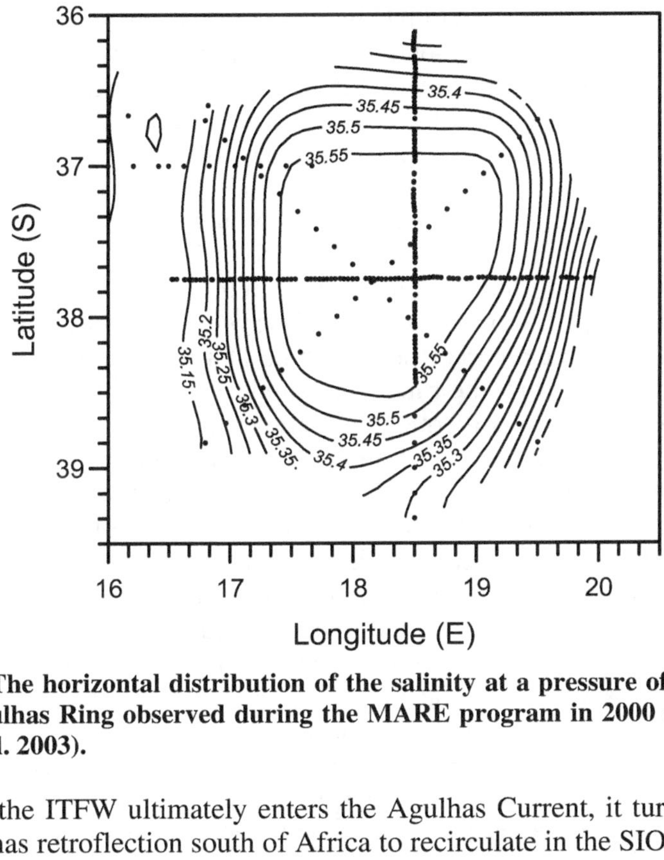

Fig. 6.7. The horizontal distribution of the salinity at a pressure of 150 dbar in an Agulhas Ring observed during the MARE program in 2000 (from van Aken et al. 2003).

When the ITFW ultimately enters the Agulhas Current, it turns east in the Agulhas retroflection south of Africa to recirculate in the SIOC. Due to instability of the retroflection occasionally large rings are shed, which drift into the Atlantic Ocean. While the differences in θ–S properties between the ring and its surroundings are not very large, the rings contain an excess of warm and saline water in the upper 500 to 700 m (Fig. 6.7; MARE in Fig. 6.2). A major part of the interocean exchange at the Agulhas retroflection, or Agulhas leakage, is contributed by these rings. They propagate into the South Atlantic Ocean with speeds ranging between 5 to 10 km/day (Schouten et al. 2000). About six rings per year are shed from the Agulhas retroflection, supporting an interocean exchange of about 7±3 Sv (de Ruijter et al. 1999). However, this transport number is quite uncertain, since it depends on the way the number of rings is estimated as well as on the decision to which depth the rings are assumed to contribute to the exchange. Existing estimates of the heat and salt fluxes, maintained by the Agulhas leakage, therefore vary within a factor 20. Therefore, these estimates do not seem to be very accurate.

The question remains where the water, brought into the Atlantic entrapped in Agulhas rings, ultimately ends. Long-lived rings have been observed to drift toward the South American continent where they become incorporated in the south-flowing Brazil Current (Schouten et al. 2000). In this way the rings do not form a direct pathway to the North Atlantic Ocean. Other rings decay or split by mutual collisions, collisions with sea mounts, and frictional effects (Schouten et al. 2000). Thereby the content of the rings, including salinified ITFW, is brought into the surrounding Atlantic water. No rings have been observed to drift directly toward the equator. Further transport of the return flow from the Indian Ocean in the South Atlantic probably will occur in the wind-driven gyre circulation, similar to the ring water ending in the Brazil Current. During this process the surface water, originating from the rings, will subduct into the thermocline of the South Atlantic gyre. Its thermohaline properties adapt to the atmospheric heat and freshwater forcing in the surface layer of the ocean. Donners (2005) found, from the analysis of a numerical simulation of the ocean circulation, that nearly all water involved in the warm return flow is modified in the South Atlantic Ocean by air–sea interaction. Ultimately the return flow crosses the equator in the wind-driven surface layer of the North Brazil Current. From there it becomes incorporated into the subtropical North Atlantic gyre, where it will flow northward in the Gulf Stream, probably recirculating several times in the subtropical and sub-arctic gyres of the North Atlantic, until it finally reaches the water mass formation regions in the arctic and polar seas. However, experimental evidence on this part of the return flow is scarce and much is founded on intuitive expectations based on mass conservation and some numerical simulations.

6.4. The cold water route

From the meridional distributions of salinity (Fig. 4.2), *AOU,* and dissolved silicate (Fig. 4.10) in the Atlantic Ocean we have seen already that two water type cores, the low-salinity AAIW and the aged UCDW with high *AOU* and silicate values, are found to extend from the Southern Ocean into the Atlantic. Both the salinity minimum of the AAIW core and the silicate maximum of the UCDW can be followed to latitudes north of the equator, where the *S* minimum and *Si* maximum coincide near 1000 dbar. These cores reflect the cold water route of the return flow toward the water mass formation regions in the northern North Atlantic Ocean can be followed northward to across the equator (Stramma et al. 2005).

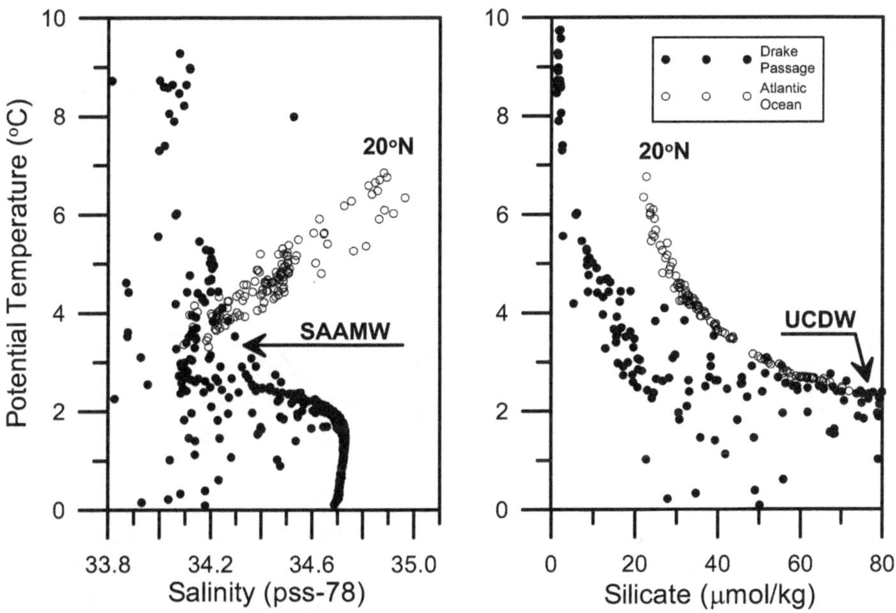

Fig. 6.8. Plots of potential temperature versus salinity (left panel) and dissolved silicate (right panel) for water samples from Drake Passage (WOCE section A21, black symbols) and from the meridional WOCE A16 section in the Atlantic Ocean (open symbols). The θ–*S* points on the A16 section were taken from the salinity minimum of the AAIW core, the θ–*Si* points from the silicate maximum of the UCDW core, both between 50°S and 20°N.

The source water type for the AAIW is the vertically homogeneous Sub-Antarctic Mode Water (SAAMW), formed by winter cooling in the Southern Ocean. The most homogeneous SAAMW is found in the southeastern Pacific sector. This SAAMW is transported to the South Atlantic through the Drake Passage, from where it flows northward in the Falkland/Malvinas Current (Piola and Gordon 1989). While further modification of the SAAMW may take place in the southwestern South Atlantic, increasing its density to the range $\gamma_\theta = 27.1$ to 27.3 kg/m^3, the θ–*S* properties of the SAAMW, which enters the Atlantic Ocean through the Drake Passage (Fig. 6.8, black dots), resemble those of the coldest most-southern

AAIW of the Atlantic Ocean (open symbols). When the SAAMW in the Falkland/Malvinas Current meets the southward flowing Brazil Current, it turns east in the South Atlantic Current. During this eastward flow it can subduct as AAIW under the thermocline of the South Atlantic subtropical gyre (Fig. 6.9). While the θ–*S* properties of the southern AAIW core strongly resemble those of the SAAMW in the Drake Passage (left panel of Fig. 6.8, open symbols), further north the AAIW core becomes more saline and warmer by mixing with the surrounding water types.

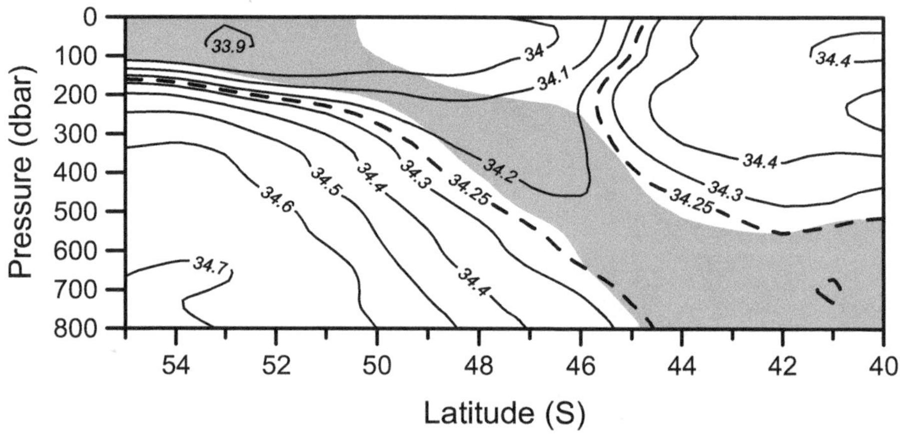

Fig. 6.9. The meridional salinity distribution along the meridional WOCE section A12 near the Greenwich meridian in the southeastern South Atlantic. The gray band represents the +/– 0.1 kg/m^3 density interval around the γ = 27.2 kg/m^3 potential isopycnal.

Also, the high-silicate UCDW core in the Atlantic Ocean appears to originate from the Drake Passage, where the θ–*Si* properties (~2.4 ºC, and 80 μmol/kg) closely resemble those of the most southern parts of the UCDW core along the WOCE A16 section (Fig. 6.8, right panel). Further north in the Atlantic Ocean the silicate concentration of the UCDW core decreases due to mixing with surrounding water types.

The vertical distance between cores of AAIW and UCDW, about 1200 m apart near 50°S and 700 m apart near 30°S, decreases while moving north to 300 m near the equator and less than 200 m at 20°N. At 50ºS the potential density anomaly of the cores of AAIW and UCDW amount to γ_θ = 27.15 and 27.65 kg/m^3, respectively. The denser parts of the UCDW are blocked from further northward spreading by the presence of the southward-moving NADW core. This causes the northward-moving UCDW core, characterized by a silicate maximum, to shallower levels and lower densities. Already at 10°S the hydrographic properties of AAIW and UCDW have converged due to mutual mixing, and the cores hardly can be discriminated any more (Fig. 6.10). If one compares the northward development of preformed phosphate in the southern hemisphere (gray symbols in the lower panel of Fig. 6.10) with the in situ phosphate which increases toward the equator, it becomes clear that the role of aging cannot be ignored for the biogeochemical tracers along the cold water route.

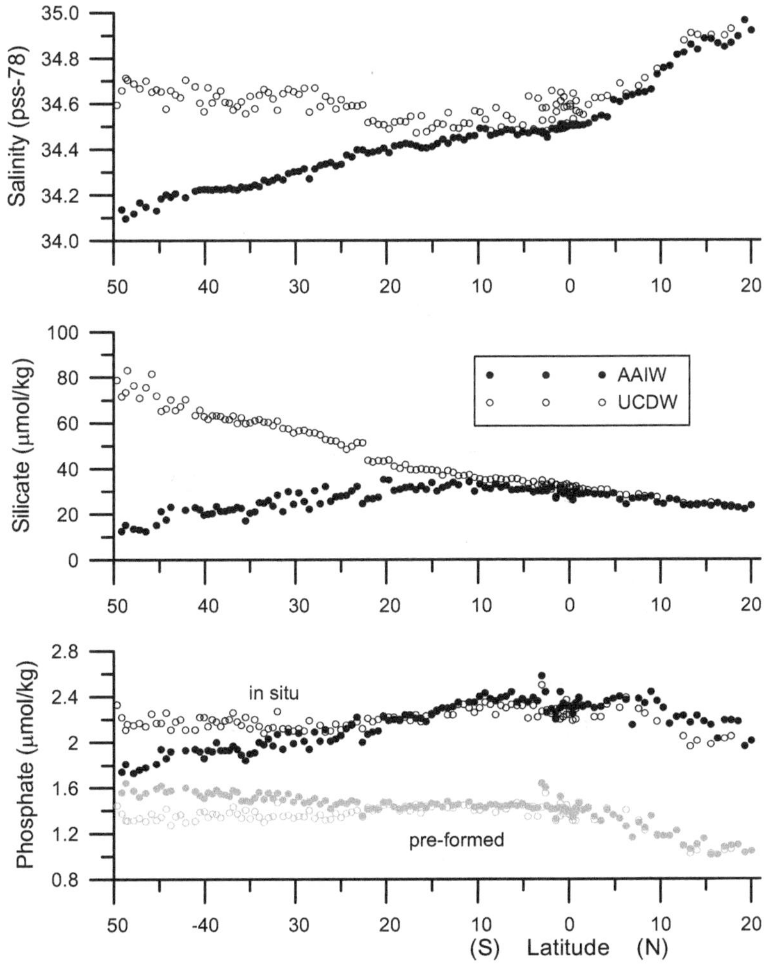

Fig. 6.10. Meridional change of the cores of AAIW (salinity minimum, filled symbols) and UCDW (silicate maximum, open symbols) along the meridional WOCE section A16 in the Atlantic Ocean: upper panel, salinity; middle panel, dissolved silicate; and lower panel, dissolved phosphate. In the lower panel the gray symbols show the preformed phosphate values.

While the lateral gradients in the basinwide cores of AAIW and UCDW are small below the subtropical gyre in the South Atlantic, the horizontal extent of the AAIW core in the North Brazil Undercurrent below the North Brazil Current has become limited (Fig. 6.11). At 750 dbar a low-salinity core is found along the continental slope near French Guyana, while at ~850 dbar a high-silicate core is located. Apparently the cores of AAIW and UCDW have nearly completely merged here. The low-salinity AAIW core has become characterized by a high silicate content. The limited horizontal extent of this core shows that the cross-equatorial transport along the cold water route is not basinwide, but takes place in a narrow boundary current along the northern continental margin of South America. This return flow enters the Caribbean between Tobago and Grenada.

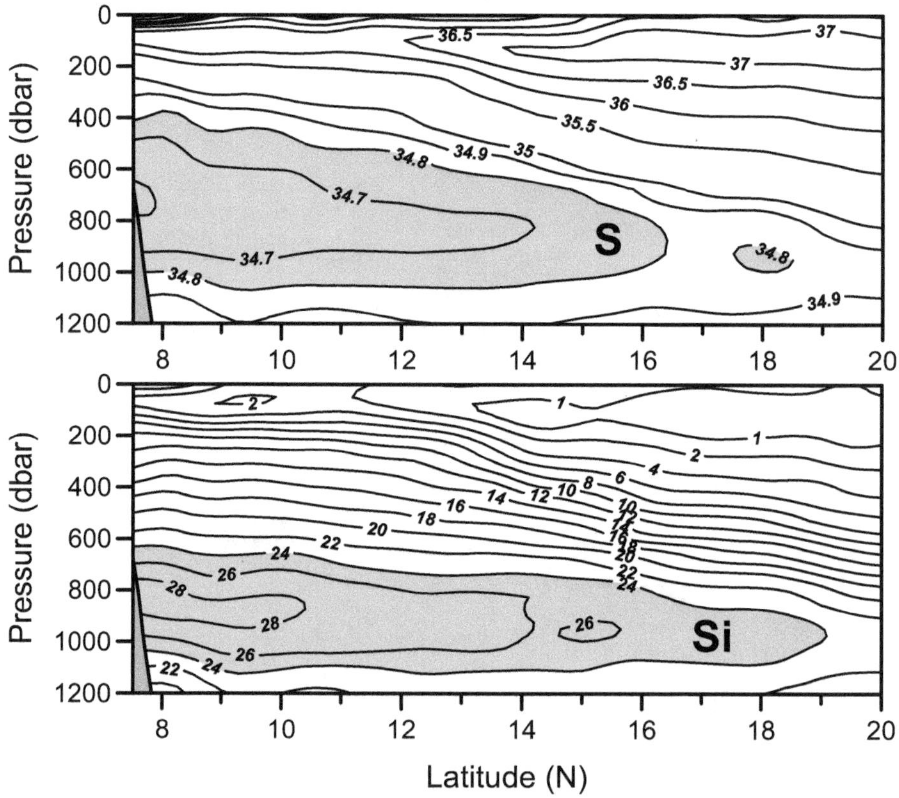

Fig. 6.11. The vertical distributions of salinity (upper panel) and dissolved silicate (lower panel) along the meridional WOCE A20 section in the tropical Atlantic Ocean near 52°W toward the continental shelf of French Guyana.

The American research vessel Knorr in 1997 surveyed the WOCE line A22 at 66°W from the Venezuelan coast to the North American continental shelf. This section crossed the path of the return flow twice: once in the Caribbean slope current along the South American continent and once in Gulf Stream region along the North American continental slope. When we plot the different hydrographic properties versus the potential density relative to 1000 dbar for the Caribbean slope current (Fig. 6.12, dots) it becomes clear that indeed the AAIW core extends from the North Brazil Current into the Caribbean Sea.

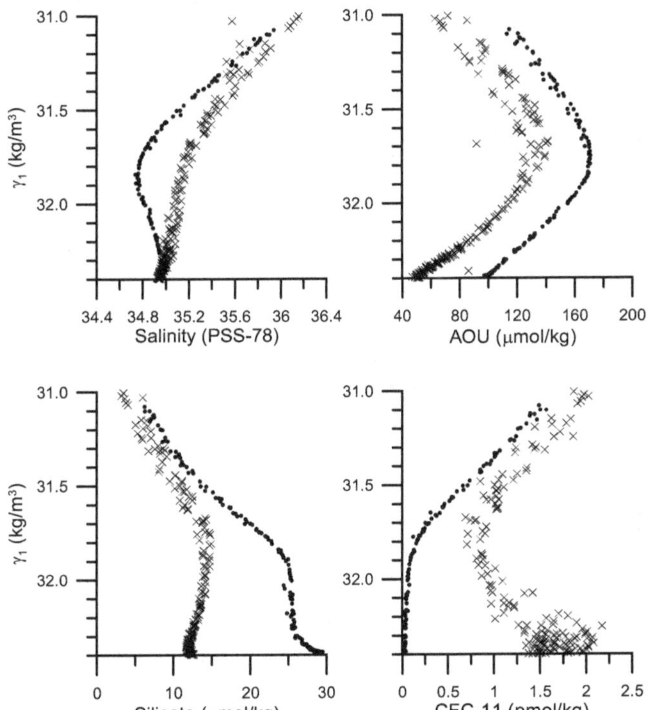

Fig. 6.12. The distribution of hydrographic properties versus γ_1, the potential density anomaly relative the 1000 dbar, measured along the WOCE A22 section at 66°W. The dots represent the data from the Caribbean slope current between the continental slope at 11°N and 15°N, while the crosses show the data points for the Gulf Stream between 36 and 41°N. The subsurface AAIW core is centered in the $\gamma_1 = 31.8$ to 31.9 μmol/kg interval.

Near a potential density anomaly of $\gamma_1 = 31.8$ to 31.9 kg/m^2 ($\theta \approx 6.3°$) a subsurface salinity minimum is encountered, which also is characterized by an *AOU* maximum, as well as an inflection point of the γ_1–silicate and

γ_1–CFC-11 curves. In the Gulf Stream region between 36 and 41°N (crosses in Fig. 6.12) the subsurface AAIW core appears to be modified by mixing with another, younger and more saline water mass. The salinity minimum of the AAIW core is modified to an inflection point, while the *AOU* maximum is still present but with lower values. The silicate in the AAIW core has been turned into a maximum, but with lower silicate values than in the Caribbean Sea, since in the Gulf Stream region it now is overlying relatively young deepwater from (sub)arctic origin. The mixing of the AAIW in the Gulf Stream region with younger water also becomes clear from the CFC-11 minimum in the AAIW density range, overlying recently ventilated high-CFC water.

6.5. Return flow into the Arctic seas

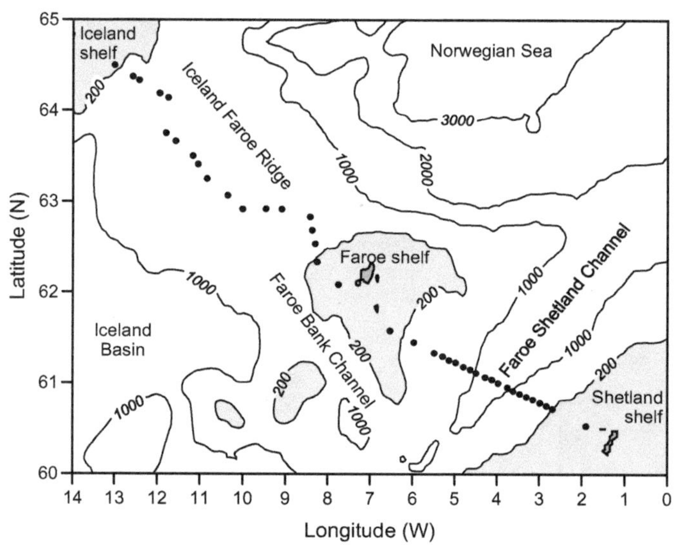

Fig. 6.13. Topography of the Iceland–Scotland Ridge. The dots show the stations of a hydrographic section, surveyed in the summer of 1983 by the Dutch research vessel Tyro.

The water, exported from the arctic seas into the deep North Atlantic Ocean, has to be replaced by low-density water from the return flow. As described above a small part of this return flow occurs via the Bering Strait. The main import of low-density water is assumed to occur across the ridge between Iceland and Scotland near the Faroe Islands (Fig. 6.13). This water originates from the Gulf Stream system and is brought, together with the underlying intermediate water, to the northeastern Atlantic Ocean via the North Atlantic Current. In the Iceland Basin remnants of the AAIW are mixed into the surface layer by convective mixing in winter and are incorporated into the return flow to the arctic seas (van Aken and Becker 1995).

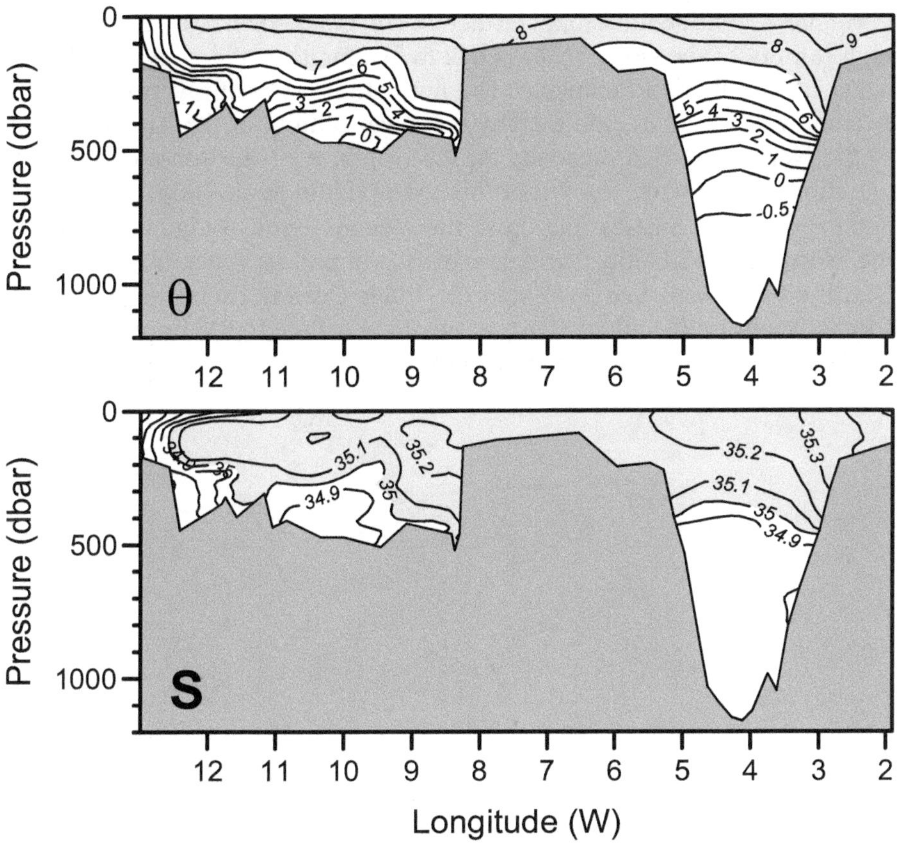

Fig. 6.14. The distribution of the potential temperature, θ, and salinity, *S*, along the hydrographic section between the continental shelves of Iceland and the Shetland islands, shown in Fig. 6.13.

Between Iceland and Scotland the warm and saline ($S > 35.0$) Atlantic return flow has two varieties. Between Iceland and the Faroe Islands warm Atlantic water of intermediate salinity ($35.0 < S < 35.2$) flows into the Norwegian Sea across the Iceland–Faroe Ridge (Fig. 6.14). This water originates directly from the North Atlantic Current in the Iceland Basin. Additional to this inflow a small flow from the Iceland Basin to the Arctic seas occurs via the Irminger Current west of Iceland. Along the Shetland shelf more saline ($S > 35.3$) warm Atlantic water flows northeastward into the Norwegian Sea (Fig. 6.14). This water originates more indirectly from the North Atlantic Current and is brought northward to the Faroe– Shetland Channel in a slope current along the European continental slope. This slope current originates from the Bay of Biscay and possibly even from the Portuguese continental slope further south (van Aken 2001, 2002). The high salinity of the slope water is due to evaporation and winter cooling of water off the Iberian Peninsula. The mutual ratio of both warm and saline Atlantic water masses, and thereby the mean salinity of the Atlantic export to the Norwegian Sea, depends on the response of the sub-arctic gyre to variations in the wind forcing of this gyre (Hátún et al. 2005). Both warm and saline water masses that cross the Iceland–Faroe Ridge contribute to the Norwegian Atlantic Current, which brings this water to the Barents Sea, the Greenland Sea gyre, and the Polar Ocean. There they are transformed again to the cold water types that contribute to the formation of the NADW, as will be described in the next chapter.

7. Formation and descent of water masses

7.1. Water mass formation

The high-density water types that contribute to the deep water masses of the THC are primarily formed by air–sea interaction in winter. During this process the sea surface loses heat and warms the atmosphere. The heat loss leads to an increase of the sea surface density. An additional increase of the surface density may be caused by the formation of sea ice. Not all the salt dissolved in the seawater will be incorporated in the newly formed sea ice. The remaining concentrated salt solution — brine — increases the surface salinity and thereby the surface density. The latter process is called "brine rejection". Due to the increase in surface density by cooling and brine rejection the upper parts of the water column can become unstable and convective mixing sets in, leading to a homogeneous water column extending downward from the sea surface.

The formation process of dense water types may take place over the continental shelf or over deep ocean basins. Examples of water type formation in shelf seas are the shallow arctic Barents Sea and the antarctic shelf regions of the Weddell Sea. There the convective mixing cannot reach very deeply, the whole water column becomes homogenized, and cooling will be fast. After reaching the freezing point ice formation with brine rejection sets in. The newly formed high-density water is not yet deepwater, since it still remains over the continental shelf. Following its formation in winter it will descend across the continental slope into the neighboring deep basins, entraining overlying less dense water on its way down and turning into deepwater. If the convective water mass formation takes place in deep basins, the convective mixing may reach larger depths, the so-called deep convection. During deep convective mixing large amounts of water from deeper levels are brought to the sea surface, which have to be cooled too. The cooling is slow because of the thermal inertia of the thick water column involved. With the slower cooling, ice formation and the following brine rejection play a lesser role compared to the relatively fast cooling of shelf water. Due to the strong density stratification in the permanent thermocline, direct exchange of water between the atmospheric modified surface layer of the ocean and the deep abyssal basins is limited to only a few special regions, mainly the Greenland Sea, Labrador Sea, western Mediterranean Sea, and occasionally the Weddell Sea gyre. These source regions are characterized by the near absence of stratification in winter, combined with a severe heat loss to the atmosphere.

In this chapter several examples of the formation of dense water types in shelf seas and deep ocean basins are presented, as well as examples of the descending motion of the dense water types into the deep abyssal ocean basins.

7.2. The Barents Sea

Warm water originating from the Gulf Stream flows via the North Atlantic Current to the Norwegian Sea (Figs. 6.14 and 7.1). From there it enters the shallow Barents Sea and it flows via the Fram Strait into the Polar Ocean (also named Arctic Ocean), or it returns southward along the east coast of Greenland, subducted under the fresher polar water in the East Greenland Current. In the arctic and polar regions air–sea interaction of this Atlantic water with the overlying atmosphere cools the ocean and warms the atmosphere.

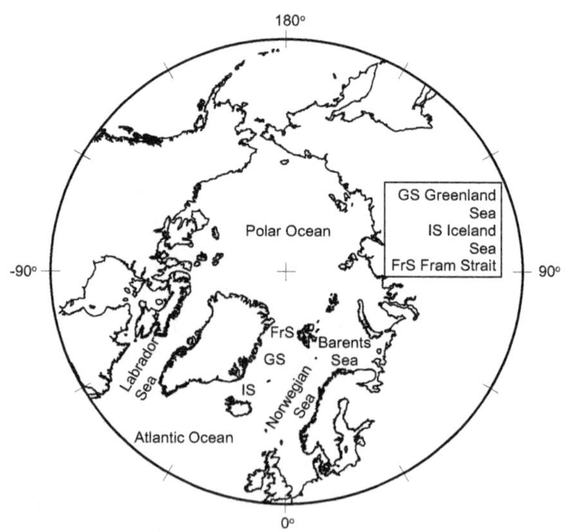

Fig. 7.1. Stereographic map of the northern polar and arctic seas.

The vast shelf areas, comprising over 30% of the Polar Ocean, are ice-free in summer and freeze over in winter. The Barents Sea is the deepest of these polar shelf seas with the most saline inflow. Relatively saline and warm Atlantic water from the Norwegian Atlantic Current flows to the east, north of Norway and Russia. While in the summer sea surface temperatures up to 10°C are observed, in winter this Atlantic water is cooled to the freezing point. Then the formation of sea ice allows brine rejection which increases the density of the underlying water. Due to adverse weather and sea ice conditions in this region most hydrographic surveys are carried out in summer and observational evidence for this process from the winter period is scarce. Our knowledge is based mainly on hydrographic surveys in spring and summer.

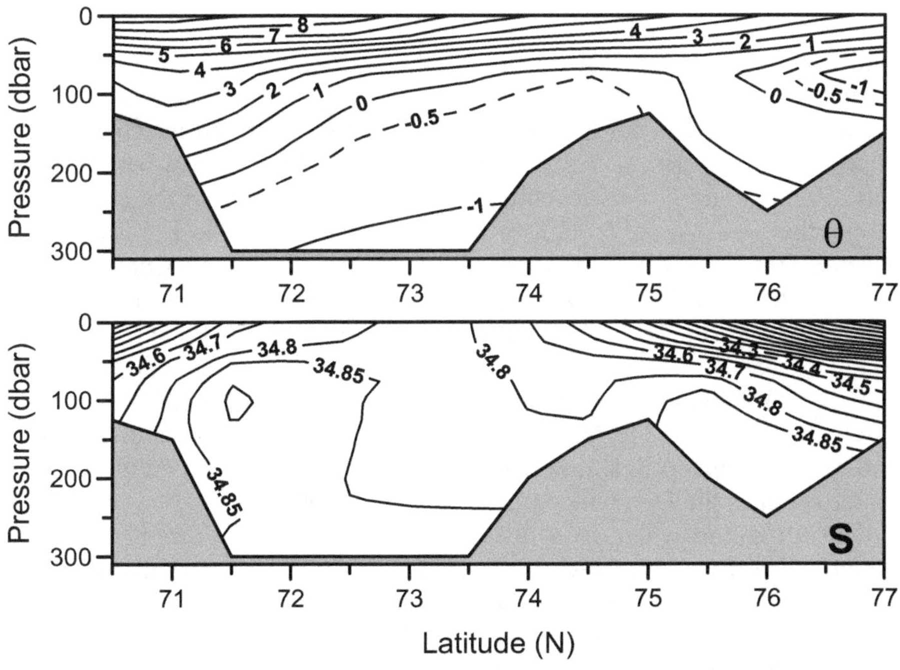

Fig. 7.2. Vertical distribution of potential temperature (upper panel) and salinity (lower panel) along a meridional section in the Barents Sea near 40°30'E. This section was surveyed in August 1990. Data from this survey were downloaded from the World Ocean Database 2001, maintained by the National Oceanographic Data Center in Maryland US (http://www.nodc.noaa.gov/OC5/WOD01/pr_wod01.html).

Along a meridional section in the Barents Sea, surveyed in August 1990 at 40°30'E, the vertical distributions of temperature and salinity (Fig. 7.2) show that the coldest water in the deeper depressions, formed in the previous winter, has the highest salinities ($S \approx 34.89$). This is the transformed Atlantic water, with temperatures still slightly above freezing point. Near-surface coastal water in the Barents Sea and Polar Surface Water from the slope region further north have lower salinities, probably due to river run-off and melting sea ice. They therefore cannot reach such high densities when cooled to the freezing point compared with the saline Atlantic water, unless brine rejection due to ice formation increases the salinity of the underlying seawater. The coldest water near the bottom ($\theta < -1.0$°C, $S = 34.89$) has a potential density anomaly, γ_θ, of over 28.05 kg/m^3.

Midtun (1985) has observed bottom temperatures near freezing point (−1.87°C) and high salinities due to brine rejection over the banks in the Barents Sea between 40 and 60°E, west of the island of Novaya Zemlya. Newly formed dense water sinks along the slope of the banks and flows into hollows and depressions of the sea bottom. Saline water ($S > 35.0$) with a potential density anomaly over 28.2 kg/m^3 has been observed in these depressions. After this high-density water is formed in winter in the shallow Barents Sea, it descends into the deep Polar Ocean between Novaya Zemlya and Fransz Joseph Land via a deep canyon in the continental slope, the Saint Anna Trough, where it contributes to the Eurasian Basin Deep Water (EBDW). Similar processes occur on the North Canadian continental shelf, where the warmer, more saline Canadian Basin Deep Water (CBDW) is formed. The cold deep water types of the Polar Ocean ultimately leave the Polar Ocean through the deep passage of the Fram Strait between Greenland and Spitsbergen. They can be followed through the Greenland Sea and the Norwegian Sea (Swift and Koltermann 1988). In the western Iceland Sea almost undiluted EBDW is occasionally observed as far south as the Denmark Strait (Buch et al. 1996).

Not all saline water from the Norwegian Atlantic Current enters the Barents Sea. Part of it flows further north in the West Spitsbergen Current and there its density has increased so much by cooling that it subducts under the fresher arctic surface waters. This water enters the Polar Ocean through the Fram Strait at subsurface levels or returns under the south-flowing East Greenland Current as subsurface return Atlantic Intermediate Water (rAtIW) (Hopkins 1991). The core of the rAtIW contains a maximum in temperature and salinity, with temperatures ranging from 0 to 3 °C and salinities greater than 34.9, The subsurface salinity maximum shows that this water mass is produced by the cooling and sinking of the saline Atlantic water.

7.3. A scheme for deep convection

Marshall and Schott (1999) have written a comprehensive overview on observations and theory of open-ocean deep convection. The deep-convection process in the ocean, driven by a surface heat loss, is intermittent and involves a hierarchy of scales. The dominant timescales and spatial scales of the source regions are sufficiently large to have the convection modified by the earth's rotation. On the scale of the convection region, O(100 km) the flow is in near-geostrophic balance and therefore is nearly non-divergent. Then simultaneous downward motion at this scale in a single convection cell is not possible, since that requires a convergent supply of water from the surrounding region, not allowed in geostrophic flow. The direct response to widespread surface cooling is at the scale of relatively small downward-moving convection cells or plumes, O(~1 km). Fluid parcels, cooled in the near-surface "thermal boundary layer," will increase in density and sink in these plumes, to be replaced by upward-floating less dense water parcels from deeper levels.

Marshall and Schott (1999) discriminate several successive phases in the seasonal development of a convective region. These can be summarized as the preconditioning, the violent convective mixing, the lateral exchange, and the restratification phase (Fig. 7.3). In the preconditioning phase weakly stratified high-density water is brought close to the surface. This occurs in subgyres with enhanced near-geostrophic cyclonic circulation, with a characteristic doming of isopycnals in its center. With the onset of winter cooling the thinned surface layer in the center of the subgyre increases further in density, approaching the density of the underlying weakly stratified water. Isopycnals from subsurface levels start to outcrop at the sea surface. Subsequent vigorous cooling events in winter lead to a further increase of the surface density until static instability develops, initiating violent convective mixing. In this phase numerous turbulent convective plumes bring dense water to greater depths, while less dense water is brought to the surface. Down-welling vertical velocities up to 10 cm/s have been observed. The individual plumes have a characteristic diameter of the order of ~1 km. Together they generate a vertically and horizontally homogenized area or "chimney" with a characteristic horizontal scale of the order of ~100 km. Caballing caused by mixing of the generally cold and less saline water in the center of the subgyre with the warmer more saline water in the surrounding rim current may add to the density increase.

In the convective mixing phase the vertical mass flux is only coherent on the scale of the convective plumes, since the near-geostrophic flow at the larger-scale of the chimney hinders a horizontal convergence of the surface flow that is required to feed the net downward motion on the scale of a chimney. A narrow near-geostrophic cyclonic rim current will develop around the chimney in response to the large horizontal density gradients at its edge. The available potential energy of the horizontal density gradients may lead to baroclinic instability of the rim current associated with eddying motion on geostrophic spatial scales. In this spreading phase the high-density water, generated in the convectively mixed chimney, is transferred laterally along the slanting isopycnals in the frontal eddies. This also will bring less dense water to the surface layers of the chimney. If the cooling event is terminated, restratification sets in at timescales of weeks to months.

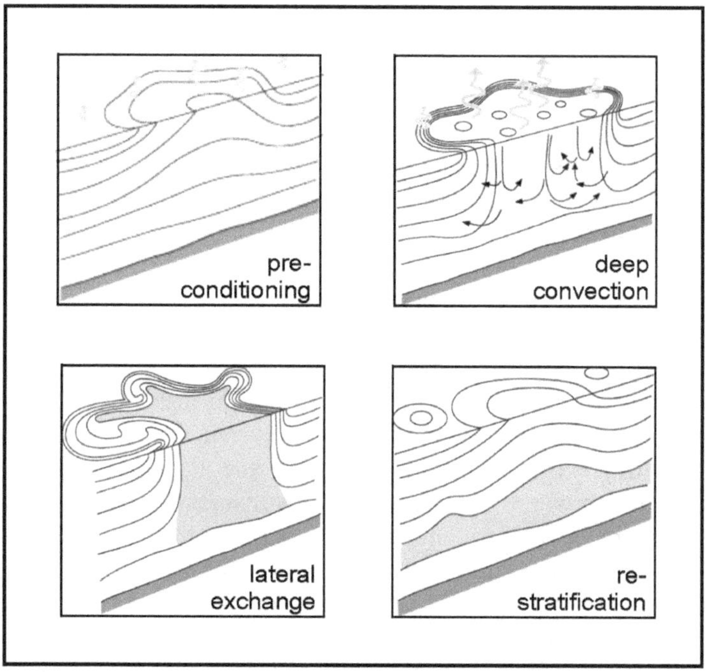

Fig. 7.3. Schematic presentation of the four phases of open-ocean deep convection: the preconditioning phase, the violent deep convection phase, the lateral spreading phase, and the restratification phase. The underlying density stratification is shown by the continuous isopycnal lines, the newly formed homogeneous water type is indicated in gray. The spreading may occur concurrently with the convection phase if the strong surface cooling continues for many days (from Marshall and Schott 1999).

The spreading may occur concurrently with the convection phase if the strong surface cooling continues for many days. In a winter with several successive strong cooling events the violent convective mixing phase may occur several times, reaching deeper and deeper isopycnals, until in spring the lateral spreading as well as the meteorologically induced re-stratification finally isolates the newly formed deepwater from further atmospheric influence.

7.4. Deep convection in the Greenland Sea

In the Greenland Sea a cyclonic surface circulation can be observed as part of the large-scale wind-driven arctic gyre in the Nordic seas (Norwegian, Greenland, and Iceland Sea). On the western side of this subgyre cold but relatively fresh low-density Polar Surface Water flows to the south in the East Greenland Current. In summer this water also spreads over the center of the Greenland Sea. East of the subgyre warm and saline water flows northward in the Norwegian Atlantic Current. North of the Island Jan Mayen the Jan Mayen Current turns east, branching from the East Greenland Current. In the center of the cyclonic subgyre, formed by these currents, the doming of the isopycnals inherent to the cyclonic circulation contributes to the preconditioning phase for convective mixing since it brings weakly stratified high-density water close to the sea surface.

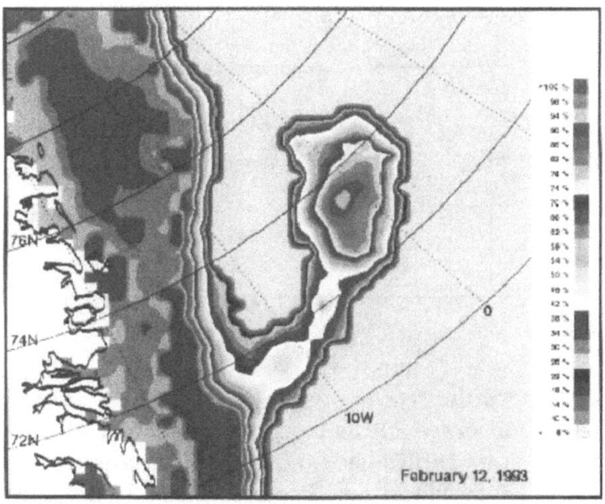

Fig. 7.4. The Is Odden in the Greenland Sea on 12 February 1993, shown by a passive microwave satellite image. The different shadings represent different percentages of ice cover. The homogeneously light gray area is the ice-free ocean. The ice-free Nordbukta is located between the Is Odden and the pack-ice of the East Greenland Current (© P. Wadhams).

In the Greenland Sea subgyre a near-permanent ice cover is observed in winter, the Is Odden (ice promontory in Norwegian, Fig. 7.4). The Is Odden consists mainly of locally formed ice, but old pack-ice from the East Greenland Current brought there by the Jan Mayen Current also has been reported. The sea ice effectively isolates the underlying water from atmospheric influences. West of the Is Odden an area is found that is regularly ice free in winter, the Nordbukta. There the thermal isolation formed by the ice cover is missing, which leads to large heat losses to the atmosphere and a fast increase in surface density. The cooling in Nordbukta destabilizes the weak density stratification and starts thermal convection. Due to the convective mixing in the Nordbukta warmer water from greater depths is brought to the sea surface, contributing to the ice-free character of the area. This process may occur successively several times during a winter, leading to a situation where deep convection may occur.

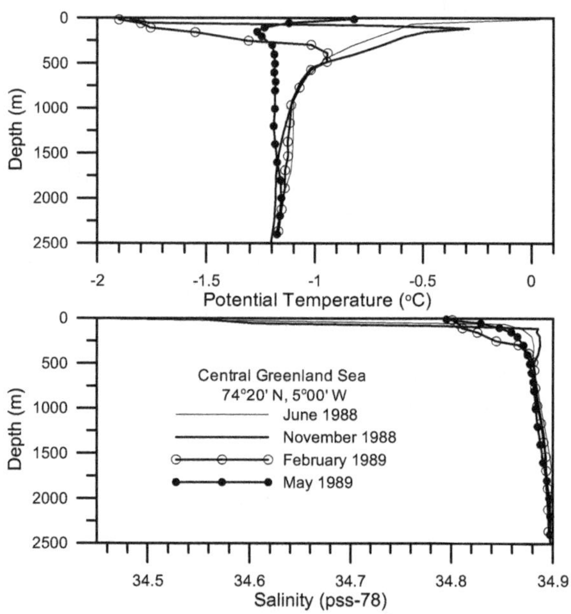

Fig. 7.5. Successive profiles from potential temperature (upper panel) and salinity (lower panel) observed in the Nordbukta in the central Greenland Sea (~74°20'N, 5°W). These profiles have been observed during successive cruises of the international Greenland Sea Project. These hydrographic data from the Greenland Sea Project were downloaded from the website http://www.ices.dk/ocean/project/data/gsp.htm.

In 1988–1989 observations for the international Greenland Sea Project were performed, initiated by the Arctic Oceans Sciences Board (see http://www.ices.dk/ocean/project/data/gsp.htm). As part of this project the water mass transformation in the central Greenland Sea was monitored by means of successive seasonal hydrographic surveys. In Fig. 7.5 the modification of the vertical stratification of temperature and salinity from June 1988 to May 1989 at a location in the Nordbukta (~74°20'N, 5°W) is shown. In June 1988 a thin, relatively warm (~0.0°C) and fresh (S = 34.46) surface layer of Polar Surface Water with a thickness of less than 50 m could be observed. This relatively freshwater spreads over the Greenland Sea from the East Greenland Current. In November 1989 the sea surface temperature was cooled to the freezing point (~–1.9°C). The maximum temperature (–0.29°C) was observed at 115 m. The surface salinity had increased to ~34.56 due to brine rejection and entrainment of more saline water from below. In February 1989 the cooling of the surface layer had already progressed to a depth of nearly 400 m and the increase of the surface salinity had continued to 34.80. In May 1989 a thick nearly homogeneous layer was found between 250 and 1750 m with a potential temperature of –1.19°C and a salinity of 34.88, a result of deep convective mixing that probably occurred in late March or early April. In May some restratification of the upper 250 m already had set in, with a sea surface temperature of –0.82°C and a sea surface salinity of 34.80.

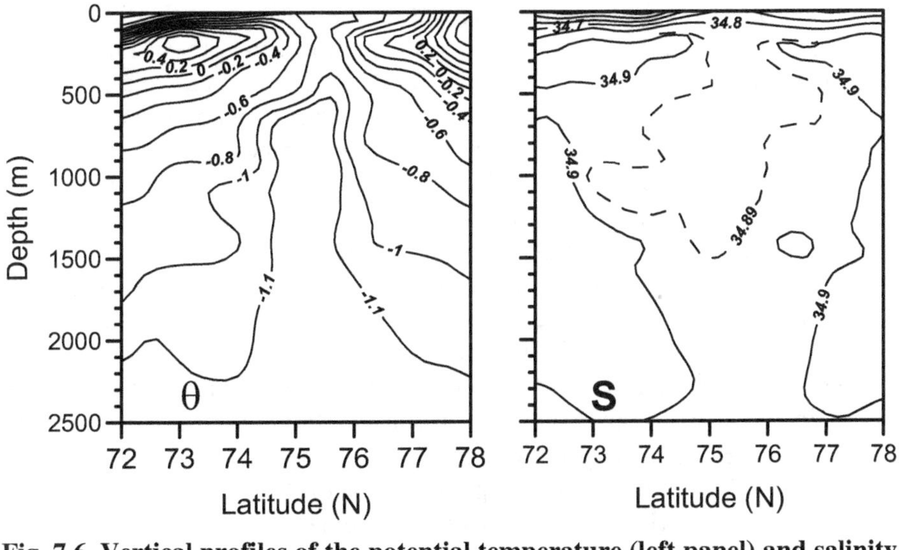

Fig. 7.6. Vertical profiles of the potential temperature (left panel) and salinity (right panel) observed during the Greenland Sea Project along a nearly meridional section through the Nordbukta in the Greenland Sea, observed in May 1989.

A nearly meridional hydrographic section through the Nordbukta, surveyed in May 1989 (Fig. 7.6), clearly shows the doming of the isothermals around the newly formed chimney and the limited size of the convective region: ~110 km between the −1.1°C isotherm from 700 to 1400 m. North of the homogenized chimney the warm and saline Atlantic water from the West Spitsbergen Current was observed, and south of the chimney the rAtIW was found below the colder and fresher surface water of the eastward-flowing Jan Mayen Current. Above 500 m the chimney has become narrower, probably because the rAtIW and the overlying fresher polar surface water had been brought over the chimney by eddies from the rim current. Chimneys in the Greenland Sea probably do not collapse immediately after the convection period in winter due to eddies on the rim current. Wadhams et al. (2004) have observed a chimney in the Greenland Sea to survive for at least 26 months. Similar to the chimney in Fig. 7.6 that chimney was capped in summer by low-density water in the upper 50 m, with further intrusion of ambient rAtIW down to 500 m. Below that level the cyclonically rotating core was left unchanged. In the following winter the low-density cap was removed by surface cooling and the chimney became open again to air–sea interaction. This cycle also was repeated the following year.

The strength of the convective mixing and the maximum convection depth that is reached in a winter depend on the initial stratification as well as on the atmospheric forcing and ice formation. Since these factors may vary from year to year, the deep convection and cold water mass production also may vary from year to year. During the Greenland Sea Project it appeared that since the 1970s deep convection had not reached the bottom layers below 2000 m depth in the Greenland Sea, and no Greenland Sea Deep Water (GSDW) was formed. However, tracer concentrations in the deep layers showed that GSDW was ventilated at previous years (Bönisch and Schlosser, 1995). Since about 1975, open-ocean convection in the Greenland Sea mainly produced intermediate water and the lighter variety of upper GSDW. This Arctic Intermediate Water (AIW) (Hopkins 1991) is generally fresher than the rAtIW formed by cooling of the saline Atlantic water in the Norwegian Atlantic Current and the West Spitsbergen Current. It is characterized by temperatures less than 2°C in the salinity range 34.7 to 34.9. The definitions for upper and lower AIW deliberately overlap in density, with AIW in the northern Greenland Sea having a density higher than the rAtIW. Elsewhere AIW always overlies rAtIW, leading to the naming of upper and lower Arctic Intermediate Water.

In the cyclonic gyre of the Iceland Sea convection cannot reach the depth range of deepwater since this sea is too shallow. Here mainly intermediate water is formed in winter. Part of that water is transported toward the Norwegian Sea near the Faroe Islands by the East Iceland Current, where it contributes to the overflow into the Atlantic Ocean, while another part is diverted toward Denmark Strait.

7.5. Norwegian Sea Deep Water

Norwegian Sea Deep Water (NSDW) already was described by Helland–Hansen and Nansen in 1909 as a nearly homogeneous water type, found below 2000 m in the Norwegian and Lofoten basins. The deepwater in the Norwegian Sea is ventilated from the Greenland Sea via a deep channel north of Jan Mayen with a sill depth slightly over 2000 m. However, at no level in the central Greenland Sea is the water as saline as NSDW ($S = 34.91$). Therefore, an additional source of saline water is required, not formed by deep convection in the Greenland Sea. Swift and Koltermann (1988), who used data from the early 1980s, proposed that NSDW is formed by the mixing of nearly equal amounts of GSDW and EBDW (Fig. 7.7). The concentration of nutrients (phosphate, nitrate, and silicate) was highest, and the concentrations of oxygen and man-made tracers (CFC-11, CFC-12, Tritium, ^{137}Cs, ^{90}Sr) was lowest in the NSDW compared to both GSDW and EBDW. This confirms that NSDW is the oldest of the water types, while EBDW and GSDW are more recently ventilated water types, both contributing to the ventilation of the deep Norwegian Sea.

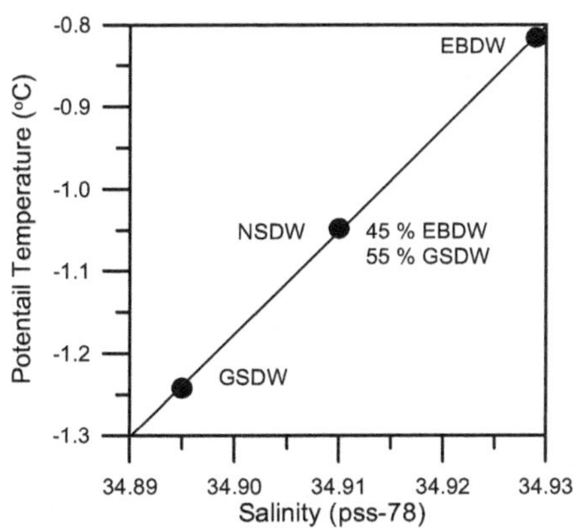

Fig. 7.7. Potential temperature–salinity diagram with the characteristic data points for Norwegian Sea Deep Water (NSDW), Greenland Sea Deep Water (GSDW), and Eurasian Basin Deep Water (EBDW) according to Swift and Koltermann (1988). Assuming linear mixing of salt and temperature, NSDW consists of 45% EBDW and 55% GSDW.

During the Greenland Sea Project it appeared that in the late 1980s deep convection in the Greenland Sea did not reach depths over 2000 m. From the study of natural and man-made tracers Bönisch and Schlosser (1995) confirmed that the formation rate of GSDW had decreased with at least a factor of 5 after 1980. The cause of this decrease is uncertain, but possibly may be ascribed to a decrease of the surface salinity in the Nordic Seas. However, there was no sign of a decrease of EBDW (and CBDW) for which brine rejection is the important mechanism to raise the salinity. The reduction of the formation of GSDW has consequences for the temperatures in the deep arctic basins. In the central Greenland Sea below 2000 m the temperature increased between 1972 and 1996 from –1.31 to –1.12°C, while between 1981 and 1997 the temperature in the Norwegian Sea at 2000 m increased from –0.97 to –0.87 (Østerhus and Gammelsrød 1999). As a consequence the potential density difference between the deep Greenland Sea and the Norwegian Sea changed sign; the deep Norwegian Sea became slightly denser than the deep central Greenland Sea in the late 1990s. This change in density difference is probably responsible for the reversal of the flow in the deep channel north of Jan Mayen, which was observed to be directed toward the Norwegian Sea in 1983–1984 and toward the Greenland Sea in 1992–1993 (Østerhus and Gammelsrød 1999). The upper boundary of the NSDW in the Norwegian Sea descended over time with several 100 m. This change will have its consequences for the composition of the dense overflow water descending over the Greenland–Scotland Ridge into the Atlantic Ocean (Hansen et al. 2001).

7.6. Exchange between the Nordic seas and the North Atlantic Ocean

The deep and bottom water, formed by convection in the Greenland Sea, together with the deepwater formed on the continental shelves of the Polar Ocean, spreads through the Nordic seas and forms NSDW. However, the southward export from the Nordic seas to the main body of the North Atlantic Ocean is restricted by the presence of the shallow (< 500 m) Greenland–Scotland Ridge (Fig. 7.8). Only at a few locations (the Faroe Bank Channel, the Iceland–Faroe Ridge, and the Denmark Strait) narrow channels with sills at depths between ~600 and 850 m allow overflow of cold high-density water into the main body of the Atlantic Ocean. These sill depths are well above the upper boundary of the deepwater in the Nordic Seas. The water crossing the sills to the Atlantic Ocean is mainly AIW and the upper parts of the NSDW.

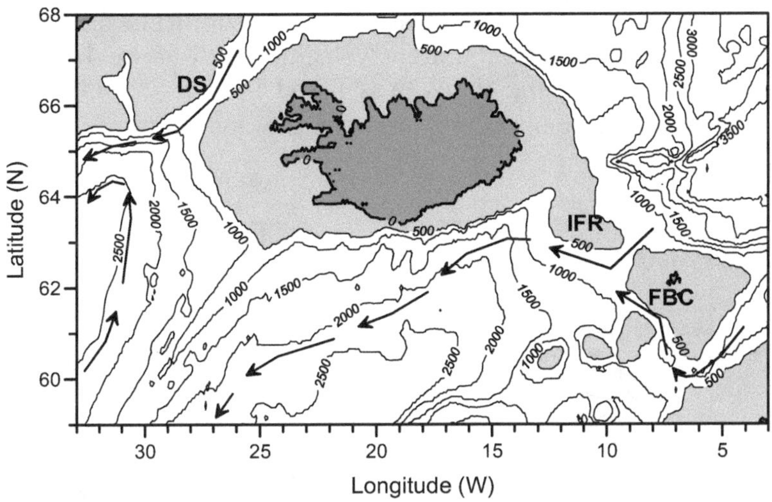

Fig. 7.8. Bathymetry of the Greenland–Scotland Ridge with the depth in meters. The areas with a depth less than 500 m in light gray. The main narrow deep connections between the Nordic Seas and the North Atlantic Ocean are the Faroe Bank Channel (FBC, ~850 m), the Iceland–Faroe Ridge (IFR, ~600 m), and the Denmark Strait (DS, ~650 m). The arrows show the main routes of high-density overflow water that passes through these gaps in the Greenland– Scotland Ridge.

Due to restrictions by the topography, the overflow is strongly determined by density differences and friction (Hansen et al. 2001). The velocities in the overflow across the sills are generally quite strong, O(1 m/s), allowing turbulent entrainment of less dense overlying water. In this way the hydrographic properties of the overflow water are strongly modified compared with the original water mass flowing from the Nordic Seas to the sills. Effectively, new water types are formed by turbulent entrainment and mixing at the sills.

The main course of overflow water from the Norwegian Sea to the North Atlantic follows the Faroe–Shetland Channel and the Faroe Bank Channel (sill depth ~850 m) with some additional flow through shallower channels on the Iceland–Faroe Ridge (sill depth ~600 m). With turbulent entrainment by the fast-flowing water, which descends from the sill into the deep Iceland Basin, warm and saline North Atlantic Central Water from the thermocline is mixed into the cold overflow water (~ –0.5°C), forming a new water type: the Iceland–Scotland Overflow Water (ISOW). South of Iceland near 17°W the transport of ISOW has been observed by Peter Saunders from the former Institute of Oceanographic Sciences in Wormly (UK) with the WOCE ACM-8 moored current meter array. The

ISOW was found to flow to the west over the south Icelandic slope, with long-term mean near-bottom current velocities well above 10 cm/s (Fig. 7.9, left panel). There the ISOW was found in a thin (~300 m) bottom layer with potential temperatures between 2.5 and 3.2°C (Fig. 7.9, right panel).

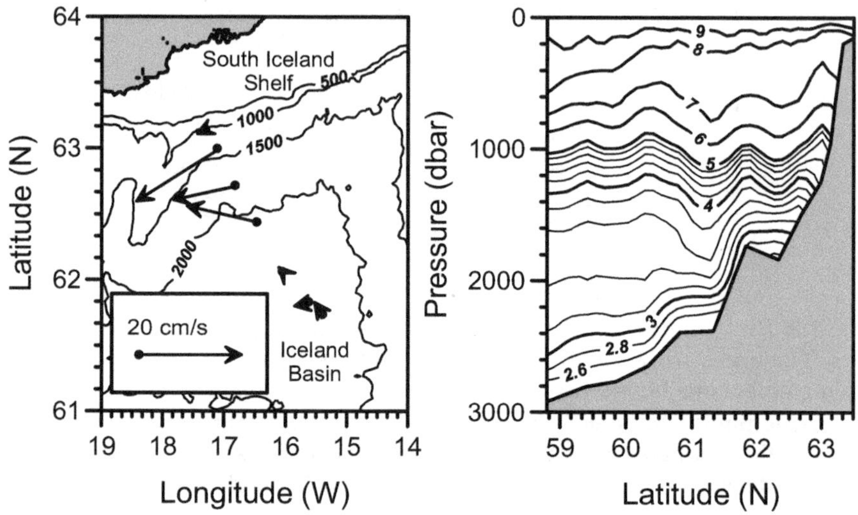

Fig. 7.9. Long-term mean current velocity vectors from the bottom layer southeast of Iceland (left panel) and the potential temperature profile along a meridional section at 17°W south of Iceland (right panel). The current meter data are from the WOCE ACM-8 moored current meter array of Peter Saunders (1996); the temperature data are from a hydrographic survey of the Dutch RV Tyro, carried out by the author in 1991.

From the current measurements Saunders (1996) derived a mean westward transport of ISOW of 3.2(±0.5) Sv. By combination of the current meter observations with observations of temperature and salinity, Saunders estimated that about 50% of this transport originated from the Norwegian Sea, while the other 50% was introduced into the ISOW by turbulent entrainment of overlying warm and saline North Atlantic Central Water (warm entrainment). Dickson and Brown (1994) estimated a smaller warm entrainment of about 25% (0.8 Sv).

Downstream from the ACM-8 array more water from the directly overlying relatively cold water is entrained (cold entrainment), decreasing the salinity of ISOW and increasing the volume transport. Part of that entrained water comes from the upper layers of the low-salinity, silica-rich, recirculating Lower Deep Water (LDW) of AABW origin; the other part comes from the slightly warmer low-salinity LSW core.

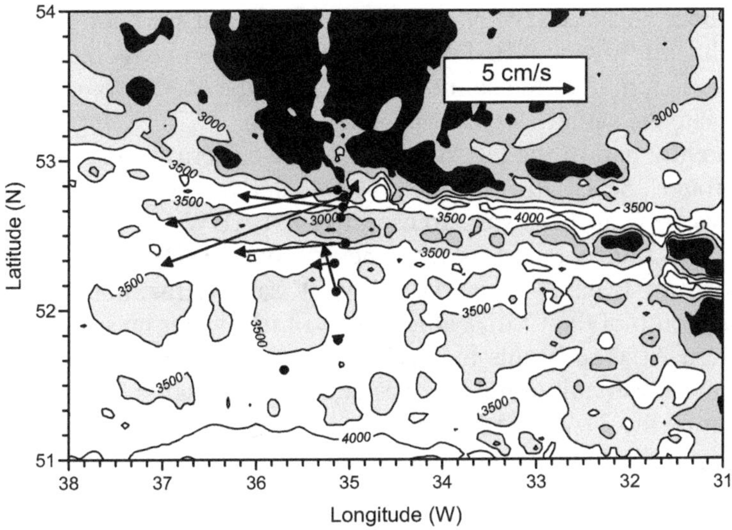

Fig. 7.10. Long-term mean current vectors in the Charlie–Gibbs Fracture Zone in the high-salinity ISOW core near a depth of 3000 m. The measurements were carried out by Peter Saunders (1994). The black shading is applied to bottom depths less than 3000 m deep.

From the Icelandic slope the ISOW core continues to follow the eastern slope of the Mid-Atlantic Ridge (MAR), here named Reykjanes Ridge, to the south. This flow reaches a gap in the MAR between 53°N and 52°N, the Charlie–Gibbs Fracture Zone (CGFZ). The CGFZ consists of two nearly zonal channels separated by a shallower ridge with peaks at depths less than 2500 m (Fig. 7.10). The sill depth in the northern channel amounts to ~3550 m and in the southern channel to ~3450 m. Peter Saunders (1994) obtained a set of current measurements in the CGFZ from a meridional array of eight current meter moorings (the black dots in Fig. 7.10). In this array, operated in the years 1988–1989, four moorings were located in the northern channel, two in the southern channel and two over the abyssal plain further south. Near the mooring array a salinity maximum ($S > 34.95$), generally referred to as Northeast Atlantic Deep Water (NEADW), was found at depths between 2500 and 3000 m, with the highest salinities in the northern channel. Most current meters near the salinity maximum showed westward velocities with a long-term mean velocity up to 7.6 cm/s (the arrows in Fig. 7.10). Even on the scale of the channels in the CGFZ the deep flow showed a definitive lateral structure. The highest velocities in both the northern and southern channel were observed over the northern slope of each channel, with near-zero velocities only 20 km south of the northern wall of the channel. Further south of the CGFZ, between about 53°N and 39°N, the core of the water mass originating from the southward flowing NEADW along the MAR can be recognized as a salinity maximum, until the saline outflow from the Mediterranean Sea starts to dominate the salinity signal (van Aken 2000).

Apparently part of the NEADW leaves the eastern Atlantic basins to the west through the CGFZ. Saunders (1994) derived from the current measurements a mean westward transport of NEADW with $S > 34.94$ to be 2.4 Sv. This transport is less than the 3.2 Sv of ISOW transported westward south of Iceland. Since the NEADW salinity maximum also can be observed south of the CGFZ, at least part of this water mass does not leave the eastern Atlantic basins but recirculates south of the CGFZ. From hydrographic observations in the western North Atlantic basins it appears that west of the MAR the NEADW recirculates in an anticlockwise, cyclonic way through the Irminger Sea.

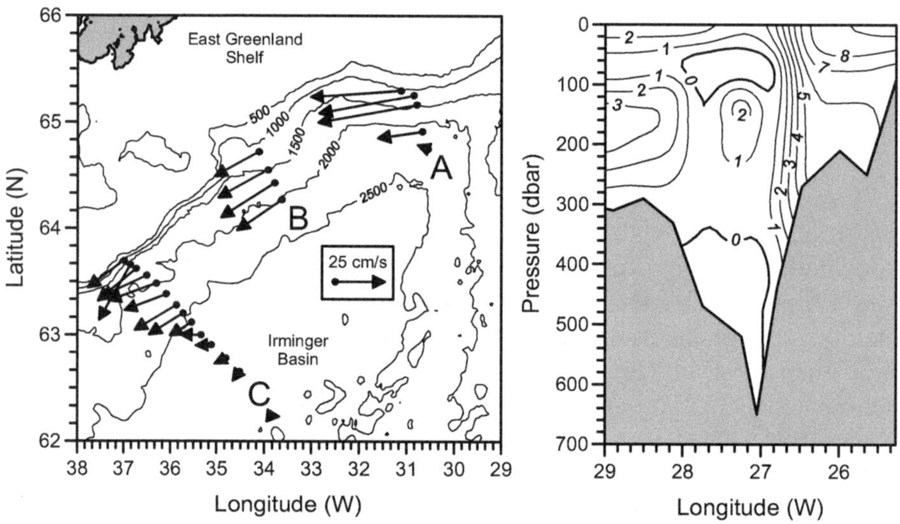

Fig. 7.11. Long-term mean current velocity vectors from the bottom layer over the East Greenland slope in the Irminger Sea (left panel) and the potential temperature profile along a near-zonal section of the potential temperature at the latitude of the sill in the Denmark Strait (right panel). The current meter data are from the WOCE ACM-8 moored current meter array of Bob Dickson from the UK and the temperature data from a hydrographic survey of the WOCE AR18 section by Svend-Aage Malmberg from Iceland.

Between Greenland and Iceland cold water is transported toward a narrow channel in the Denmark Strait (sill depth ≈ 650 m). Current observations suggest that the main transport of AIW and the upper deepwater from the Greenland Sea (GSDW, EBDW, and CBDW) occurs via the Iceland Sea where a final water mass transformation toward the Denmark Strait Overflow Water (DSOW) may take place (Jonsson and Valdimarsson 2004). A reasonable estimate of the mean cold overflow through the Denmark Strait, based on several data sources, is 2.9 Sv (Dickson and Brown 1994). On the Icelandic side of the Denmark Strait warm Atlantic water from the Irminger Current flows into the Nordic seas, while over the Greenland continental shelf a core of relatively warm and saline rAtIW at ~160 m depth flows southward under the colder and fresher Polar Surface Water from the East Greenland Current (Fig. 7.11, right panel). South of the sill in the Denmark Strait the cold overflow water descends over the East Greenland continental slope in a thin (~300 m) bottom layer, and DSOW is diluted by turbulent entrainment and mixing with less dense overlying water. Near Cape Farewell at the southern tip of Greenland the DSOW core has reached a depth of about 3000 m.

Bob Dickson from the Fisheries Laboratory in Lowestoft (UK) has carried out long-term current measurements over the East Greenland continental slope in the Irminger Sea from 1986 to 1999, the WOCE ACM-8 current meter mooring array (Dickson and Brown 1994). The array consisted of three sections (A, B and C) perpendicular to the continental slope (the black dots in Fig. 7.11, left panel). The purpose of this array was to determine the mean transport of DSOW in the deep slope current and the variability of this transport. Relatively high mean near-bottom velocities, of about 50 cm/s, were observed in the DSOW core at mooring section A, closest to the Denmark Strait (Fig. 7.11, arrows in left panel). These velocities decreased downstream to about 29 cm/s at section B and 25 cm/s at section C. The deep transport across mooring section A was estimated to be 5.1 Sv, increasing to 10.7 Sv at section C (Dickson and Brown 1994). The increase of the transport of 2.2 Sv between the sill and mooring section A is assumed to be caused mainly by turbulent entrainment of overlying less dense water. The mean near-bottom temperature in the fast-flowing DSOW core of the slope current at section A was 1.1°C, lower than the temperature of the ISOW core south of Iceland. This difference is caused by the lower temperatures of the entrained water near the Denmark Strait, compared to the warm thermocline water in the Faroe Bank Channel. At section C the deep-water transport had increased by about 100% by incorporation of recirculating water in the deep slope current. This includes the 2.5 Sv contribution of the recirculating NEADW from the CGFZ. Near Cape Farewell, at the southern tip of Greenland, the deep slope current had increased further to 13.3 Sv (Dickson and Brown 1994).

7.7. Convection in the Labrador Sea

Already in 1935 Wüst noted that the Labrador Sea was the source region of the salinity minimum near $\sigma_\theta = 27.8$ in the North Atlantic Ocean. Then the formation mechanism of this LSW was not known. Sverdrup et al. (1942) stated in their classic textbook that both the intermediate and deep water types in the Labrador Sea were formed in winter by deep convection. In April 1962, the Woods Hole Oceanographic Institution surveyed the hydrography of the Labrador Sea in early spring with the Danish vessel Erika Dan. During that survey no large homogeneous water mass was found, implying that in the preceding winter deep convection was absent (Lazier 1973).

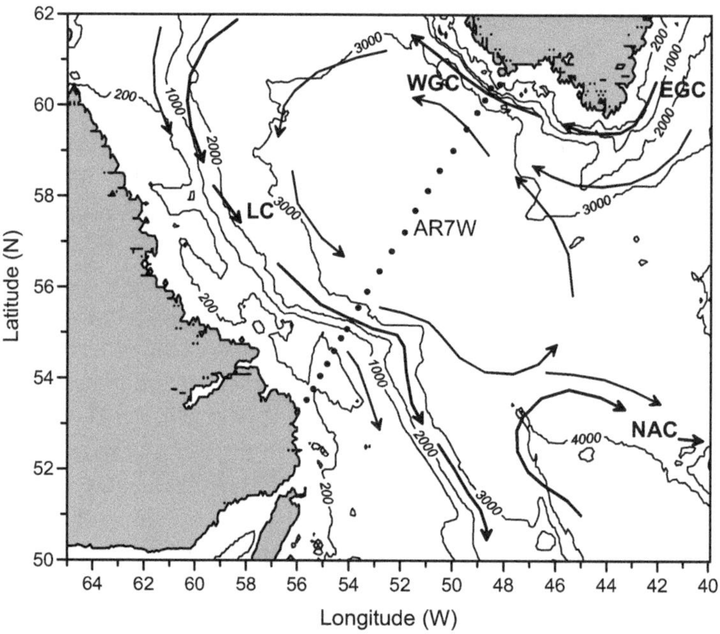

Fig. 7.12. Scheme of the cyclonic surface circulation in the Labrador Sea (arrows) that shows the East Greenland Current (ECG), West Greenland Current (WGC), Labrador Current (LC), and North Atlantic Current (NAC). The circulation forms part of the larger subarctic gyre circulation in the northern North Atlantic Ocean. The dots show the hydrographic stations occupied during the survey of WOCE hydrographic section AR7W by the Canadian oceanographer John Lazier with the research vessel Dawson.

In March 1976, late winter, the Canadian research ship Hudson found the first direct evidence of deep convection in the Labrador Sea (Clarke and Gascard 1983). A near homogeneous water column was encountered in the Labrador Sea, extending to depths of over 2000 m. However, 2 years later convection in winter reached less deep, indicative of interannual variability of the formation of LSW. During the WOCE a hydrographic section between Labrador and south Greenland, the AR7W hydrographic section, was surveyed annually from 1990 to 1998. This section ran through the region where deep convection was observed in 1976. The late 1980s and early 1990s (1987 to 1994) appeared to be a period of strong convective formation of LSW. Here we will use the survey from July 1990 with the Canadian vessel CSS Dawson as an example of the hydrographic situation in the Labrador Sea (the dots in Fig. 7.12).

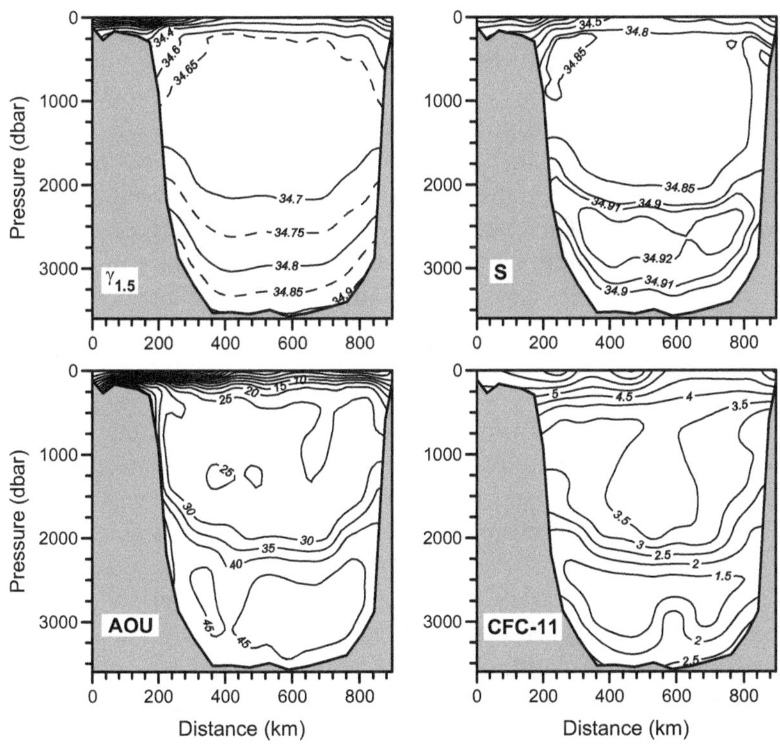

Fig. 7.13. Vertical distribution of the potential density relative to 1500 dbar, $\gamma_{1.5}$ (kg/m^3), salinity, *AOU* (µmol/kg), and CFC-11 concentration (pmol/kg) along the WOCE AR7W section in the Labrador Sea, surveyed in June 1990. The Labrador shelf and slope are located on the left side of the section and the Greenland slope on the right side.

The surface circulation in the Labrador Sea (arrows in Fig. 7.12) forms a subgyre of the subarctic cyclonic gyre. Due to this cyclonic circulation the isopycnals dome upward, bringing high-density water with low temperature and salinity stratification to the surface in the center of the Labrador Sea (Fig. 7.13, upper panels). The resulting density gradients sustain a northwestward flow in the West Greenland Current (WGC) along the continental slope of Greenland and a southeastward flow in the Labrador Current (LC) along the Labrador slope. A branch of the WGC crosses the northern Labrador Sea westward, feeding the LC. In the WGC relatively fresh surface water is observed, originating from the East Greenland Current (EGC). Also over the shelf and slope off Labrador fresh surface water is encountered (Fig. 7.13, upper right panel). It has been discussed in the previous chapter that part of that water comes from the Bering Strait through-flow. Both in the WGC and in the LC relatively saline water is encountered below the fresher surface layer. That is the saline water from the Irminger Current that is brought to the Labrador Sea at sub-surface levels in the EGC. In deep layers below 1500 dbar, the horizontal density gradients are reversed compared to the upper parts of the water column. These density gradients sustain a deep, downward increasing, cyclonic geostrophic flow which transports DSOW and NEADW along the slopes of the Labrador Sea. The combination of a baroclinic surface circulation as well as a deep baroclinic circulation, both with the same cyclonic rotation sense, results in a finite barotropic contribution to the cyclonic flow along the boundaries of the Labrador Sea.

In the center of the Labrador Sea the density as well as the salinity are nearly homogeneous between 500 and 2000 m (Fig. 7.13, top panels). This is a remnant of the chimney where deep convective mixing, driven by a surface heat loss to the atmosphere, had homogenized the water column in the previous winter. Restratification driven by eddies generated in the WGC and LC, the rim current of the Labrador Sea chimney, apparently has brought less saline surface water back to the center of the Irminger Sea. According to Pickart et al. (2002) a more saline variety of LSW, upper LSW is formed closer to the Labrador slope by admixture of the saline subsurface core, originating from the Irminger Sea. This water may leave the Labrador Sea relatively fast in the deep-reaching LC over the continental slope.

This flow and convection scheme is confirmed by the distribution of several tracers, e.g., salinity *AOU* and CFC-11 (Fig. 7.13, lower panels). Below the chimney that reached to about 2000 dbar, the saline core of the NEADW overlying the fresher DSOW is found. The water from the chimney in the center of the Labrador Sea is characterized by low salinities and *AOU* values and high CFC-11 concentrations. The underlying NEADW core with the salinity maximum has relatively high *AOU* values and low

CFC-11 concentrations, while in the fresher DSOW again lower *AOU* and higher CFC-11 values are encountered.

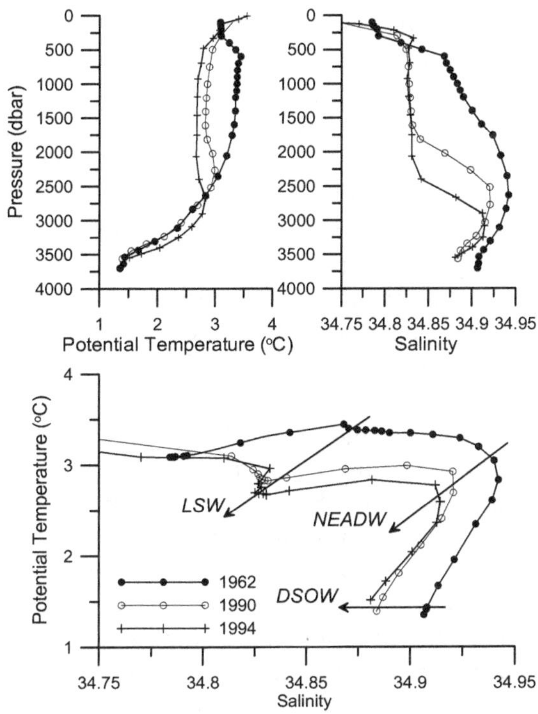

Fig. 7.14. Profiles of potential temperature (upper left panel), salinity (upper right panel), and θ–*S* diagram in the center of the Labrador Sea for the years 1962 (black dots), 1990 (open circles), and 1994 (crosses). The location of these profiles is halfway the AR7W section. In the θ–*S* diagram arrows indicate the position and trend of the main intermediate and deep water types.

It was already stated above that the formation of LSW has an interannual variability. This is shown by profiles of potential temperature and salinity as well as in the θ–*S* diagram for different years (Fig. 7.14). The data were obtained from the survey by the Danish Erika Dan in 1962 and the surveys of the WOCE AR7W section in 1990 and 1994. While in 1962, the potential temperature was nearly homogeneous between 700 and 1500 dbar, the salinity still showed a finite stratification in that pressure interval. After 1962, the water below 400 m remained stratified while the temperature slowly increased. This lasted until in the severe winter of 1972–1973 the surface water was sufficiently cooled and deep convection started again, decreasing the temperature and salinity of the deeper water.

In the early 1980s convection stagnated again. In 1990, both θ and S were nearly homogeneous from 500 to 1700 dbar, a result of deep-reaching convection in the previous winters. Winter convection in the Labrador Sea continued in the following winters and reached the deepest level in the winter of 1993–1994. In early June 1994, a nearly homogeneous layer was encountered between 500 and 2400 dbar. The temperature and salinity of the LSW core was significantly lower in the early 1990s compared to 1962 because of the downward convective mixing of cold and fresh surface water. The decrease of temperature and salinity in the LSW layer continued from 1990 to 1994. Apparently more low-salinity surface water was brought downward by this deep convection between 1990 and 1994, while the progressive heat loss to the atmosphere was responsible for the temperature decrease. The upper parts of the high-salinity NEADW core were eroded by the progressive deep convection, leading to a decrease of the salinity of the NEADW core from 1990 to 1994. Note here that also the θ–S properties of the DSOW changed over the years, especially between 1962 and 1990. This reflects a long-term trend of decreasing salinities in the overflow water from the Nordic Seas (Dickson et al. 2002).

Interannual to interdecadal variability in the formation rate and properties of LSW and of other convectively formed water types may have different causes, since it will depend on the initial stratification before the winter convection starts, varying surface properties, differences in preconditioning, and difference in wintertime heat loss to the atmosphere due to variations in atmospheric forcing. The initial stratification is related to the intensity of winter convection in the previous winters.

7.8. Bottom water formation in the Southern Ocean

In the Southern Ocean AABW is formed by several processes, involving interaction between ocean, atmosphere, sea ice, and shelf ice (Gammelsrød et al. 1994). The water mass formation occurs in the Weddell Sea, the Ross Sea, and along the Adelie coast.

In the Weddell Sea the main circulation is cyclonic (arrows in the top panel of Fig. 7.15). The WOCE S04A section, surveyed in 1996 (Fig. 7.15), shows that the hydrographic properties the deep incoming water along the eastern continental slope hardly differ from the center of the Weddell gyre. However, the slightly enhanced CFC-11 levels indicate that even there the deepwater already is influenced by relatively recent ventilation. In the top 1000 m a warm and saline water mass is found that, according to the high *AOU* and low CFC-11 values, is hardly ventilated. This water type is named Warm Deep Water (WDW), a remnant of the LCDW of the Southern Ocean, recirculating in the Weddell Sea, brought close to the surface by the cyclonic circulation.

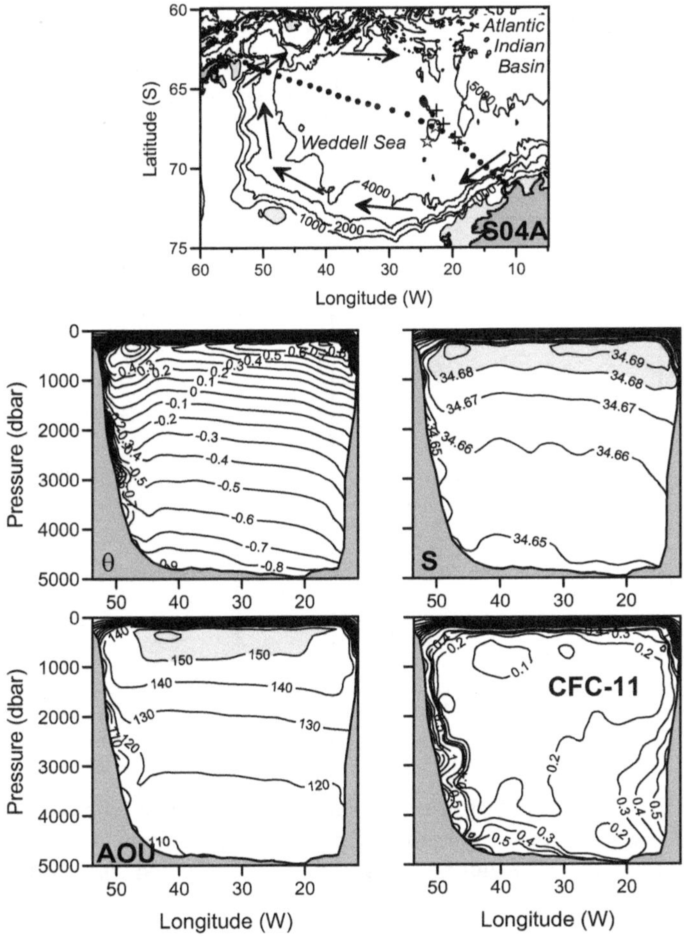

Fig. 7.15. Vertical distribution of potential temperature, θ (°C), salinity, *S*, *AOU* (μmol/kg), and CFC-11 (pmol/kg) along the near-zonal WOCE S04A section through the Weddell Sea, surveyed in April 1996 by the German ice breaking research vessel Polarstern (black dots in top panel). The crosses and stars in the top panel indicate hydrographic stations occupied in, respectively, 1973 and 1977.

Off the Antarctic Peninsula near 50°W, a cold (θ < 1.0°C) and relatively fresh (*S* < 34.65) water mass is encountered in the Weddell Sea near the bottom along the western slope of the basin. The relatively low *AOU* values (< 100 µmol/kg) and high CFC-11 values (>2.5 pmol/kg) are characteristic for recent contact with the atmosphere. This bottom water is formed in winter over the continental shelf of the Weddell Sea and descends along the continental slope to greater depths. During the descent along the continental slope the new high-density bottom water entrains overlying less dense water from the interior of the Weddell Sea and changes its characteristics accordingly. This newly formed AABW from the Weddell Sea is brought to the Atlantic–Indian Basin of the Southern Ocean with the cyclonic circulation in the Weddell Sea.

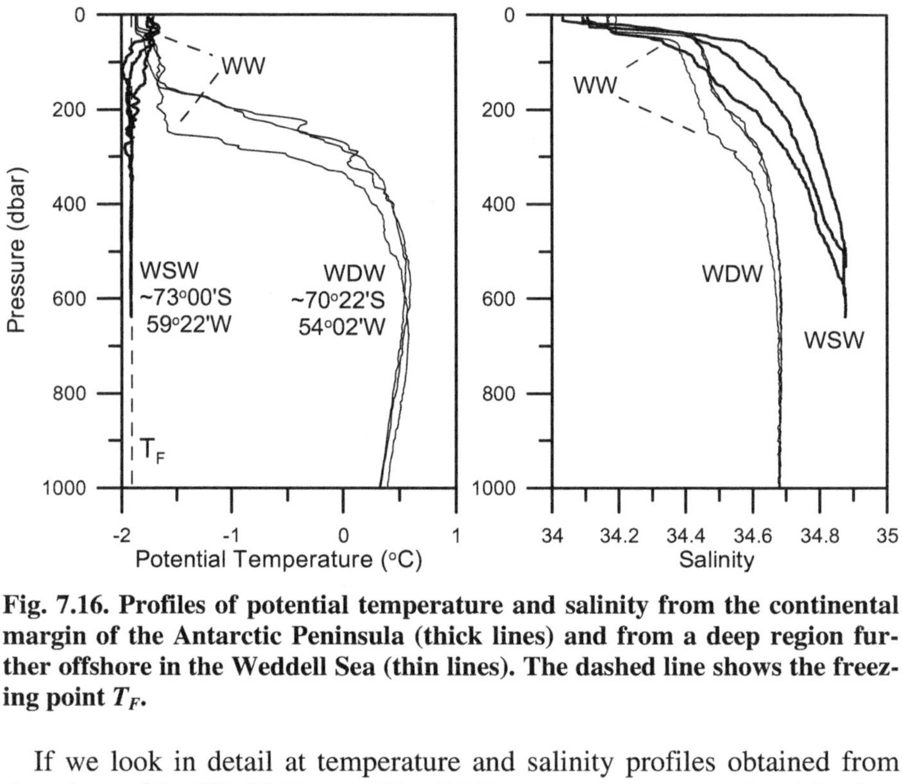

Fig. 7.16. Profiles of potential temperature and salinity from the continental margin of the Antarctic Peninsula (thick lines) and from a deep region further offshore in the Weddell Sea (thin lines). The dashed line shows the freezing point T_F.

If we look in detail at temperature and salinity profiles obtained from the edges of the Weddell gyre (Fig. 7.16) we can learn more about the formation process. Over the continental shelf regularly ice-free zones or polynyas are formed in winter, when the prevailing winds force the sea ice offshore. If such a polynya is formed along the coast or ice shelf, the sea ice cover no longer isolates the underlying water. By air–sea interaction the surface water loses large quantities of heat to the atmosphere. The Winter Water (WW) produced in this way is relatively fresh. When the decreasing temperature comes to the freezing point T_F ice formation starts, and due to the accompanying brine rejection the salinity of the underlying water starts to increase. The most saline water produced in this process is found near the Antarctic Peninsula, the Western Shelf Water (WSW), characterized by a salinity well above 34.7 (Fig. 7.16). The salinity is definitely higher than the salinity of the WDW in the deeper part of the Weddell Sea, further offshore of the Antarctic Peninsula.

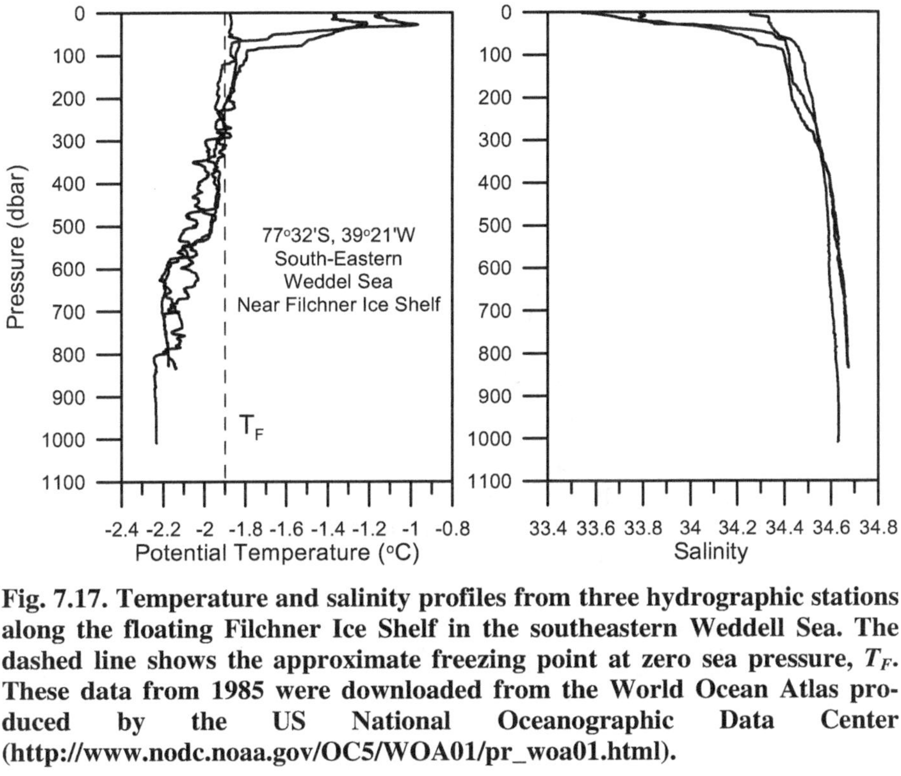

Fig. 7.17. Temperature and salinity profiles from three hydrographic stations along the floating Filchner Ice Shelf in the southeastern Weddell Sea. The dashed line shows the approximate freezing point at zero sea pressure, T_F. These data from 1985 were downloaded from the World Ocean Atlas produced by the US National Oceanographic Data Center (http://www.nodc.noaa.gov/OC5/WOA01/pr_woa01.html).

Along the antarctic continental slope the Weddell Sea is bordered by floating ice shelves that bring land ice from the continent to the ocean: the Ronne Ice Shelf in the west and the Filchner Ice Shelf in the east. Below this very cold ice mass the warmer seawater will induce ice melt. This will cool the seawater locally to the freezing point. Since this process occurs at a pressure of several hundreds dbar, the local freezing point will be lower than the freezing point at zero sea pressure, T_F (Fofonoff and Millard 1983). This pressure-induced lowering of the freezing point can be recognized from temperature profiles, collected in 1985 over the continental slope along the Filchner Ice Shelf (Fig. 7.17). These profiles show temperatures below the freezing point at zero pressure (–1.9°C) at subsurface levels, down to potential temperatures of about −2.2°C. This "super-cooled" water is known as Ice Shelf Water (ISW).

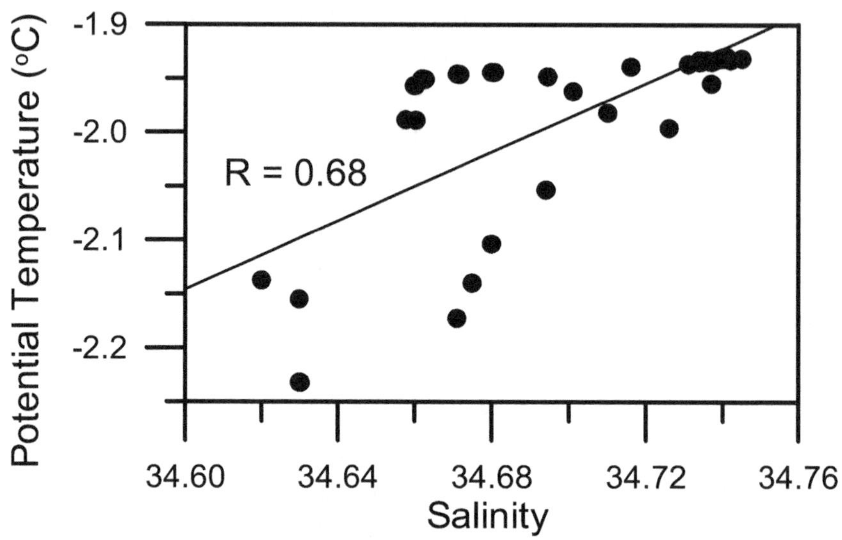

Fig. 7.18. Potential temperature–salinity diagram of Ice Shelf Water observed at the bottom near the Filchner Ice Shelf in the western Weddell Sea at pressures between 413 and 1133 dbar. The linear regression line shows a positive correlation between θ and *S* in the ISW of 0.68.

The salinity of the coldest ISW is between 34.6 and 34.7, lower than the salinity of the WSW. It can be expected that when more heat from the ISW is used to melt the lower side of the ice shelf, the ISW becomes more diluted with the fresh melt water from the ice shelf. Assuming a more or less homogeneous origin of the water that is transformed to ISW, e.g., WSW, this agrees with observations on ISW at the ocean bottom, where a positive correlation of R = 0.68 was found between potential temperature and salinity (Fig. 7.18).

Gammelsrød et al. (1994) have proposed a mixing scheme for the water mass formation which ultimately leads to the Weddell Sea Bottom Water (WSBW), the local variety of AABW (Fig. 7.19). According to this scheme WW becomes more saline when sea ice formation accompanied by brine rejection takes place. In this way WSW is formed over the deep continental shelf of the western Weddell Sea. This water type that reaches depths of over 500 m can flow under the floating ice shelves, where it is cooled and freshened by melting at the lower side of the ice shelf until the ISW properties are reached. Gammelsrød et al. (1994) have estimated from CFC-11 measurements that this water remains about 10 years under the ice shelf which has a typical melting velocity of 0.1 m/yr. The salinity decrease from WSW to ISW is of the order of 0.2, in agreement with dilution of a water layer of 175 m seawater with 1 m of melt water. The resulting ISW mixes with entrained overlying WDW when it descends along the continental slope into the abyssal basin of the Weddell Sea and WSBW is formed.

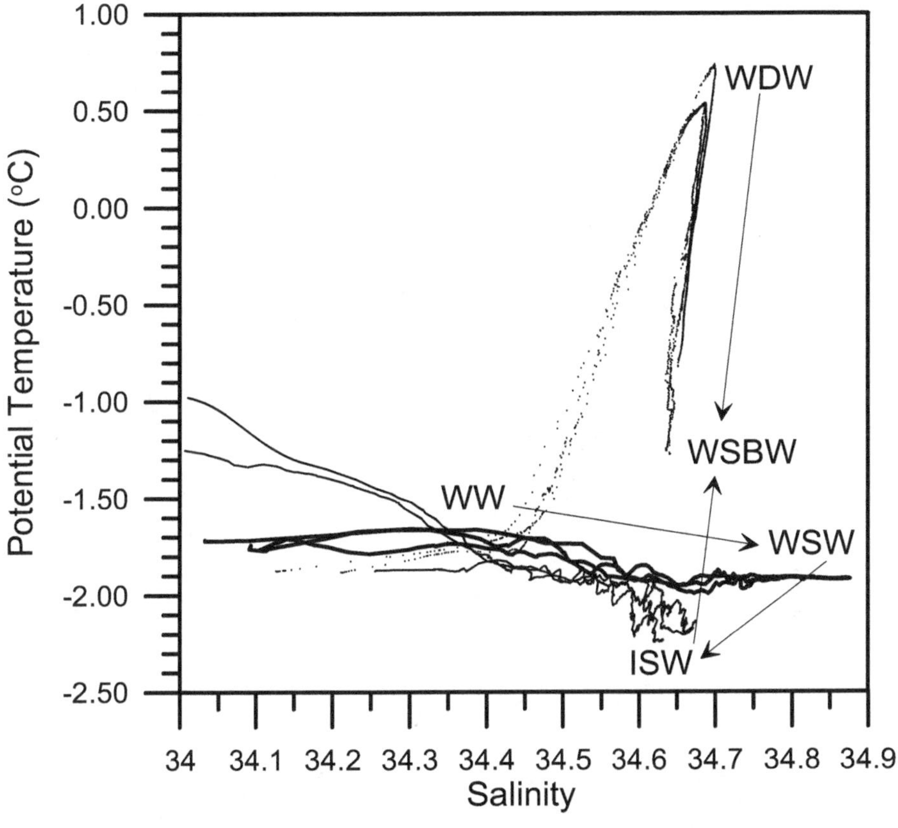

Fig. 7.19. Potential temperature–salinity diagram of the different water types that contribute to the formation of Weddell Sea Bottom Water (WSBW). The dotted line presents stations along the WOCE S04A section from the western continental rise and slope between 4500 and 2000 m depth. The thick line presents the hydrographic stations from the continental margin of the Antarctic Peninsula, shown in Fig. 7.16. The thin lines are from the hydrographic stations along the Filchner Ice Shelf shown in Fig. 7.17. The arrows indicate the direction of the water mass transformations, proposed by Gammelsrød et al. (1994).

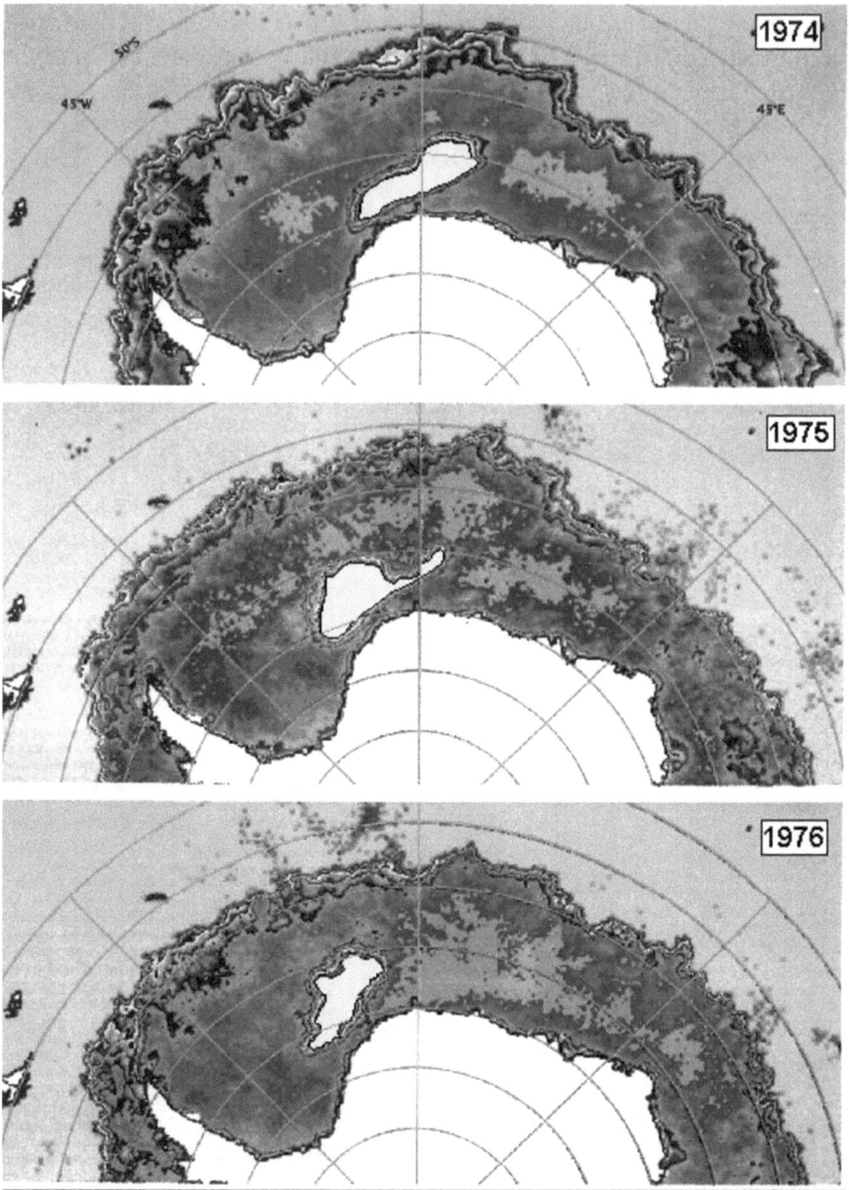

Fig. 7.20. The Weddell Polynya, the large open ocean ice-free region in the ice cover of the Weddell Sea for the austral winter month September in the years 1974 (top), 1975, and 1976 (bottom), observed with an electrically scanning microwave radiometer (SMMR) on board the Nimbus 5 satellite. The white region is the Antarctic land mass. The light gray is the polynya and the darker region the surrounding ice cover (from Gordon and Comiso, 1988).

In winter the ice cover around Antarctica stretches well north of 60°S. However, satellite observations, using passive microwave sensors, have shown that occasionally a vast region, up to 350,000 km^2, is completely free of ice, a polynya (Gordon and Comiso 1988). This happened in three successive years: 1974, 1975, and 1976 (Fig. 7.20). In the preceding austral winter of 1973, no polynya was observed. Such a huge open ocean polynya in winter allows large heat losses of the ocean to the atmosphere, similar to the ice-free Is Odden in the Greenland Sea. Because of these observations a cooperative US–USSR research program was initiated in the early 1980s. However, after 1976 the so called Weddell Polynya as a permanent winter feature was no longer observed. Only in 1994, the Weddell Polynya was observed as a short-term transient feature. Further east, near the Maud Rise at ~2°E and in the Cosmonaut Sea near ~40°E, occasionally open ocean polynyas are observed in winter (Comiso and Gordon 1987). In the summer the open ocean polynyas disappear with the seasonal melting of the sea ice.

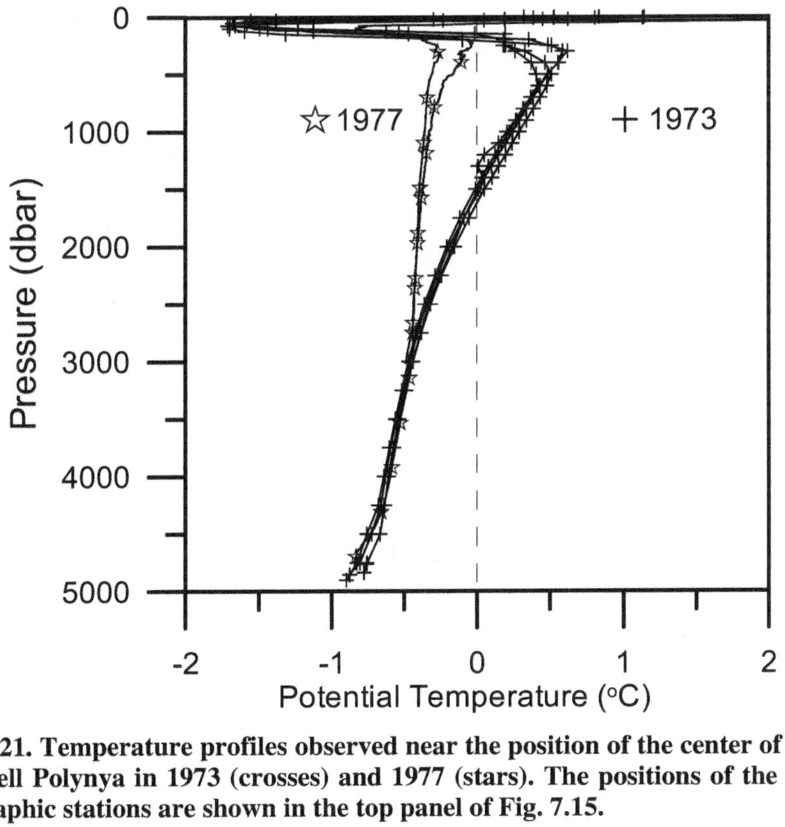

Fig. 7.21. Temperature profiles observed near the position of the center of the Weddell Polynya in 1973 (crosses) and 1977 (stars). The positions of the hydrographic stations are shown in the top panel of Fig. 7.15.

It happened that just before the emergence of the Weddell Polynya, in the summer of 1973–1974 as well as in the summer of 1976–1977 after the last polynya winter, hydrographic observations were carried out at stations in the region of the polynya, respectively, the crosses and stars in the top panel of Fig. 7.15. The temperature profiles from these hydrographic stations show the large influence of the presence of the polynya on the underlying stratification. In 1973, the temperature stratification was pronounced over the whole water column. A temperature maximum of over 0.5°C, connected with the warm WDW core was observed (Fig. 7.21), similar to the hydrographic survey of 1996, shown in Fig. 7.15. In 1977, temperatures above zero had disappeared completely. Apart from seasonal stratification in the upper 200 dbar, the water column was nearly homogeneous to a pressure of nearly 3000 dbar; only some restratification influenced the layer between 200 and 700 dbar slightly. This complete removal of the warm WDW core has supplied large amounts of heat to the atmosphere. The transient character of the open ocean polynyas in the Southern Ocean makes it difficult to quantitatively assess the long-term importance of these features for the water mass conversion and AABW formation in the Southern Ocean with only 33 years of satellite observations (1973 to 2006).

The WOCE section S04P through the Ross Sea, surveyed in 1992 by the Russian research vessel Akademik Ioffe, shows a distribution of hydrographic parameters, with similarities with the western Weddell Sea (Fig. 7.22). The bottom water west of 140°W had a potential temperature below zero, with a near-bottom minimum in salinity and *AOU* and a maximum in CFC-11. This is characteristic for recently ventilated bottom water. Along the continental slope at the western margin of the Ross Sea lower temperatures and higher *AOU* and CFC-11 values can be observed, characteristic for even younger high-density water, formed over the continental shelf in winter and descending along the continental slope into the abyssal basin. The temperature distribution reflects the structure of the density distribution. This distribution agrees with a deep geostrophic outflow of bottom water from the Ross Sea into the deep Pacific-Antarctic Basin near 140°W, as well as a fast northwestward geostrophic outflow of the younger bottom water along the continental slope. The latter transport may possibly bring bottom water from the Ross Sea along the antarctic continental margin toward the Australian–Antarctic Basin. There it is added to the bottom water formed over the continental shelf of the Adelie coast.

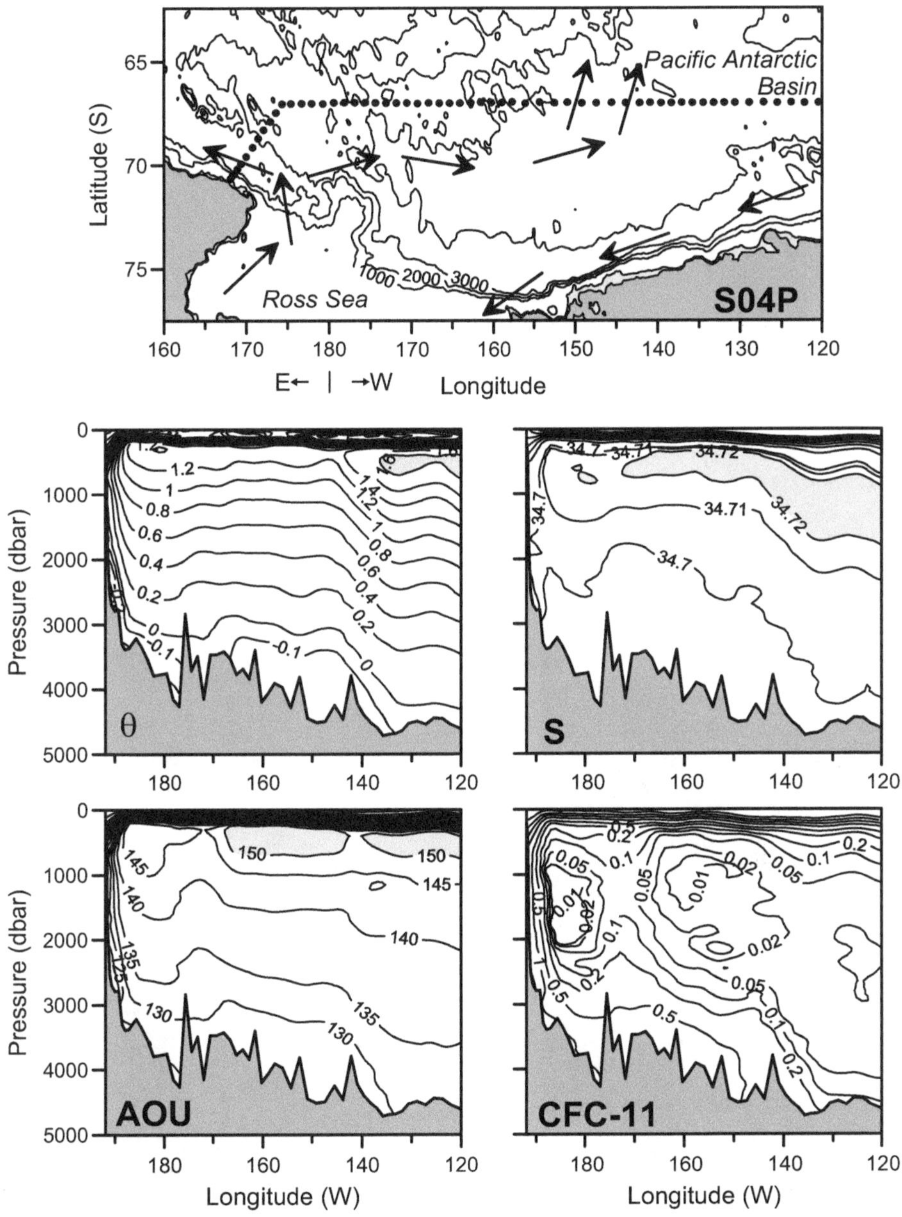

Fig. 7.22. Vertical distribution of potential temperature, θ (°C), salinity *S*, *AOU* (μmol/kg), and CFC-11 (pmol/kg) along the near-zonal WOCE S04P section through the Ross Sea, surveyed in February and March 1992, by the Russian research vessel Akademik Ioffe (black dots in top panel). The arrows in the top panel indicate the prevailing direction of the deep circulation.

8. Dynamics of the THC

8.1. Meridional overturning circulation

In Chapter 3 it was stated that in major parts of the ocean, outside the thin turbulent boundaries near the sea surface and the bottom, the dynamics of the ocean circulation are well described by the geostrophic balance. There the pressure gradient balances the Coriolis force. This even applies in a reasonable approximation to the deep western boundary currents where friction modifies the flow. The pressure gradient, or at least its vertical change, is related to the density distribution and can be derived from observations of temperature and salinity according to Eq. (3.15). In order to maintain a mean meridional flow in the ocean, like in the THC, a zonal pressure gradient is required that will vary with depth, supporting, e.g., in the North Atlantic Ocean an equatorward flow in the NADW core and a poleward flow near the surface.

However, on an earth that is completely water covered, with a constant water depth, the geostrophic balance does not allow a mean meridional transport at any level, as is illustrated in Fig. 8.1 (top panel). The zonally averaged meridional geostrophic velocity at a certain depth level, $\bar{v}$, is described by

$$f\bar{v} = \frac{1}{L}\int_{x(0)}^{x(360)} \frac{\partial P}{\partial x}dx = \frac{P(360^o) - P(0^o)}{L} = 0 \quad . \tag{8.1}$$

There the integral of the pressure gradient along the circumference L at a parallel around the earth equals zero, since the pressure is continuous and cyclic, repeating itself every 360°. However, if the isobar in the ocean is interrupted at least once by topography, the zonal integral of Eq. (8.1) in the ocean may have a value different from zero, as illustrated in the bottom panel of Fig. 8.1. There the pressure on the oceans on average decreases to the east, allowing a mean southward geostrophic velocity in the northern hemisphere. This shows that the presence of zonal boundaries, in the form of the continents or subsurface topography like mid-ocean ridges is essential for the maintenance of a geostrophic thermohaline circulation in the world ocean. As we will discuss later, the absence of zonal boundaries in the upper 1.5 km of the Southern Ocean at the latitudes of the Drake Passage creates a special problem. The finite zonal integral of the zonal pressure gradient, resulting when the ocean topography intersects isobaric surfaces, is referred to as topographic form stress.

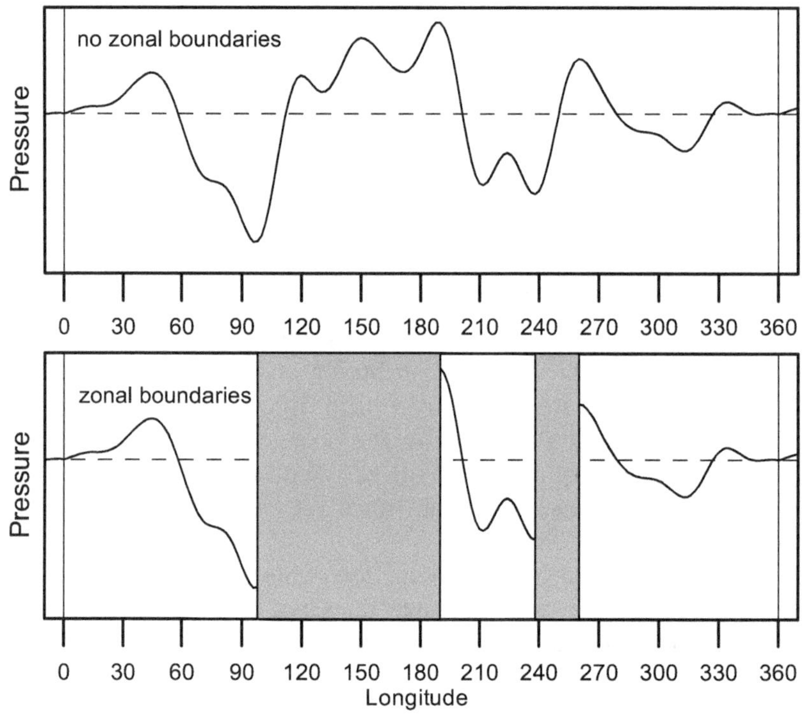

Fig. 8.1. Examples of the zonal pressure distribution along a parallel on a water-covered earth (top panel) and an ocean interrupted by continents (the gray blocks in the bottom panel).

In Chapter 3 it was shown that a zonal stress due to bottom friction will maintain a meridional transport in the bottom Ekman layer. In the ACC the bottom velocity is in general directed eastward, with a typical value of about 2 cm/s (Gille 2003). According to Eq. (3.21) such a bottom velocity will cause a bottom friction of the order of 0.4 10^{-3} Pa. At 60°S, this zonal friction will support a total meridional transport in the bottom Ekman layer according to Eq. (3.22) of ~0.06 Sv (1 Sv = 10^6 m^3/s). Even with a mean bottom velocity of 10 cm/s the meridional Ekman transport in the bottom layer of the Southern Ocean will be only 1.6 Sv. Moreover, this is a southward Ekman transport, opposite to the mean northward meridional transport of AABW. Apparently bottom friction is not able to support meridional transports with a magnitude of the order of 10 to 20 Sv, and the deep transport in the THC has to be described as a near-geostrophic flow in ocean basins, in balance with the topographic form stress, bound in the zonal direction by topography.

A characteristic of the meridionally overturning THC is that on large spatial scales the direction of the (geostrophic) velocity reverses vertically. Near the sea surface a return flow is directed toward the formation region of high-density deepwater, while at abyssal depths the mean flow is directed away from the source region of deepwater. One can question what the relation is between the mean vertical derivative of the deep flow and the large-scale density gradients. According to Eq. (3.17), the meridional geostrophic velocity, v, is:

$$v = \frac{1}{f}\left.\frac{\partial z}{\partial x}\right|_P \quad . \tag{8.2}$$

Following Eq. (3.15), with pressure as a vertical coordinate, the vertical derivative of v, or the current shear, is then:

$$\frac{\partial v}{\partial P} = \frac{1}{f}\frac{\partial^2 z}{\partial P \partial x} = -\frac{1}{f}\frac{\partial V}{\partial x} \approx \frac{1}{f}\frac{1}{\rho_0^2}\frac{\partial \rho}{\partial x} \quad , \tag{8.3}$$

where the capital V represents the specific volume and ρ_0 is a constant reference density, approximately equal to the density of seawater. According to Eq. (8.3) the vertical change of the meridional velocity is proportional to the zonal gradients of the specific volume or the density. A meridional density gradient, e.g., between subtropical and polar latitudes, apparently does not contribute to the vertical current shear that is characteristic for the THC. It only contributes to the current shear in the zonal flow. In this sense the meridional differences in density do not drive the THC.

Because of the meridionally overturning character of the THC, with a basinwide mean transport close to zero, the zonally averaged meridional velocity, $\overline{v}$, will equal zero at least at one pressure level, P_0: the basinwide level of no motion. If the velocity at a certain level is prescribed but does not equal zero, we generally name it the reference velocity, v_r, at the reference level, $P_r(x)$. One can determine $v(P)$ by integration of Eq. (8.3) from P_0 or P_r to P. The choice of P_0 or P_r and v_r should meet several physical constraints, at least that of mass conservation. The net meridional geostrophic transport of mass across a zonal section with width L in an ocean equals M_{gy}, according to

$$M_{gy} = \int_0^L \int_{-H(x)}^0 \rho v dz dx = -\frac{1}{g} \int_0^L \int_0^{P_{max}(x)} v dP dx \quad , \tag{8.4}$$

where $H(x)$ is the depth of the ocean at the zonal coordinate x and $P_{max}(x)$ the bottom pressure at that location. The meridional geostrophic volume flux V_{gy} can be approximated by M_{gy}/ρ_0, ρ_0 being a constant reference density. For the real ocean the value of M_{gy} is determined by interocean exchange like the through-flow through the Bering Strait or the Indonesian seas and by the Ekman transport in the wind-driven surface layer and the frictional bottom layer as well as by the hydrological constraints of evaporation, precipitation, and river runoff. In most practical cases the hydrological constraints and the Ekman transport in the bottom layer are ignored.

An illustration of the zonally averaged THC can be obtained from WOCE sections A03 and A10 in the Atlantic Ocean, already presented in Chapter 4 (Figs. 4.1, 4.14, and 4.15). For simplicity we will assume that the zonal cross section of these sections is rectangular with width L, ignoring further all effects of the topography. The near-bottom T and S values near the eastern and western ends of the sections are projected on the vertical sides of the rectangular cross sections. It is easy to derive for that situation from Eq. (8.3) the integral geostrophic balance,

$$L\frac{\partial \overline{v}}{\partial P} = -\frac{1}{f}[V(L) - V(0)] \quad . \tag{8.5}$$

This equation shows that the zonal mean velocity times the width of the ocean, or the vertical derivative of the volume transport in a rectangular ocean, only depends on the difference in specific volume or density across the ocean, not on the width of the ocean.

Along the A03 section the specific volume is higher near the European continent than near North America (Fig. 8.2, left panel). Only in the upper 200 dbar this gradient is clearly reversed because of the presence of low-salinity coastal water over the American continental shelf, while near 2000 dbar the mean gradient is close to zero. This large-scale zonal gradient maintains a northward geostrophic return flow near the sea surface and a southward flow of NADW at large depths, according to Eq. (8.5). This overturning circulation also is maintained in the South Atlantic Ocean. Since f changes sign from the northern (A03) to the southern (A10) hemisphere it can be expected that the zonal gradient in V or ρ in the upper part of the ocean at A10 is reversed compared with A03. This is confirmed by the observations along both sections for the upper 2500 dbar (Fig. 8.2, left panel). However, below 2500 dbar the horizontal specific volume gradient at the A10 section changes sign, indicating that below that level the vertical velocity shear also changes sign, corresponding with the northward flow of AABW in the near-bottom layer.

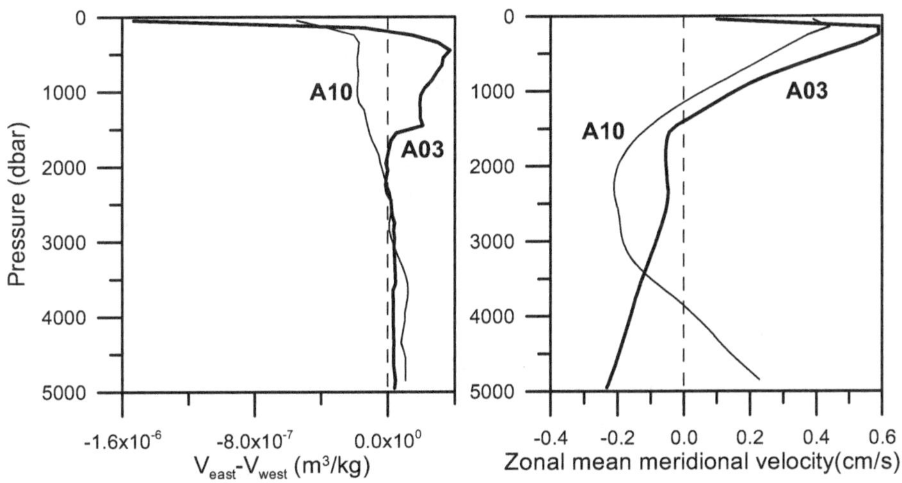

Fig. 8.2. The difference between the specific volume on the eastern and western side of the WOCE A03 and A10 sections in the Atlantic Ocean (left panel) and the corresponding zonal mean of the meridional velocity (right panel). For the calculation of $\overline{v}$ it is assumed that the wind-driven Ekman transport, limited to the upper 100 m, together with the overturning circulation are responsible for a net mass transport across the sections equal to zero. The thick lines represent the A03 section, the thin lines the A10 section.

If we assume that the total net mass transport across both rectangular sections, maintained by the geostrophic flow and the wind-driven Ekman transport, equals zero, we need a single mean reference velocity $\overline{v}_r$ at some reference pressure P_r to fulfill this constraint. The resulting profiles of the zonally averaged meridional velocity $\overline{v}$ of the overturning flow are shown in the right panel of Fig. 8.2. At section A03 in the North Atlantic the meridional overturning consists of a northward flowing return flow above 1500 dbar and a southward flowing NADW layer below that level. The magnitude of the southward flow of NADW in this highly simplified inverse model equals 24 Sv. At section A10 a three-layer system of the overturning THC fits the density observations, with a northward return flow in the upper 1200 dbar, a southward flow of NADW between 1200 and 3900 dbar, and northward-flowing AABW below 3900 dbar. Here the net southward transport of NADW amounts to 23 Sv. The mean magnitude of the zonally averaged meridional velocity per transport layer involved in the overturning circulation at both sections varies between 1.1 and 2.3 mm/s.

Here we have used a simplified rectangular ocean model to estimate in an inverse way the thermohaline overturning from the large-scale characteristics of the observed density field. Although it gives qualitatively the right results, the rectangular ocean model contains several errors because of its high degree of simplification. The width of the real ocean decreases at increasing pressure levels. This is due partly to the presence of a finite slope at the continental boundaries and partly by the presence of the MAR and other topographic features. The presence of the MAR, ignored in our rectangular ocean, may cause an overestimate of the overturning transport. Because of the zonally averaged character of the model, any contributions from the correlation between the water depth and the density gradient are ignored. The latter is responsible for the strong influence of the presence of fresh coastal water over the North American shelf on the small contribution of the upper 200 dbar to the near-surface return flow at section A03. A method to estimate the meridional overturning circulation from basinwide density gradients in a more realistic topography is applied in the monitoring of the meridionally overturning circulation by a research group from the National Oceanography Centre in Southampton as part of the Rapid Climate Change program (http://www.noc.soton.ac.uk/rapid/rapid.php). Their analysis of the zonal density distributions, determined for five hydrographic surveys of the Atlantic Ocean near 25°N, has suggested a decreasing trend in the strength of the THC from 22.9 Sv in 1957 to 14,8 Sv in 2004 (Bryden et al. 2005). However, this trend is only marginally significant, and only if the short-term variability of the THC can be ignored.

In more advanced inverse models the ocean topography will be more realistic and additional constraints may be prescribed, e.g., conservation of heat, salt, potential vorticity, and dissolved substances. Often continuity of the mass transport between two successive sections in different density intervals is added. With additional simplifications the constraints may lead to an overdetermined set of equations describing the reference velocity v_r at a reference pressure P_r which can be estimated by a linear least-squares fit. However, in most cases the set of equations resulting from the constraints is underdetermined with less equations than unknowns, and sometimes it also is inconsistent. In order to solve such sets of equations in an optimal way, inverse problem theory is used (Tarantola 1987), for which specific oceanographic methods have been developed (Fiadeiro and Veronis 1984; Wunsch 1996). The optimal solution for such underdetermined inverse problems is the solution that minimizes some specified penalty function, e.g., the summed squares of the reference velocities between successive hydrographic stations.

Several estimates exist of the magnitude of the volume transport of the meridional overturning circulation, all based on inverse modeling efforts. Macdonald and Wunsch (1996) estimated with a global inverse model the overturning circulation, with a surface layer and a deep and bottom water layer separated by the 3.5 °C isothermal surface. In the lower layer of the Atlantic Ocean a net volume transport of 17(±5) Sv NADW minus AABW entered the Southern Ocean at 30°S. Near that latitude 18 Sv cold deep and bottom water entered the Indian Ocean northward, while only 3 Sv flowed into the Pacific Ocean near 28°S. The latter transport is quite small, because also the southward return flow of the aged PDW was contained in the cold deep and bottom layer. Only these 3 Sv that entered the Pacific were converted to water in the warmer surface layer. The main part of the ITFW, flowing from the Pacific to the Indian Ocean, probably originated from the inflow of warmer intermediate and thermocline water ($\theta > 3.5$°C) from the Southern Ocean.

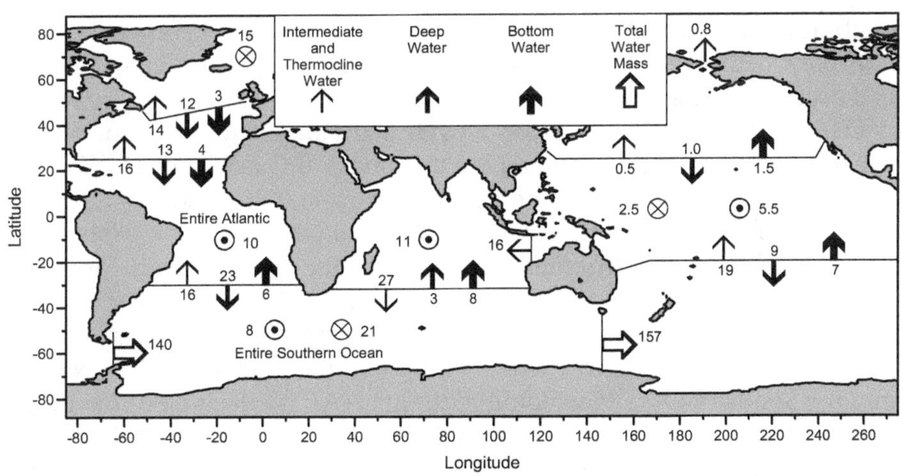

Fig. 8.3. Volume transports in Sv (10^6 m^3/s) of the thermohaline overturning circulation for three layers, derived by Ganachaud and Wunsch (2000) by means of an inverse model, using hydrographic section data collected during WOCE as well as current meter observations, mass conservation, climatological wind fields that drive the surface Ekman transport, and biogeochemical tracer constraints.

Ganachaud and Wunsch (2000) discriminated three density layers in their inverse model that used the global data, collected during WOCE (Fig. 8.3). They derived a net production of NADW of 15(±12) Sv in the North

Atlantic Ocean, and a net production of 21(±6) Sv AABW in the Southern Ocean. In the Indian Ocean a net amount of 11 Sv of AABW and LCDW was converted to NIDW and shallower water masses. A net amount of about 7(±2) Sv AABW and LCDW entered the Pacific Ocean, while within the accuracy of the estimates, a nearly equal amount of 9(±3) Sv PDW returned to the Southern Ocean. The ITFW originated completely from the intermediate and thermocline water in the Pacific Ocean which entered the Pacific from the Southern Ocean.

Sloyan and Rintoul (2001) used an inverse model of the Southern Ocean with observations mainly from the pre-WOCE period. The northern boundary of their model was located between 30 and 40°S. They arrived at an import of 17 Sv of NADW from the South Atlantic. The total formation rate of AABW was estimated to be 25 Sv, of which 6 Sv flowed into the South Atlantic, 10 Sv into the Indian Ocean, and 8 Sv into the Pacific Ocean. Additionally 18 Sv LCDW entered the Indian Ocean and 18 Sv entered the South Pacific. The return flow from the Indian Ocean as NIDW and from the Pacific as aged PDW nearly completely compensated the deep inflow of AABW and LCDW. Only in the Indian Ocean about 3 Sv of deepwater was converted to intermediate water. However, it is questionable whether that number significantly differs from zero, given the limited accuracy of the inverse model estimates.

Overall the inverse model of Sloyan and Rintoul (2001) shows a reduced export of NADW compared to the model of Ganachaud and Wunsch (2000), and a stronger ventilation of the deep and bottom waters in the Indian and Pacific oceans. However, in these and other inverse models of the THC the authors use different definitions of the water masses for which they calculate the transports. This makes it difficult to compare the results of different inversions quantitatively. Qualitatively the results of the different inverse models agree, with an inflow of the order of ~20 Sv NADW into the Southern Ocean including returning AABW, flushing of the Indian and Pacific Ocean by AABW and LCDW, and a return of aged NIDW and PDW. According the inverse model results export of aged deepwater to the Southern Ocean seems to be an order of magnitude more important than upwelling into the intermediate layers and the thermocline in the Indian and Pacific basins. The Indonesian through-flow appears to be only loosely connected with the THC. This qualitative circulation scheme roughly agrees with the interpretation of hydrographic data, presented in previous chapters.

8.2. Upwelling and divergence of the abyssal circulation

The sources of deepwater in the ocean are presently located at high latitudes either in the northern or the southern hemisphere. From there the deepwater spreads toward the equator while part or all of it leaves the deep layer by upwelling to shallower levels. The deep and bottom water in the North Pacific Ocean and the northern Indian Ocean are flushed by a deep flow across the equator. We will consider here what basic type horizontal structure is connected with the abyssal circulation, following the ideas developed in the 1950s by Henry Stommel.

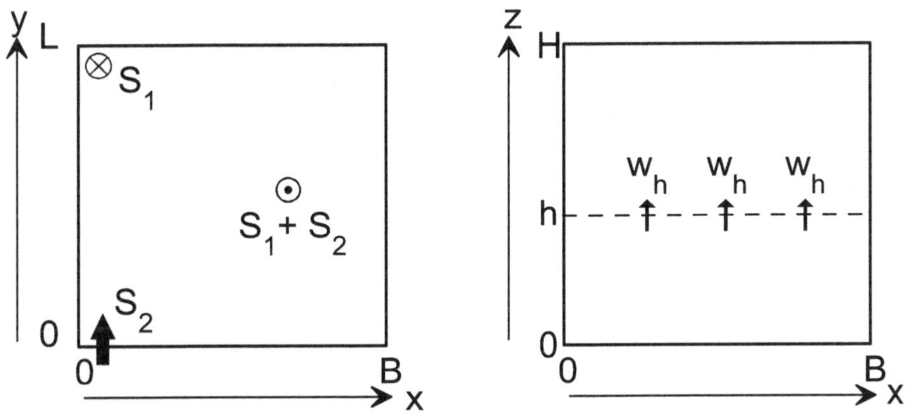

Fig. 8.4. The schematic northern single-hemispheric rectangular ocean used to study the horizontal structure of the deep flow below the level $z = h$. The symbol in the top of the left panel shows the position of a northern source S_1 of deepwater, not unlike the source of NADW in the North Atlantic. The arrow near the bottom of the left panel shows an inflow S_2 across the equator. The vertical velocity at the top of the deep layer at $z = h$ equals w_h.

For simplicity we use here a flat-bottomed single-hemispheric rectangular ocean with fixed boundaries at $x = 0$, B, $y = 0$, L. The northern boundary of the abyssal ocean is located at $y = L$, e.g., the latitude of the Greenland–Scotland Ridge in the North Atlantic Ocean. The deep flow is assumed to be restricted to the bottom layer below the level $z = h$. (Fig. 8.4). In the northwestern corner of this ocean basin a source of deepwater can be located, mimicking, e.g., the source of NADW in the North Atlantic. That source supplies the deepwater to the ocean basin with a rate of S_1 m^3/s. A deep cross-equatorial import of S_2 at $y = 0$ is indicated by the arrow in the southwestern corner. The sum of both deepwater imports, $S_1 + S_2$, equals S. For the North Atlantic Ocean S_2 is negative, representing the net cross-equatorial export of NADW to the South Atlantic Ocean. In the North Pacific Ocean S_1 equals zero, since in that ocean basin no deep-water is formed. There the positive S_2 represents the import of deep and bottom water from the South Pacific. It is easy to adapt this model to the southern hemisphere. Then the results are mirrored relative to the equator.

Over the whole basin large-scale upwelling occurs with a vertical velocity w_h at the level $z = h$ above the bottom ($z = 0$). Following Stommel (1958) and Stommel and Arons (1960) we assume, because of lack of detailed knowledge, that in zeroth order w_h is constant and does not depend on latitude y or longitude x, so that $S = w_hBL$; S_1 may be zero and S_2 may be either positive or negative, as long as S is positive. In order to describe the mass conservation in the deep ocean and its related flow structure, we will use the continuity equation (3.30), repeated here:

$$\frac{\partial u}{\partial x} + \frac{\partial v}{\partial y} + \frac{\partial w}{\partial z} = 0 \tag{8.6}$$

The parameters u, v, and w are the velocity components in, respectively, the x, y, and z direction. The vertical velocity w equals zero at the flat bottom ($z = 0$). By vertical integration of Eq. (8.6) to the level $z = h$ we can derive a simple relation between the horizontal divergence of the abyssal volume transport density (U_A,V_A) and the upwelling velocity w_h at the level $z = h$:

$$\int_0^h \left(\frac{\partial u}{\partial x} + \frac{\partial v}{\partial y} \right) dz = \frac{\partial U_A}{\partial x} + \frac{\partial V_A}{\partial y} = -w_h \quad . \tag{8.7}$$

Integration of Eq. (8.7) from the western to the eastern ocean boundary, assuming that U_A equals zero at these closed side walls, results in

$$\int_0^B \frac{\partial V_A}{\partial y} dx = -\int_0^B w_h dx = -\frac{S}{L} \quad . \tag{8.8}$$

Since S and w_h are positive (deep upwelling!) the meridional derivative of the volume flux is negative. The mean total southward flow will decrease linearly to the south or the mean total northward flow will decrease to the north, in each case feeding the warm upper branch of the THC by upwelling.

The magnitude of the mean value of v is rather small. With a meridional volume flux of NADW of 15 Sv (1 Sv = 10^6 m^3/s) over a layer of 2000 m thickness in the subtropical North Atlantic ($B \approx 7000$ km) v will be of the order of 1 mm/s. This magnitude agrees with the typical magnitude of the mean meridional geostrophic velocity across the WOCE A03 section, presented in the previous section (Fig. 8.2). If with a global mean deepwater production of the order of 40 Sv (NADW + AABW) upwelling occurs in about 80% of the ocean ($2.9 \cdot 10^{14}$ m^2 of the $3.6 \cdot 10^{14}$ m^2), the mean upwelling velocity will be 1.4 10^{-7} m/s or about 4.3 m/year, several orders of magnitude smaller than in the wind-driven coastal upwelling regions.

8.3. Geostrophic flow in the abyssal ocean

Stommel (1958) and Stommel and Arons (1960) assumed that, except for boundary currents, the dynamics in the deep ocean interior is well approximated with the geostrophic balance. While the wind-driven surface circulation can be described as a quasigeostrophic flow with a horizontal structure that responds to the vertical motion imposed by wind-driven Ekman pumping, the horizontal structure of the deep abyssal circulation should supply a convergent mass flow for the vertical upwelling motion at middepth, imposed by the overturning circulation. Note here that the prescription of upwelling at middepth is a boundary condition, not the result of the dynamics of the deep flow. The geostrophic flow in the homogeneous bottom layer is described by

$$\begin{aligned} fu_g &= -\frac{1}{\rho_0}\frac{\partial P}{\partial y} \\ fv_g &= \frac{1}{\rho_0}\frac{\partial P}{\partial x} \end{aligned} \quad , \tag{8.9}$$

where u_g and v_g are the geostrophic velocity components, f is the Coriolis parameter, ρ_0 a reference density, and P the pressure. By differentiating these equations to, respectively, x and y and subsequent addition we get rid of the pressure term, leaving a relation between u_g and v_g:

$$f\left(\frac{\partial u_g}{\partial x} + \frac{\partial v_g}{\partial y}\right) + v_g \frac{\partial f}{\partial y} = 0 \quad . \tag{8.10}$$

Stommel and Arons (1960) solved these equations in a homogeneous bottom layer on a spherical earth. They assumed homogeneity of the density ensures that the geostrophic velocity components do not depend on depth in the bottom layer and they are locally barotropic. We will assume here that the length L of the basin is large enough to account for a significant meridional change of f to allow the β–plane approximation, $f = \beta y$ with

$\partial f/\partial y = \beta \approx$ constant. Also in this approximation f equals zero at the equator ($y = 0$). With the continuity equation (8.6), Eq. (8.10) reduces to

$$f\frac{\partial w}{\partial z} = \beta y\frac{\partial w}{\partial z} = \beta v_g \quad . \tag{8.11}$$

Equation (8.11) represents the so-called vorticity balance for a rotation-free flow. The change of the vertical component of the planetary vorticity of a water particle, f (f = twice the earth's angular velocity around a locally vertical rotation axis), due to a change in latitude of a water particle (βv_g or advection of planetary vorticity) requires stretching/compression of the water column or vortex stretching, described by $f\partial w/\partial z$. Integrating Eq. (8.11) vertically over the bottom layer h results in

$$v_g = y\frac{w_h}{h} \quad . \tag{8.12}$$

Since w_h is positive, the abyssal flow in the ocean interior is poleward, both on the northern ($y > 0$, $v_g > 0$) and the southern ($y < 0$, $v_g < 0$) hemispheres. Because we have used the geostrophic balance, cross-equatorial flow is not allowed in the geostrophic ocean interior, which is confirmed by Eq. (8.12) for $y = 0$. For the present deep circulation with deep and bottom water production at high (arctic and antarctic) latitudes and cross-equatorial flow of NADW and AABW, an additional nongeostrophic flow system is required next to the geostrophic ocean interior to bring the cold deep water masses toward and across the equator. Note that for the derivation of Eq. (8.12) we have not yet used the assumption that w_h is constant. If this upwelling velocity is locally zero, then v_g is zero and the abyssal flow there will be exclusively zonal.

From Eq. (8.12) we can derive an expression for the divergence of the meridional geostrophic transport in the deep ocean:

$$h\frac{\partial v_g}{\partial y} = w_h \quad . \tag{8.13}$$

Integrated from the western to the eastern ocean boundary (8.13) becomes

$$\frac{\partial}{\partial y}\int_0^B h v_g dx = \int_0^B w_h dx = +\frac{S}{L} \quad . \tag{8.14}$$

The deep poleward flow not only is directed to the pole, it also increases in the poleward direction since w_h is positive. This result seems rather counterintuitive for a circulation in ocean basins with either a polar source S_1 in arctic or antarctic regions or a near-equatorial input S_2. Comparison of Eq. (8.14) with Eq. (8.8) shows that the divergence of the geostrophic poleward transport is equal in magnitude to the divergence of the total abyssal transport, but with an opposite sign.

Integration of Eq. (8.6) for the geostrophic velocity over the vertical between the ocean bottom and the level $z = h$ and substitution of Eq. (8.13) for the divergence of the meridional geostrophic velocity gives an expression for the divergence of the zonal geostrophic velocity,

$$h\frac{\partial u_g}{\partial x} = -h\frac{\partial v_g}{\partial y} - w_h = -2w_h \quad , \tag{8.15}$$

that agrees with an eastward flow decreasing linearly to the east. Assuming as boundary condition for Eq. (8.15) that the zonal velocity, u_g, at the eastern boundary of the ocean equals zero, this results in the following expression for u_g:

$$u_g = (B - x)\frac{2w_h}{h} \quad . \tag{8.16}$$

Streamlines, corresponding to the solution for the deep geostrophic abyssal flow, given by Eqs. (8.12) and (8.16) are shown in Fig. 8.5.

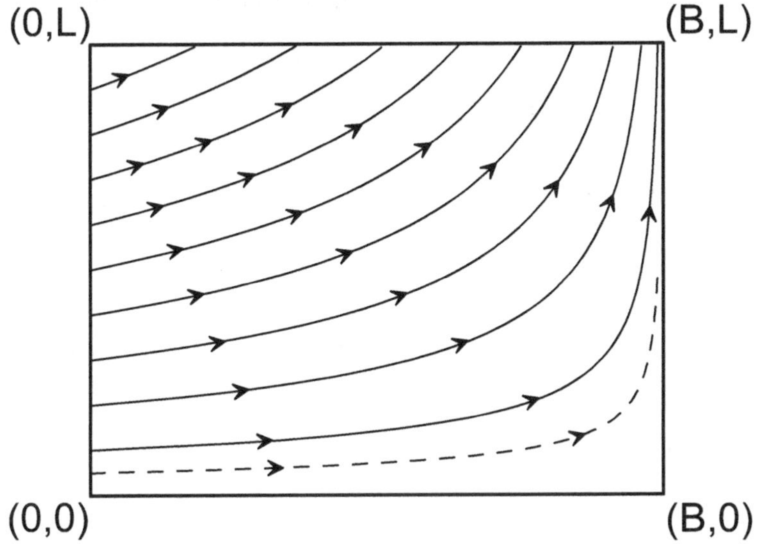

Fig. 8.5. Streamlines corresponding to an interior geostrophic abyssal flow in the northern hemisphere with a constant deep upwelling velocity w_h at a level h above the flat bottom, given by Eqs. (8.12) and (8.16).

This solution for the geostrophic interior flow has velocity components different from zero, perpendicular to the western and northern boundaries. Continuity of the mass transport requires that at those boundaries additional nongeostrophic boundary currents are present, which are discussed in the next section.

8.4. Deep boundary currents

The abyssal geostrophic circulation, derived above, requires an input of water from the western boundary of the basin. This water can be supplied by a deep western boundary current (DWBC) governed by a non-geostrophic balance. The DWBC is present in a narrow strip with width εB ($\varepsilon << 1$) along the western boundary of the ocean basin. In this boundary layer the velocity v_{BC} is nongeostrophic, with a volume flux V_{BC} equal to

$$V_{BC} = \int_0^{\varepsilon B} \int_o^h v_{BC} \, dz \, dx \quad . \tag{8.17}$$

The zonally integrated meridional volume flux density, as used in Eq. (8.8), equals the sum of the integrated meridional volume flux density in the geostrophic ocean interior and the nongeostrophic volume flux in the DWBC, according to

$$\int_0^B \frac{\partial V_A}{\partial y} dx = \int_0^B h \frac{\partial v_g}{\partial y} dx + \frac{\partial}{\partial y} V_{BC} \quad . \tag{8.18}$$

Substitution of Eqs. (8.8) and (8.14) into Eq. (8.18) then gives

$$\frac{\partial}{\partial y} V_{BC} = -\frac{2S}{L} \quad . \tag{8.19}$$

So the volume flux in the DWBC changes twice as fast in meridional direction as the total meridional mass flux in Eq. (8.8). The solution of Eq. (8.19) is

$$V_{BC} = V_0 - y \frac{2S}{L} \quad , \tag{8.20}$$

where V_0 is determined by the boundary conditions, given by the source terms S_1 and S_2.

When in a single-hemispheric ocean basin a single deepwater source is present in the northwestern corner, V_{BC} is given by $V_{BC} = -y \cdot 2S_1/L$. That implies that near the source region of deepwater the DWBC transports twice as much water as given by the deepwater source, while an equal amount of recirculating deepwater is supplied from the geostrophic interior regime and brought to the source region by a deep northern boundary current (DNBC) (Fig. 8.6, left panel). If in a similar single-hemispheric ocean basin the deepwater in the DWBC is supplied by a cross-equatorial transport S_2, the transport of the DWBC is given by $V_{BC} = S_2 - y2S_2/L$. This implies that halfway the DWBC the flow direction reverses, since in its northern parts the DWBC transports recirculating water originating from the DNBC: a southward flow. The westward-flowing DNBC in the northwestern corner has a magnitude S_2, similar to the cross-equatorial transport (Fig. 8.6, right panel). The DNBC is fed by the interior geostrophic flow. It is easy to derive that, since v_g does not depend on x, the westward transport in the DNBC increases linearly to the west.

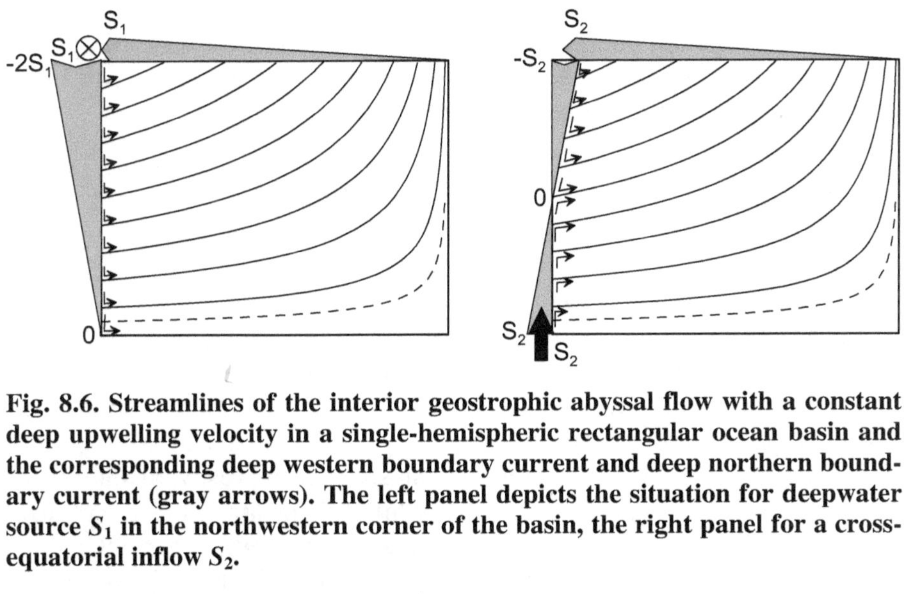

Fig. 8.6. Streamlines of the interior geostrophic abyssal flow with a constant deep upwelling velocity in a single-hemispheric rectangular ocean basin and the corresponding deep western boundary current and deep northern boundary current (gray arrows). The left panel depicts the situation for deepwater source S_1 in the northwestern corner of the basin, the right panel for a cross-equatorial inflow S_2.

If the DWBC is in complete geostrophic balance, Eq. (8.11) also will apply to that part of the abyssal circulation, reducing the total net upwelling in the boundary current and the ocean interior combined to zero. The specific dynamics of a westward intensified boundary current prevents this occurrence. Some qualitative insight in the dynamics of the DWBC can be gained from a simple approximation of the balance of forces, whereby the friction, τ, is assumed to have only a meridional component, opposite to the mean flow direction. That implies that for the meridional velocity component, v_{BC}, the geostrophic balance applies, while the Coriolis force due to the zonal component, u_{BC}, is balanced by both the friction and the pressure gradient:

$$
\begin{aligned}
fv_{BC} &= -\frac{1}{\rho_0}\frac{\partial P}{\partial x} \\
fu_{BC} &= \frac{1}{\rho_0}\frac{\partial P}{\partial y} + \frac{\tau}{\rho_0}
\end{aligned}
\quad . \tag{8.21}
$$

This is a simplification of the balance of forces sketched in Fig. 3.9. We will not go into detail whether τ is determined by lateral friction or bottom friction, but we will assume that the quadratic parameterization of τ in the y direction according to Eq. (3.21) can be applied:

$$
\frac{\tau}{\rho_0} = -Cv_{BC}\left|v_{BC}\right| \quad , \tag{8.22}
$$

where C is a friction coefficient. By differentiation of the Eqs. (8.21) to y and x, respectively, and then adding the results, the pressure term is removed and we get the frictional form of Eq. (8.11):

$$
f\left(\frac{\partial u_{BC}}{\partial x} + \frac{\partial v_{BC}}{\partial y}\right) = -f\frac{\partial w_{BC}}{\partial z} = -v_{BC}\beta - v_{BC}2C\frac{\partial\left|v_{BC}\right|}{\partial x} \quad . \tag{8.23}
$$

The torque exerted by the friction adds to the vorticity budget. Therefore, vortex stretching is not the only possible way to close this budget, like in Eq. (8.11). We will first assume that the left-hand side of Eq. (8.23) which represents the effects of vortex stretching on the vorticity can be ignored because of absence of deep upwelling in the DWBC. Then the frictional torque should balance the planetary vorticity advection, $v_{BC}\beta$. The sketch of the current and lateral current shear in a westward intensified boundary current (Fig. 8.7) shows that while β is positive, the x derivative of the absolute velocity is always negative in a westward intensified boundary current. In that case both terms on the right-hand side of Eq. (8.23) have opposite signs, independent of the direction of the boundary current. If the width of the DWBC, εB, does not alter, the frictional torque will increase faster in magnitude than the vorticity advection $v_{BC}\beta$ when v_{BC} increases in magnitude. Therefore, these terms may completely compensate when $2C\partial|v_{BC}|/\partial x = -\beta$, generating a dynamic vorticity balance in the westward intensified DWBC.

In general these opposite terms will reduce the necessity of a divergence term, the left-hand side of Eq. (8.23), in the boundary current, and thereby it can reduce any upwelling velocity w_h to zero if both remaining terms balance completely in the DWBC. In that case the existence of a westward intensified boundary current will not reduce the overall mean upwelling in the total ocean basin, boundary current plus geostrophic interior.

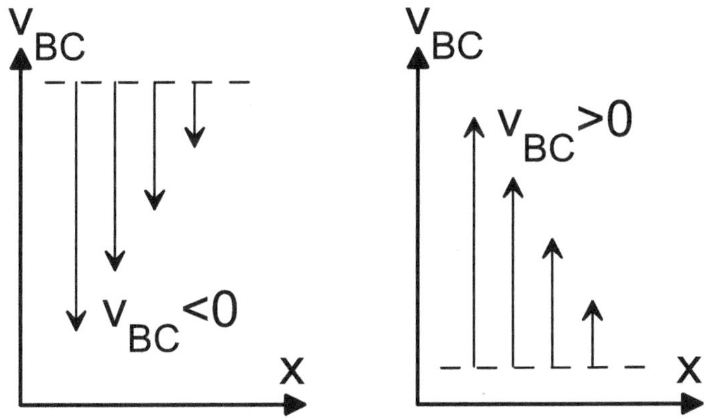

Fig. 8.7. Scheme of the lateral current shear in a westward intensified western boundary current with southward (left panel) or northward (right panel) flow. In both cases the *x* derivative of the meridional velocity has a sign opposite to the sign of the velocity in the boundary current, v_{BC}.

The extremely simple model, presented above for the description of the horizontal structure of the cold branch of the thermohaline circulation, is not intended as a realistic description of the actual ocean. It does not cope with the compartmentalization of the deep ocean basins by the presence of meridionally extended midocean ridges and zonal ridges between subbasins, with only narrow sills allowing a connection between these basins. The model also does not cope with the present situation with deepwater formation both at arctic and antarctic latitudes which leads to a vertical stratification of the deep flow with, e.g., opposite flow directions of NADW and AABW in the Atlantic Ocean. However, this model contains the essentials of the abyssal circulation; upwelling connected with meridional advection of planetary vorticity and the existence of a convergent DWBC which feeds the interior flow.

Stommel (1958) summarized the results on the interior geostrophic abyssal flow and the deep boundary currents in a simple sketch (Fig. 8.8). Sources of deepwater (black dots) are located in the North Atlantic Ocean (S_1) and the Weddell Sea (S_2). The poleward geostrophic flow in the interior of the ocean basins supports the deep upwelling. Deep boundary currents supply water to the geostrophic interior circulation. The Southern Ocean forms the main pathway for interocean exchange of deepwater. Despite the shortcomings of the simple model it has some qualitative predictions on the overall structure of the deep circulation which can be tested.

After Stommel proposed this abyssal circulation scheme, the search for the DWBC was open. Its existence in the North Atlantic Ocean was confirmed soon by Swallow and Worthington (1957, 1961). In Section 8.6 observational evidence will be presented.

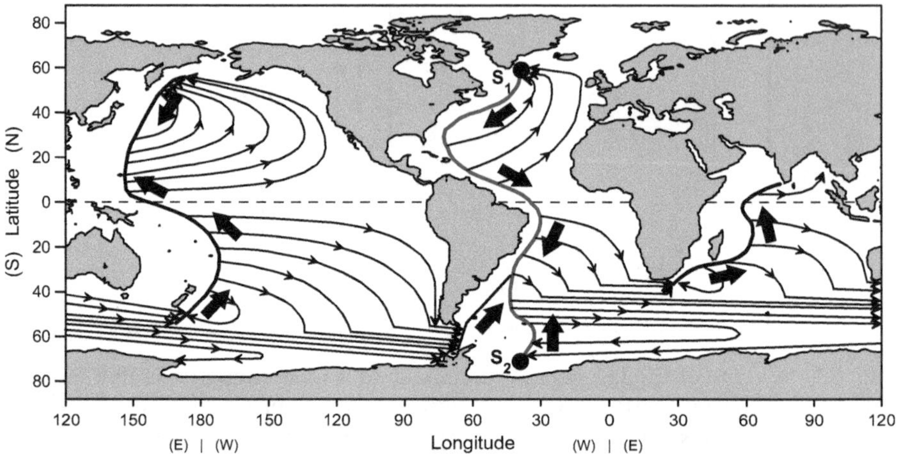

Fig. 8.8. Schematic sketch of the abyssal circulation in the world ocean based on a geostrophic interior flow, generating the required upwelling and the existence of deep boundary currents. The symbol with S_1 represents the source of NADW in the North Atlantic, the symbols with S_2 the source of AABW in the Weddell Sea. The ocean is assumed to be of constant depth (after Stommel 1958).

8.5. Topographic influence on the abyssal circulation

The Stommel–Arons model, discussed in Section 8.3, uses the simplification that the ocean bottom is horizontal. This implies that w is zero at the bottom. Here we will allow variation in the bottom depth, assuming that the bottom depth varies as a function of location. The thickness of the bottom layer below the level $z = h$ equals H when the bottom is then located at $z = h-H$. Then Eq. (8.11) still can be derived for a geostrophic abyssal circulation, connecting the stretching of the water column to the poleward velocity component. In a more general form, not a β–plane approximation, Eq. (8.11) is written as

$$f\frac{\partial w}{\partial z} = v_g \frac{\partial f}{\partial y} \quad . \tag{8.24}$$

For the vertical integration we now have to use the correct boundary condition for a sloping bottom, similar to Eq. (3.23), that is:

$$w_B = -u\frac{\partial H}{\partial x} - v\frac{\partial H}{\partial y} \quad . \tag{8.25}$$

Under the same condition as before, that the geostrophic velocity below the level h does not change with depth, we can integrate Eq. (8.25) vertically between $z = h–H$ and $z = h$, resulting in

$$f\int_{h-H}^{h}\frac{\partial w}{\partial z}dz = fw_h - fw_B = fw_h + u_g f\frac{\partial H}{\partial x} + v_g f\frac{\partial H}{\partial y} = v_g H\frac{\partial f}{\partial y} \quad . \tag{8.26}$$

Division of Eq. (8.26) by the square of the layer thickness and some rearranging of the terms leads to

$$\frac{fw_h}{H^2} = -\frac{u_g f}{H^2}\frac{\partial H}{\partial x} - \frac{v_g f}{H^2}\frac{\partial H}{\partial y} + \frac{v_g}{H}\frac{\partial f}{\partial y} = u_g\frac{\partial f/H}{\partial x} + v_g\frac{\partial f/H}{\partial y} \quad , \tag{8.27}$$

or, in vector notation,

$$\frac{fw_h}{H^2} = \vec{v}_g \cdot \nabla_h \frac{f}{H} \quad . \tag{8.28}$$

The parameter f/H is named the planetary potential vorticity of the abyssal layer. According to Eq. (8.28) no upwelling will occur as long as the geostrophic velocity is parallel to the contours of the planetary potential vorticity. When on the horizontal scale of a circulation system f does not change significantly, a nondivergent deep flow ($w_h = 0$) will follow the depth contours of the ocean. Model experiments have shown that the topographic effect, described by Eq. (8.28), may under certain conditions alter the structure of the abyssal flow, allowing eastern boundary currents (e.g., Stephens and Marshall 2000). That is probably the mechanism that maintains a deep eastern boundary current as is observed in the flow of LDW in the northern North Atlantic Ocean (McCartney 1992). When a water parcel leaves the relatively thin (εB) DWBC along the western continental slope in an upwelling ocean ($w_h > 0$), H generally increases. To balance the increase in layer depth, the velocity then will have a poleward component so that the contours of f/H are crossed in a direction where f/H also increases.

Another effect of the topography is that it may force a strong variation in the vertical turbulent diffusivity K_v, and thereby, as will be described in Chapter 9, it also will enforce a variable upwelling velocity, w_h. An increase of the turbulent motion, and therefore of K_v, may be due to the generation and reflection of internal waves over sloping or undulating topography. When the internal waves break at critical bottom slopes or focal points of reflected wave beams, such waves provide an energy source for the turbulent mixing (Polzin et al. 1997; Ledwell et al. 2000). It was remarked above that the derivation of Eq. (8.12) is also possible, when w_h varies with the location in an ocean basin. So within a basin also the upwelling velocity may vary, being high where K_v is high and low where K_v is low. According to Eq. (8.12) such a geographic variation of w_h within an ocean basin also will induce a variation of the interior geostrophic circulation compared with the simple Stommel–Arons model with a constant upwelling velocity.

8.6. Observational evidence for the abyssal circulation scheme

As stated before, the solution for an ocean basin in the southern hemisphere is similar to that sketched in Fig. 8.5 but mirrored relative to the equator ($y = 0$). Figure 8.5 shows that the most aged deepwater can be found in the northeastern part of the ocean basins at the northern hemisphere and in the southeastern part of the basins on the southern hemisphere. The *AOU* maximum of the aged deepwater in the eastern parts of WOCE section P03 in the North Pacific Ocean (Fig. 5.18) and sections I05P (Fig. 5.10) and P06 (Fig. 5.16) in, respectively, the southern Indian and Pacific ocean basins agree with this aspect of the abyssal circulation.

During WOCE a total of 140 subsurface floats were released in the Brazil Basin in order to find out whether the Stommel–Arons scheme for the deep circulation could be confirmed. The results did not conform to that circulation scheme (Hogg 2001). At the level of the NADW as well as at the level of the AABW the deep motion was mainly zonal, suggesting that the deep upwelling velocity w_h in the Brazil Basin was extremely small, according to Eq. (8.12). However, low-frequency interannual variability of the abyssal zonal flow was observed from the float motion, suggesting that the experiment that took 2.5 years probably did not last long enough to establish a reliable long-term mean circulation. Tracer studies have the advantage that the large-scale tracer distributions are integrated long-term results of time-varying advection and mixing. Lagrangian velocity measurements with floats of the very slow abyssal circulation in the ocean, with variability on timescales from months to several years, require large numbers of float years to obtain a stable long-term average circulation.

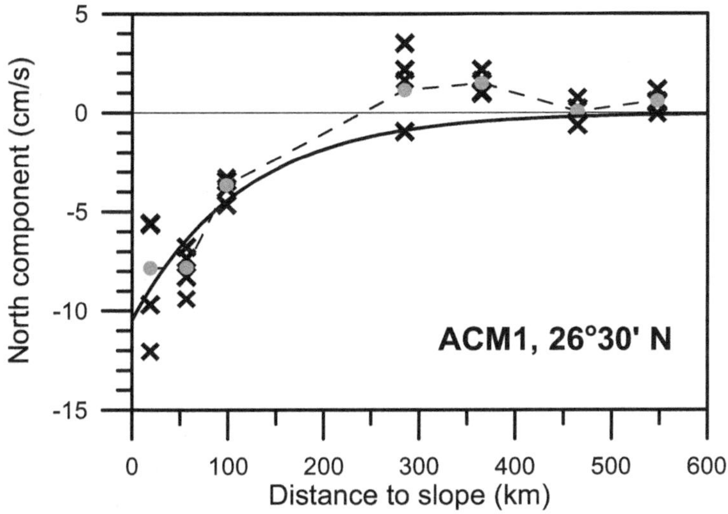

Fig. 8.9. The long-term mean northward component of the velocity, measured with WOCE current meter mooring array ACM1 near 26°30' N as a function of the distance to the continental slope off Abaco Island, one of the Bahamas. The observations were made at nominal depths of 1200, 2000, 3000, and 4000 m (crosses). The vertical mean is shown as gray dots, connected with a dashed line. The thick line gives a least-squares fit of an exponential function, decaying eastward. The current meter data were downloaded from website of the Oregon State University Buoy Group, http://kepler.oce.orst.edu/ .

However, the existence of a DWBC is a well-established fact. It has been observed with current meter arrays and can be derived from tracer distributions. From 1988 to 1997, arrays of current meter moorings have been maintained in the North Atlantic at 26°30'N near the slope off Abaco Island, the WOCE ACM1 array (Lee et al. 1996).

The observations show a significant, narrow, southward flow in the western boundary along the slope of Abaco Island. The decay of the vertically averaged current speed (gray dots in Fig. 8.9) can be modeled adequately with an exponential curve, with an e-folding length of 115 km (thick line in Fig. 8.9). This length scale is definitely smaller than the width of the Atlantic Ocean at the latitude of the mooring array (~7000 km). The typical long-term mean southward velocity component is about 10 cm/s at the continental slope. The volume transport estimate of the southward DWBC below 1000 m, within 250 km from the slope, based on these observations amounts to about 35 Sv. This is about twice the estimate of the meridional transport of NADW, formed at higher latitudes (Schmitz 1995; Macdonald and Wunsch 1996; Ganachaud and Wunsch

2000). It is not clear yet whether the observed slow northward transport between 200 and 450 km from the slope is a permanent direct recirculation within the western boundary itself or whether it is due to the statistical uncertainties introduced by the mesoscale and interannual variability of the deep circulation. Over half of the 20 available current meter records for $x > 200$ km had a mean northward velocity component that did not differ significantly from zero. Anyway, even for the mean transport between the slope and 445 km, the estimated mean meridional transport amounts to 26 Sv southward flow of deepwater in the DWBC, still 12 Sv larger than the net southward NADW transport reported by Schmitz (1995), suggesting a significant abyssal recirculation in the interior of the Atlantic Ocean, in agreement with the Stommel–Arons model.

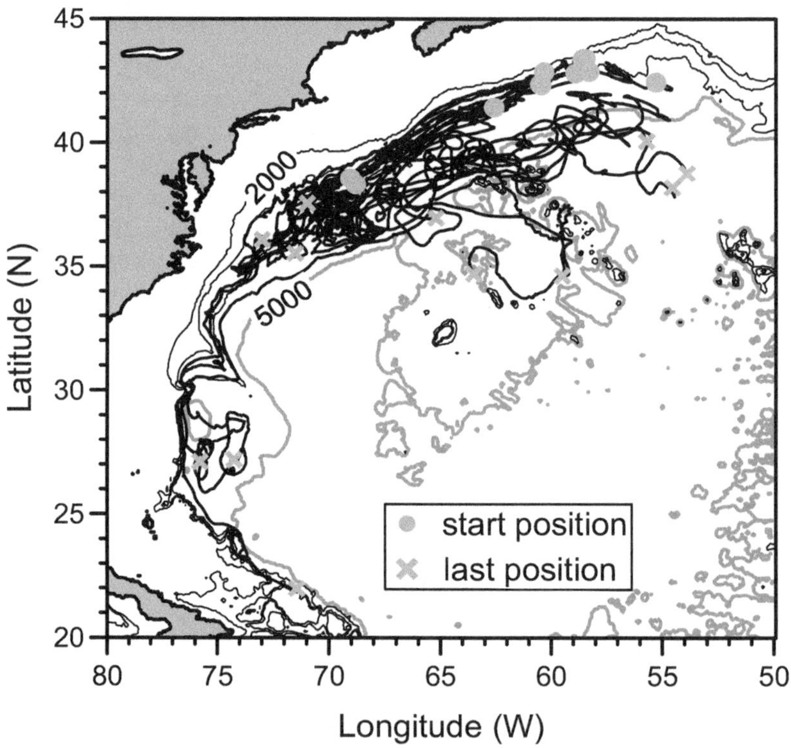

Fig. 8.10. Tracks of subsurface floats deployed during the Boundary Current Experiment from 1994 to 1997. The 13 floats were ballasted to follow the deep circulation at a pressure of 3000 dbar. The gray dots indicate the start position, while the crosses show the last reported position of the floats. The 2000, 3000, and 4000 m isobaths are shown as thin lines, the 5000 m isobath as a thick gray line. The float data were downloaded from http://woce.nodc.noaa.gov/woce_v3/wocedata_1/woce-ssf/welcome.htm .

During the Boundary Current Experiment (BOUNCE) from 1994 to 1997 the southward flow of deepwater from the Nordic seas in the DWBC of the North Atlantic was studied by means of deep isobaric floats (Bower and Hunt 2000a,b). These floats were ballasted for 3000 dbar, near the level of a CFC maximum characteristic for the NADW core. The position of the free-floating sub-surface floats was acoustically tracked. The tracks of the successive drifter positions show the Lagrangian motion of the surrounding water. Tracks of 13 of these floats are shown in Fig. 8.10. The experiment clearly showed the Langrangian character of the DWBC, with water flowing in equatorial directions for considerable periods. All but one of the floats followed the continental slope for some time, ranging from 45 days to 21 months. Of these 12 floats 11 were observed ultimately to be swept out of the DWBC into the abyssal ocean. There they started some meandering course, indicative for the presence of mesoscale eddies at the level of the NADW core. These eddies may be an active agent for the diffusive exchange of tracers between the DWBC and the ocean interior. The mean Lagrangian along-slope velocity between 35 and 42°N is estimated to be 6.1 cm/s, a value with a magnitude comparable to the mean Eulerian velocities measured by the ACM1 current meter array (Fig. 8.9). The floats that left the DWBC current between 35 and 38°N circulated in a northeastern direction, crossing the isobaths at a small angle.

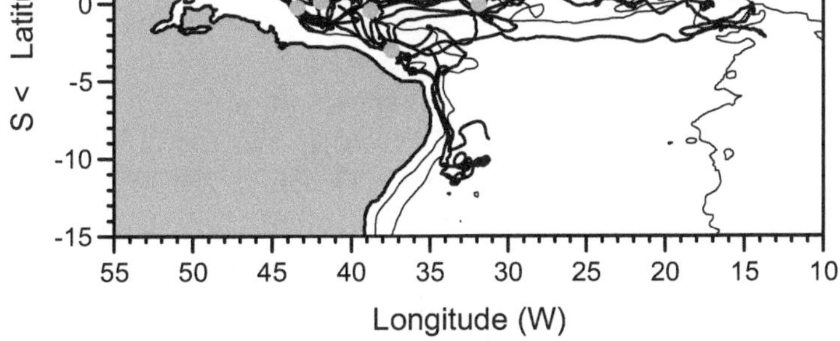

Fig. 8.11. Float tracks of sub-surface floats deployed from 1989 to 1992 in the Tropical Atlantic Ocean. The floats were ballasted to follow the circulation at a depth of 1800 m. The dots indicate the deployment positions, the thin lines the 2000 and 4000 m isobaths. The data were downloaded from http://woce.nodc.noaa.gov/woce_v3/wocedata_1/woce-ssf/welcome.htm .

From 1989 to 1992 scientists from the Woods Hole Oceanographic Institution deployed sub-surface floats in the tropical Atlantic at a depth of 1800 m, in the upper layers of the NADW (Richardson and Fratantoni 1999). The floats showed a complicated drift pattern (Fig. 8.11). At 11°N three floats slowly drifted westward toward the DWBC, but with a lot of eddy motion. Two other floats drifted around between 5° and ~10°N before they entered the DWBC. From there they drifted directly in a narrow DWBC from ~7°N to 10°~S, following the Brazilian continental slope until they were stalled by mesoscale eddy motion. Other floats left the DWBC to flow eastward in the tropical region up to longitudes near 15°W and then returned westward again. These equatorial excursions took 2 to 3 years. A fourth group of floats drifted westward in the equatorial current system until they were incorporated in the southward-flowing DWBC. Mean eastward-flowing jets were observed near 2°N and 2°S bounding a mean westward jet centered on the equator (1°S to 1°N).

Apparently there existed a considerable exchange between the DWBC and the upper deepwater in the abyssal basins, although some floats (and water) in the DWBC managed to cross the equator directly with an average velocity of about 8 cm/s. The zonal current system in the tropical band does not belong to the geostrophic abyssal flow of the Stommel–Arons model but probably forms part of a nongeostrophic equatorial current system. The southern jet at 2°S coincides with a CFC-rich plume centered south of the equator. The CFC plume is inferred to have been advected by the southern jet across the Atlantic to its eastern boundary (the high-CFC-11 core at ~1800 dbar in Fig. 4.13). Over much of the region a considerable eddy motion strongly influenced the float paths.

In tracer distributions the DWBC in the North Atlantic is most easily observed when we use tracers that show the "age" contrast between the relatively young water of the DWBC and the "aged" water in the ocean interior. Tracers that can show that age contrast are the natural tracer *AOU* and the manmade CFCs. Figure 8.12 shows the distribution of *AOU* and two different CFCs, CFC-11, and CFC-12 in the first 1500 km from the western boundary of the north Atlantic Ocean at 24°30' N from WOCE section A05. Relatively young water (low *AOU*, high CFC-11 and CFC-12) is observed along the western boundary (Fig. 8.12). A vertical succession of different cores of deepwater can be recognized in the DWBC, with different regions of origin. A low-*AOU* core ($AOU < 60$ µmol/kg) between 1600 and 2100 dbar and relatively high CFC-11 and CFC-12 values (> 0.4 and > 0.2 pmol/kg, respectively) indicate the presence of deepwater, mainly LSW, originating from the Labrador Sea (at $\Theta \approx 3.5$°C). A secondary low-*AOU* core and high-CFC-11 and CFC-12 cores are observed between 3000 and 4000 dbar (at $\Theta \approx 2.2°C$) and represent the contribution to the NADW of water entering the North Atlantic through the Denmark Strait (DSOW). The apparently older water in between these cores (2100 to 3000 dbar) originates from overflow of arctic water across the sills on the Iceland–Scotland Ridge (NEADW). This water is aged more than the water from the Denmark Strait because of mixing with older recirculating water in the Iceland Basin (LDW). Near 1200 m a high-CFC core is found over the continental slope. The core of this water, with a temperature range of 4 to 5°C, can be traced back toward the Labrador Sea (Pickart 1992). However, it is not dense enough to have been formed in the central Labrador Sea, even during anomalous conditions. Based on its density and tracer characteristics, its area of formation appears to be the southern Labrador Sea, inshore of the North Atlantic Current. This water type nowadays is referred to as Upper Labrador Sea Water (ULSW) in contrast to the already known classic LSW (Smethie et al. 2000). Fine and Molinari (1988) have shown that the water types in the DWBC of the North Atlantic Ocean can be followed further downstream to at least the latitude of Barbados (13°N).

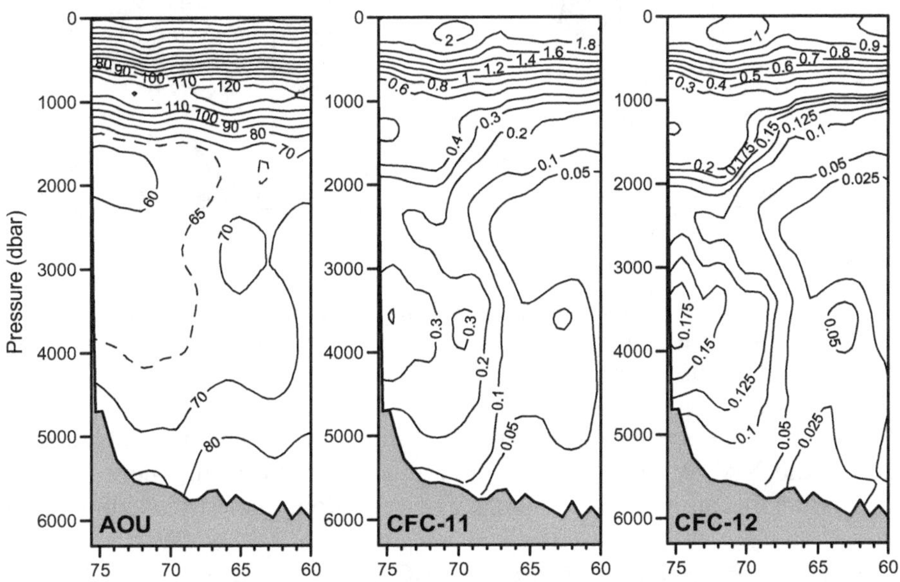

Fig. 8.12. The vertical distribution of *AOU* (μmol/kg, left), CFC-11 (pmol/kg, center), and CFC-12 (pmol/kg, right) along the zonal WOCE A05 section in the North Atlantic Ocean near 24°30' N.

The width of the DWBC, suggested by the tracer distributions in Fig. 8.12, is definitely larger than the width derived from the current measurements only 200 km further north, depicted in Fig. 8.9. This is caused by the specific information present in tracer distributions. The spatial distribution of tracers is built up during many years, advected by a temporally varying flow field. Spatial and temporal variations in the flow field act as a diffusive process, broadening the tracer distribution relative to the mean flow in the DWBC, observed with current meters. Next to these variations the eddy motion in the ocean also acts as a turbulent diffusive process. As derived above, the zonal flow into the interior will bring water from the core of the convergent DWBC into the ocean interior. These processes contribute to a widening of the cores in the tracer distribution relative to the width of the mean flow field in the DWBC.

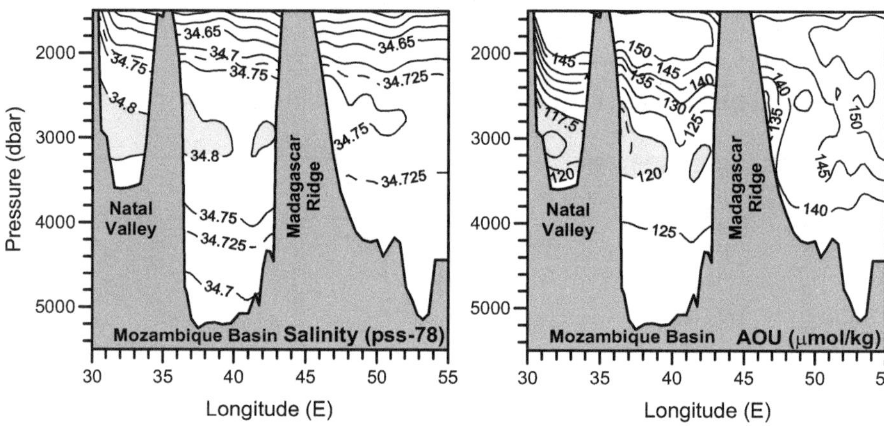

Fig. 8.13. The vertical distribution of salinity (left) and apparent oxygen utilization (right) below 1500 dbar in the southwestern Indian Ocean along WOCE section I05 near 32°S.

Also in other ocean basins a DWBC can be found. The Indian Ocean receives the NADW directly from the Atlantic Ocean, from where it enters the Indian Ocean in a deep boundary current along the African continental slope (see Fig. 5.2). A hydrographic section near 32°S (WOCE section I05) shows the characteristics of DWBC below the strong south flowing Agulhas Current (Fig. 8.13). The NADW in the Natal Valley and Mozambique Basin is characterized as a deep salinity maximum, because of the relative high salinity in its North Atlantic region of origin. This water is also "young" compared to the surrounding water masses, reflected by a low-*AOU* core coinciding with the high-salinity core along the continental slope near 3000 dbar. Across the Madagascar Ridge (the rightmost ridge in

Fig. 8.13) the amount of NADW is much less, since this ridge acts as a topographic barrier for the eastward extension of the NADW. At pressures of over 4000 dbar, below the NADW core, we can observe low-salinity water in the Mozambique Basin with an *AOU* that is only slightly higher than in the NADW core. This is AABW, bottom water of antarctic origin.

In the southwestern Pacific Ocean, against the slope of Kermadec Ridge northeast of New Zealand, another DWBC can be observed in tracer distributions (Fig. 8.14). It contains two water mass cores. A LCDW core of saline water ($S > 34.72$) with a relative low dissolved silicate content ($Si < 110$ μmol/kg) overlies a less saline AABW core ($S < 34.71$) near the bottom, with higher silicate concentrations ($Si > 120$ μmol/kg). The LCDW core still contains remnants of the high-salinity NADW formed in the relatively saline North Atlantic, while the less saline bottom layer reflects the low salinity of bottom water of antarctic origin. The dissolved silicate concentration reflects the same contrast. North Atlantic water is characterized by low silicate concentrations, while these concentrations are high in water from the antarctic region. Both in the South Pacific and in the South Indian Ocean the tracer maxima, characteristic for the LCDW in the DWBC, reach quite far east into the ocean, O(1500 km), similar to the DWBC the North Atlantic (Fig. 8.12).

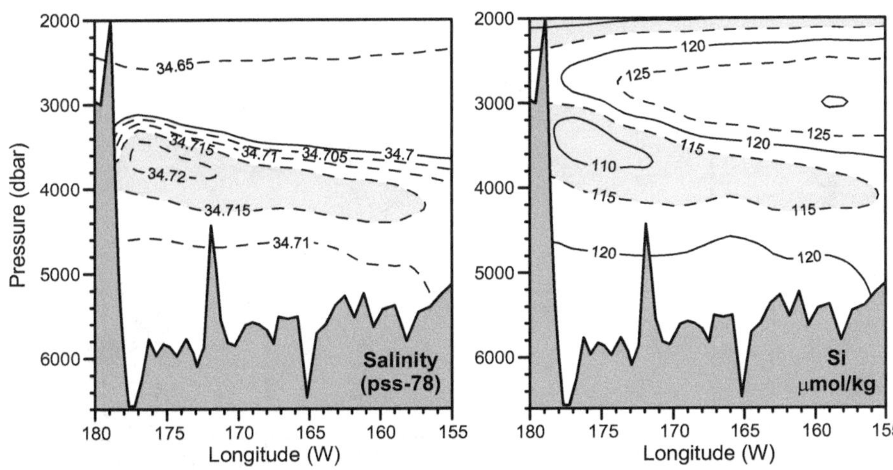

Fig. 8.14. The vertical distribution of salinity (left) and dissolved silicate (right)below 2000 dbar in the southwestern Pacific Ocean along WOCE section P06 near 32°S.

Hogg (2001) has noted that estimates of the meridional volume transport in the Pacific DWBC over the continental slope showed a convergent flow and a reversal to a southward transport at the latitudes of the Japanese islands (Fig. 8.15). Whitworth et al. (1999) derived from the WOCE mooring line over the Chatham Rise northeast of New Zealand at 32.5°S a 22 months averaged northward transport of 15.8 ± 1.4 Sv. For the latitude of ~10°S, at the Samoan Passage, Roemmich et al. (1996) and Rudnick (1997) reported a northward transport of bottom water with a mean transport of 10.6 ± 1.7 Sv. Near 36°N, Hallock and Teague (1996) reported a southward flow below 2000 m from WOCE current meter mooring line PCM-7 over the Japanese continental slope. They stated that one can derive a southward transport from the 2 years means velocities in this DWBC of 1.7 ± 1.7 Sv. Further north, at WOCE mooring line PCM-6 near 42°N, Owens and Warren (2001) also found a deep southward transport, here of about 4 Sv. However, the precise value of this transport is uncertain because of the small horizontal extent of the DWBC over the continental slope, indicating a standard error of the order of the magnitude of this transport. Plotted versus the latitude (black dots in Fig. 8.15) these transports agree with a convergent boundary current with a reversal of the flow direction near 24°N, as can be expected from the flat-bottom, constant upwelling Stommel–Arons model (Owens and Warren 2001). This flow reversal was already suggested by Stommel in 1958 (Fig. 8.8).

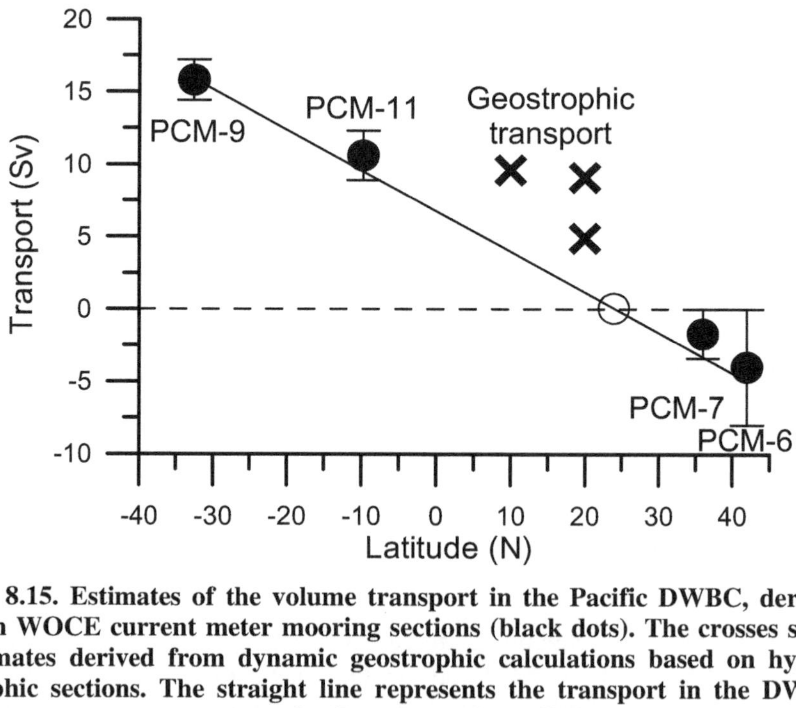

Fig. 8.15. Estimates of the volume transport in the Pacific DWBC, derived from WOCE current meter mooring sections (black dots). The crosses show estimates derived from dynamic geostrophic calculations based on hydrographic sections. The straight line represents the transport in the DWBC with constant convergence and a flow reversal at 24°N (open symbol).

Hogg (2001) also mentioned some transport estimates derived from geostrophic calculations using the data from hydrographic sections (crosses in Fig. 8.15). However, such dynamic estimates are quite uncertain, because of the problem of defining a reliable level of no motion. The current meter observations mentioned above often show a columnar structure of the flow without a specific flow reversal in the vertical. This uncertainty is reflected in the relatively large magnitude of the geostrophic estimates, compared with a linear fit to the current meter derived transports.

According to the Stommel–Arons model of the abyssal circulation in the northern hemisphere water from the ocean interior reaches the northern boundary where it feeds a westward-flowing DNBC. The potential density distribution near the northern boundaries of both the northern Atlantic Ocean and the northern Pacific Ocean shows upward sloping isopycnals (Fig. 8.16). The resulting northward density gradients support a downward intensified, westward-flowing near-geostrophic DNBC on the northern side of both oceans. The width of the DNBC is of the order of 100 km in both cases.

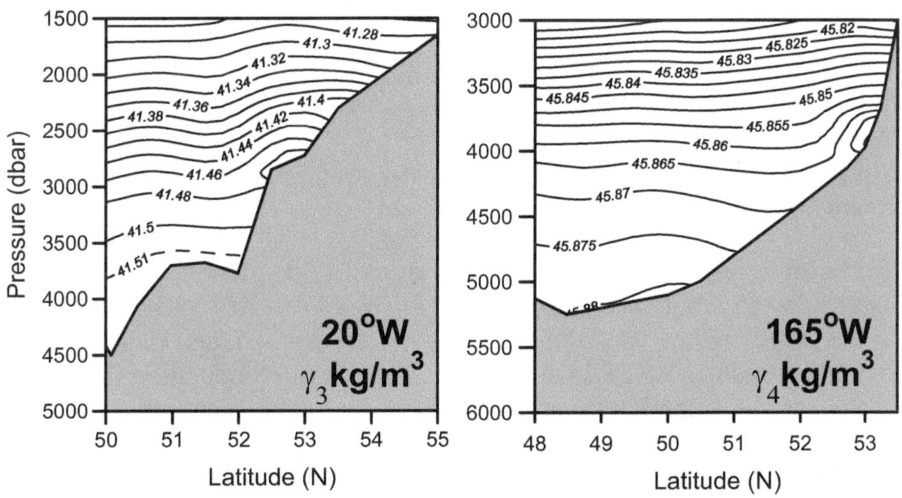

Fig. 8.16. Potential density anomaly γ_3 along WOCE section A16 near the slope of the Rockall–Hatton Plateau in the North Atlantic Ocean at 20°W (left panel) and γ_4 along WOCE section P15 near the continental slope of the Alaskan Peninsula in the North Pacific Ocean at 165°W (right panel). In both cases the meridional density gradient near the slope due to the upward sloping isopycnals agrees with a downward intensified westward deep northern boundary current.

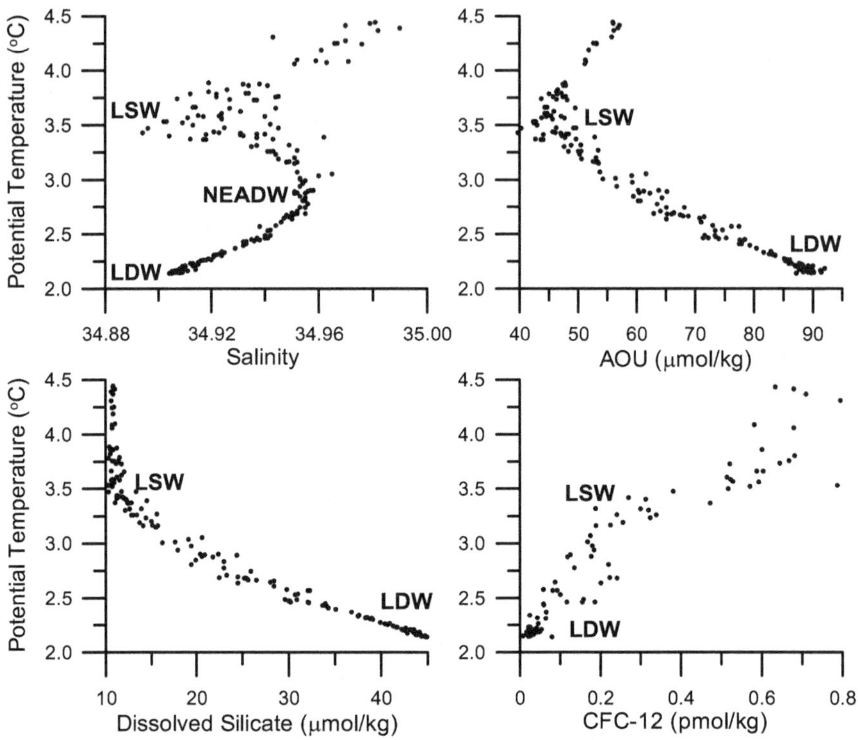

Fig. 8.17. Property–property plots of the deepwater mass in the DNBC in the North Atlantic Ocean, south of Hatton Bank.

The analysis of hydrographic properties of the DNBC in the northeastern North Atlantic Ocean (Fig. 8.17) shows that the water transported in the DNBC below 3000 dbar, south of the Rockall–Hatton Plateau, is the relatively fresh LDW. That water type consists of recirculating deepwater in the eastern Atlantic Ocean that contains reasonable amounts of AABW that entered the eastern Atlantic basins near the equator (McCartney 1992). LDW shows its partly antarctic origin in its low salinity and the high concentrations of dissolved silicate. The great age of this water is reflected in the high *AOU* values and near-zero CFC concentrations. Between 2500 and 3000 dbar the DNBC consists of recirculating aged NEADW, the eastern Atlantic variety of the NADW that is formed by the input of relatively saline ISOW (van Aken 2000). The DNBC flows around the Hatton Bank into the Iceland Basin where it is strengthened by influx of ISOW from the Norwegian Sea. Further west, the Atlantic DNBC flows to the southwest along the Reykjanes Ridge (Fig. 7.8). The flow of deepwater through the Charlie–Gibbs Fracture Zone (Fig. 7.10) and subsequent cyclonic recirculation of NEADW in the Irminger Sea can be seen as a continuation of the DNBC in the western North Atlantic Ocean. Ultimately the DNBC meets the DSOW near the Denmark Strait (Fig. 7.11) and from there the NEADW contributes to the DWBC along the East Greenland continental slope (Dickson and Brown 1994). Current measurements at 175°W and 180°W in the North Pacific Ocean have confirmed the existence of the westward-flowing DNBC in this ocean basin (Warren and Owens 1985, 1988; Onishi and Ohtani 1999).

The deep westward flow along the antarctic continent, especially in the Weddell Sea (Fig. 7.15) and the Ross Sea (Fig. 7.22) can be seen as the southern hemispheric counterparts of the DNBC, in this case a deep southern boundary current (DSBC). In both southern subbasins the DSBC adds to the equatorward flow of newly formed AABW, as can be expected from the abyssal circulation scheme shown in Fig. 8.6 (left panel), mirrored at the equator.

8.7. Wind-driven deep upwelling in the Southern Ocean

Drake Passage between 56 and 62°S connects the Pacific and Atlantic oceans and provides for a continuous zonal band of water above the sill depth of the Scotia Ridge, ~1500 m. As discussed in Section 8.1 the mean zonal pressure gradient in this uninterrupted band of water in the southern ocean will be zero, and therefore any mean meridional geostrophic transport at these latitudes will be confined to depths below 1500 m where the zonal form stress will balance the Coriolis force of the mean meridional flow. The water mass directly below the sill of the Scotia Ridge in the Atlantic Ocean is the NADW core that can flow southward in a geostrophic balance to form the LCDW core in the Southern Ocean.

Toggweiler and Samuels (1995) pointed out that the surface winds in the southern circumpolar belt exert an eastward wind stress that drives an equatorward Ekman transport in the near-surface layer. The eastward wind stress has a maximum at latitudes north of Cape Horn, the southern tip of South America (Fig. 8.18, left panel). In the ocean that wind stress is balanced by the Coriolis force of the directly wind-driven Ekman transport according to Eq. (3.22).

The northward increasing zonal wind stress over the Southern Ocean will lead to a divergent northward Ekman transport (Fig. 8.18, right panel). Since mass conservation requires that the meridional mass transport across that latitude belt in the Southern Ocean has a zonal mean equal to zero (apart from precipitation effects), a convergent southward geostrophic return flow is required from which the divergent Ekman transport draws up deepwater. In a zonally bound ocean the upwelling can come from any depth, brought south by a southward geostrophic transport supported by the finite topographic form stress in a bound ocean. In the Southern Ocean, where the latitude belt of the Drake Passage is located, a zonally averaged mean meridional geostrophic flow only can exist below 1500 m, and it is there where upwelling LCDW and UCDW is found to ascend toward the surface (figs. 5.3, 5.4, and 5.5). The dynamics of the upwelling of deepwater in the Southern Ocean is in principle equal to the deep upwelling in the ocean interior, described by Eq. (8.11). The divergent Ekman transport only determines a boundary condition for the underlying water mass. The resulting geostrophic adjustment of the deep density distribution then will force a southward transport of the deep water mass that drives upwelling water according to Eq. (8.11) into the divergent wind-driven Ekman transport.

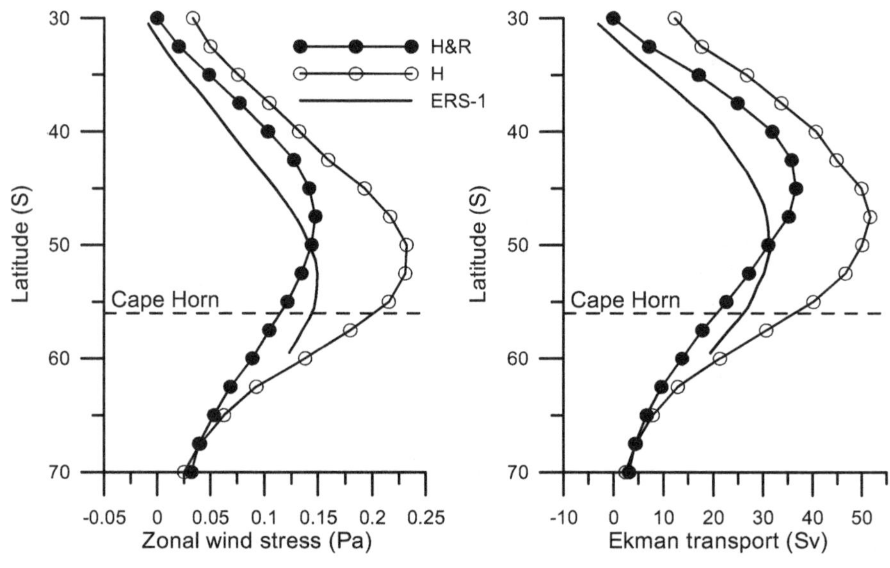

Fig. 8.18. Zonally averaged zonal wind stress over the oceans, τ_x, for different climatologies (left panel) and the mean northward Ekman transport, driven by that stress (right panel). H&R is the climatology of Hellerman and Rosenstein (1983), based on ship observations between 1870 and 1976; H is the Hellerman (1967) climatology, based on wind rose data from the Marine Climatic Atlases and Pilot Chart Data of the US Navy; ERS-1 is the climatology for August 1991 to May 1996, derived from scatterometer measurements of the European ERS-1 satellite. Those data were obtained from IFREMER, Plouzané (France) (http://www.ifremer.fr/cersat/en/index.htm). The dashed line shows the latitude of Cape Horn, the southern tip of South America.

If we assume that the geostrophic transport of deep and bottom water completely compensates the wind-driven Ekman transport at the latitude of Cape Horn (56°S), we arrive at a total poleward geostrophic transport of between 21 and 36 Sv for the different wind stress climatologies shown in Fig. 8.18. The wind stress derived from the ERS-1 scatterometer data is probably the most reliable in this remote part of the world ocean, resulting in a mean poleward geostrophic transport of 26 Sv at the latitude of Cape Horn. Of this transport 6 Sv can be attributed to the Ekman transport in the Atlantic sector of the Southern Ocean. From the analysis of model simulations of the geostrophic transport that compensates the northward Ekman transport, Toggweiler and Samuels (1995) found that an eastward stress in the ACC, or lateral friction, was generated due to mesoscale eddy motion south of the latitude of Cape Horn. This stress was so large that it enabled a shallow (< 1500 m) compensating southward flow where the Coriolis force of that meridional motion was balanced by the lateral friction of the ACC. The volume transport in this compensating meridional flow was 36% of the northward Ekman transport. If we apply the same ratio to our estimate of the Ekman transport, based on the ERS-1 data, we come to a shallow southward recirculation of 9 Sv and a net deep poleward transport across 56°S of 17 Sv. In the model estimates of Toggweiler and Samuels (1995) the eddies were not explicitly modeled but their effect was approximated by means of an arbitrary large lateral friction coefficient compensating for the lack of spatial resolution, which makes such quantitative estimates not perfectly reliable. But the results at least show that understanding the effects of mesoscale eddies on subsurface transport, either shallow or deep, and preferentially explicit modeling of these eddies in ocean general circulation models also is required when the effects of the wind stress on the THC is studied.

In model simulations with their coarse resolution numerical model Toggweiler and Samuels (1995) found a strong dependence of the intensity of the North Atlantic overturning THC on the wind stress over the southern Ocean. From a similar model simulation Sijp and England (2004) concluded that the open connection through the Drake Passage is essential for the existence of the THC with a North Atlantic cold deepwater source, while in contrast Klinger et al. (2004) found from similar simulations that also with a closed Drake Passage wind over the Southern Ocean strongly determines the North Atlantic overturning THC. All three model simulations showed the deep upwelling poleward flow in the Southern Ocean, compensating for the wind-driven equatorward eddy transport, but apparently details in the model formulation and boundary conditions were more important for the outcome of the model simulations on a global scale.

9. Deep upwelling and mixing

9.1. Profiles of conservative tracers

It has become clear from the oceanic tracer distributions, presented in Chapter 5, that in the deep Pacific and Indian oceans deep and bottom waters move upward to at least the levels of the aged deepwater like NIDW and the upper PDW, the so-called deep upwelling. In the Atlantic Ocean the deep upwelling is less clear. However, the existence of a DWBC and a DNBC in both the Pacific and Atlantic Ocean, in agreement with the Stommel–Arons model, indicate that deep upwelling of poleward flowing deepwater probably occurs in the Atlantic Ocean too. In the abyssal basins, where global upwelling occurs, a balance has developed between vertical advection of heat and salt, due to the upwelling velocity and the vertical turbulent diffusion of heat and salt. We follow in this chapter the reasoning of Munk (1966) and Munk and Wunsch (1998) with regard to this advective–diffusive balance. For any conservative tracer C, e.g., salinity or potential temperature, the conservation equation for a stationary distribution can be written according to (4.1), with the source term, Q, set to zero. This equation is here repeated as

$$u\frac{\partial C}{\partial x}+v\frac{\partial C}{\partial y}+w\frac{\partial C}{\partial z}=\nabla_h\cdot(K_h\nabla_h C)+\frac{\partial}{\partial z}\left(K_v\frac{\partial C}{\partial z}\right)\quad . \tag{9.1}$$

The vertical (diapycnal) turbulent diffusivity, K_v, is known to vary strongly with occasionally extremely high values (> 10^{-1} m^2/s) in the benthic boundary layer, which may lead to fully mixed bottom layers with a thickness of the order of 100 m (Munk and Wunsch 1998). Enhanced values of K_v also are observed over steep or irregular topography, but over nearly all of the pelagic ocean low values of only $K_v \approx 10^{-5}$ m^2/s have been derived from tracer experiments and microstructure observations (Polzin et al. 1997). The deep ocean seems to mix primarily at its topographic boundaries, and the results of that locally intensified turbulent mixing is redistributed laterally in an ocean basin by isopycnal mixing and advection at sub-basin scales (Garrett, 1979). The net effect of these processes for the dynamics of the mean vertical stratification of an ocean basin, encountered by the upwelling branch of the THC, can be obtained by averaging Eq. (9.1) over the whole basin surface A. Hereby we assume that w is not laterally correlated with C (since such "turbulent" correlations are modeled by K), that K_v is not laterally correlated with $\partial C/\partial z$, and that horizontal diffusion over the open boundary of the basin can be neglected, relative to advection. The horizontal integration of Eq. (9.1) over the basin surface A, using also the continuity equation (3.30) then leads to

$$A\frac{\mathrm{d}}{\mathrm{d}z}\left(wC - K\frac{\mathrm{d}C}{\mathrm{d}z}\right) = q(z)C(z) \quad , \tag{9.2}$$

where the variables $w(z)$ and $C(z)$ and the parameter $K(z)$ are now basin-wide averages of w, C, and K_v. The variable $q(z)$ represents the inflow at level z of water with trace concentration $C(z)$ over the lateral boundary of the basin. Horizontal integration of the continuity equation (3.30) separately over the basin surface A gives a relation between q and w:

$$q(z) = A\frac{\mathrm{d}w}{\mathrm{d}z} \quad . \tag{9.3}$$

Substitution of Eq. (9.3) into Eq. (9.2) and rearranging of terms then results in

$$K\frac{\mathrm{d}^2C}{\mathrm{d}z^2} - \left(w - \frac{\mathrm{d}K}{\mathrm{d}z}\right)\frac{\mathrm{d}C}{\mathrm{d}z} = 0 \quad . \tag{9.4}$$

For simplicity, we will assume here that the effective diffusion coefficient K and the mean upwelling velocity w are in first approximation not depth dependent. For any conservative tracer C the advective–diffusive balance in the upwelling regions then can be written (with z positive upwards) as

$$K\frac{\mathrm{d}^2C}{\mathrm{d}z^2} - w\frac{\mathrm{d}C}{\mathrm{d}z} = 0 \quad . \tag{9.5}$$

In advection–diffusion problems like Eq. (9.5), a characteristic vertical length scale L can be defined as $L = K/w$. The condition that w is approximately constant implies that $\mathrm{d}w/\mathrm{d}z \ll w/L$. With the length scale L Eq. (9.5) can be rewritten to

$$\frac{\mathrm{d}^2C}{\mathrm{d}z^2} - \frac{1}{L}\frac{\mathrm{d}C}{\mathrm{d}z} = 0 \quad . \tag{9.6}$$

From Eq. (9.6) it is easily derived that the vertical gradient dC/dz increases exponentially upward, with an e-folding scale L. An approximately exponential form of the vertical density gradient has been confirmed by observations in many ocean regions. With a global upwelling velocity of $w \approx 1 \cdot 10^{-7}$ m/s (~25 Sv deep and bottom water production) the molecular thermal diffusion coefficient of the order of $K = 10^{-7}$ m/s^2 leads to a characteristic length scale of the thermal stratification of about 1 m. The permanent thermocline has a much larger scale, of the order of 300 to 1000 m, indicative of the importance of turbulent diffusion in the ocean interior represented by a much larger turbulent K. Without deep turbulent mixing the deep ocean will turn into a pool of homogeneous stagnant cold water. Then equilibrium is maintained by near-surface molecular diffusion, with a surface intensified overturning circulation driven by weak shallow convection with a characteristic depth of only a few meters. The strength of the THC and the associated meridional heat and freshwater fluxes in the real ocean may well be determined by the power available for turbulent mixing and not primarily by the high-latitude density forcing due to the heat loss to the atmosphere. Knowledge of the turbulent mixing is essential for the understanding and accurate modeling of the THC.

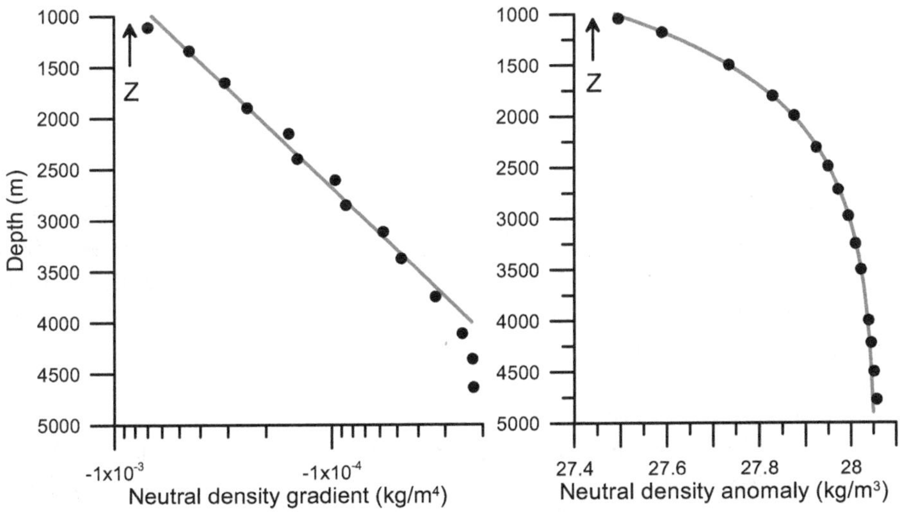

Fig. 9.1. The mean vertical profile of the vertical density gradient, corrected for adiabatic compression effects (left panel) and the vertical profile of the neutral density with $L = 885$ m (right panel) for regionally averaged data from the northeastern Pacific Ocean, WOCE section P16 between 25° and 30°N. The full lines show the fits with a function according to Eq. (9.7).

In order to derive $C(z)$ from Eq. (9.6) we have to add two boundary conditions. Here we assume as boundary conditions $C = C_0$ at a certain depth level where $z = 0$ and $C = C_1$ at a level $z = z_1$; z is defined upward positive. Then the solution of Eq. (9.6) can be derived as:

$$C(z) - C_0 = (C_1 - C_0)\frac{(e^{z/L} - 1)}{(e^{z_1/L} - 1)} \quad . \tag{9.7}$$

We can compare these results with observations from WOCE section P16 in the northeastern Pacific Ocean at 152ºW (Fig. 5.12). The North Pacific is an appropriate semienclosed region for this type of studies, since it is relatively isolated from the South Pacific by the equator while only a Southern Ocean source of deep and bottom water exists. We have averaged the hydrographic profiles between 25 and 30ºN from section P16. From these profiles the vertical density gradient was calculated, corrected for adiabatic compression effects, $\mathrm{d}\rho/\mathrm{d}z|_{ad}$ (Fig. 9.1, left panel). This observed gradient $\mathrm{d}\rho/\mathrm{d}z|_{ad}$ is close to exponential between depths of 1000 and 4000 m (and negative). When we fit an exponential function to the data points by a least-squares method in this depth interval, an e-folding length $L = 885$ m is found. By integration of the adiabatic density gradient from the surface downward a neutral density $\rho_{n.}$ can be calculated. The related neutral density anomaly γ_n is defined as:

$$\gamma_n(z) = \gamma(0) + \int_0^z \frac{\mathrm{d}\rho}{\mathrm{d}z}\bigg|_{ad.} \mathrm{d}z \quad . \tag{9.8}$$

With the parameter L already derived from the observed density gradient, it is easy to find a least-squares fit for the data points between 1000 and 4000 m depth. The resulting fit of the theoretical curve with the neutral density data is quite good, with an RMS value of the residuals of only 0.002 kg/m^3 in the depth interval from 1000 to 4000 m (Fig. 5.1, right panel).

For similar relatively isolated semienclosed deep basins like the northern Indian Ocean or the Angola Basin in the southeastern Atlantic Ocean Eq. (9.7) also can be used to model the density stratification over a depth range of approximately 3000 m (Fig. 9.2, left and middle panel). For these basins the estimated advective–diffusive length scale is also of the order of 1000 m, being 892 and 716 m, respectively.

However, in the Brazil Basin the alternating north- and south-flowing cores of AAIW, NADW, and AABW make it questionable whether the assumption that w is nearly constant will be met, since there the horizontal volume influx q certainly does not equal zero at intermediate levels. Then the vertical advective mass flux may be divergent or convergent, allowing horizontal recirculation of upwelled water in the overlying water mass core. Analysis of the adiabatic density gradient shows that between 700 and 1600 m, above the UCDW-related silicate and *AOU* maxima, the stratification is characterized by a value of L of 938 m (Fig. 9.2, right panel). The layer between the UCDW core at 1600 m and the NADW core below 2500 m is characterized by a lower value of L, 515 m. If K can be considered to have similar values in both layers, this suggests an upward increase of w of nearly 100% across the UCDW core. Part of the upwelled NADW may be incorporated into the UCDW and brought back to the north. Another explanation may be a lower value of K between 1600 and 2500 m relative to the overlying layer. In the thin depth interval between 4200 and 4700 m, between the AABW and the NADW cores, the density stratification also can be modeled according to Eq. (9.6) with L =539 m. The NADW core itself is characterized by a minimum of the vertical density gradient between 2500 and 3600 m. As expected, Eqs. (9.6 and 9.7) certainly cannot be applied for the whole water column in the Brazil Basin.

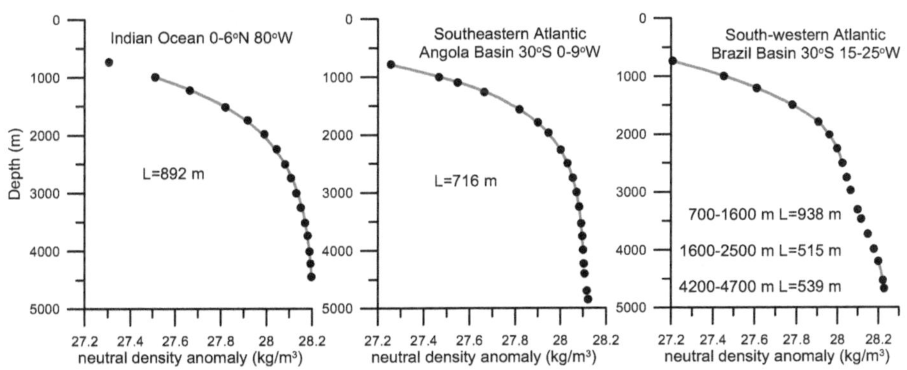

Fig. 9.2. Profiles of the neutral potential density for different ocean basins. The dots represent the regionally averaged profile, while the full lines represent the model results according to Eq. (9.6) as described in the text.

9.2. Profiles of a tracer with first-order decay: radiocarbon

The advective–diffusive length scale L can be estimated from the application of Eq. (9.6), but not the values of K or w separately. In order to obtain their value, a clock tracer with its own characteristic timescale has to be used, e.g., ^{14}C or radiocarbon. If we want to describe the vertical distribution of the radiocarbon concentration in the ocean, $[^{14}C]$, due to the advective–diffusive balance in the global upwelling we have to add a decay term to Eq. (9.5):

$$K\frac{d^2}{dz^2}\left[{}^{14}C\right]-w\frac{d}{dz}\left[{}^{14}C\right]-\frac{\left[{}^{14}C\right]}{\tau}=0 \quad . \tag{9.9}$$

Next to the length scale L we can introduce a decay length λ, defined as $\lambda = w\tau$, where τ is the ${}^{14}C$ decay time. Then Eq. (9.9) can be rewritten as

$$\frac{d^2}{dz^2}\left[{}^{14}C\right]-\frac{1}{L}\frac{d}{dz}\left[{}^{14}C\right]-\frac{1}{L\lambda}\left[{}^{14}C\right]=0 \quad . \tag{9.10}$$

The general solution of Eq. (9.10) is the difference between two exponential functions:

$$\left[{}^{14}C\right]=A\cdot e^{\frac{1}{2}(1+\sqrt{1+4L/\lambda})\frac{z}{L}}-B\cdot e^{\frac{1}{2}(1-\sqrt{1+4L/\lambda})\frac{z}{L}} \quad , \tag{9.11}$$

where the integration constants A and B are determined from the boundary conditions.

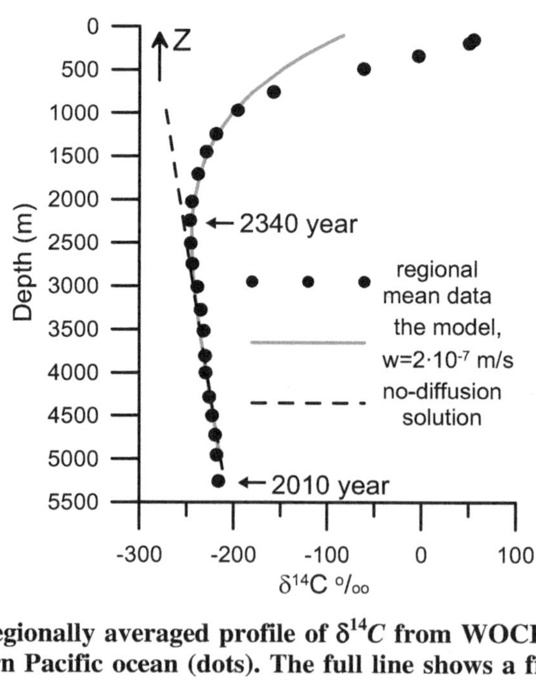

Fig. 9.3. The regionally averaged profile of $\delta^{14}C$ from WOCE section P16 in the northeastern Pacific ocean (dots). The full line shows a fit with the solution of (9.11) assuming that $w = 2\ 10^{-7}$ m/s. The dashed line gives the no-diffusion solution according to Eq. (9.13). The radiocarbon ages, indicated near 2250 and 5250 m, are not corrected for the reservoir effect.

Unfortunately it is not possible to determine w from an optimal fit, since fitting of Eq. (9.11) to the data presents, in mathematical terms, an ill-posed problem. Any value for w, from $0.5\ 10^{-7}$ to 10^{-2} m/s, gives a nearly perfect fit to the data from most ocean basins. Radiocarbon data from the northeastern Pacific Ocean are shown in Fig. 9.3. The full line gives a least-squares fit of Eq. (9.11) to the data points between 1000 and 5000 m, assuming that $w = 2\ 10^{-7}$ m/s (~6.3 m/year) and $L = 885$ m, derived from the density stratification.

Some constraints for w still can be derived from the $\delta^{14}C$ data by ignoring the diffusion term in Eqs. (9.9) or (9.10). Below 2500 m the δ^{14}C profile appears to be nearly linear. This suggests the possibility that at these levels the diffusion term in Eq. (9.10) can be ignored relative to the advection and radioactive decay terms. Then Eq. (9.10) reduces to

$$\frac{\mathrm{d}}{\mathrm{d}z}\left[{}^{14}C\right] = -\frac{\left[{}^{14}C\right]}{\lambda} \quad . \tag{9.12}$$

The solution of this simple advection–decay balance is an exponential function:

$$\left[{}^{14}C\right] = \left[{}^{14}C_0\right] e^{-\frac{z}{\lambda}} \quad . \tag{9.13}$$

Expressed in δ^{14}C, the exponential curve from Eq. (9.13) gives quite a good fit for the data points below 2250 m (the dashed line in Fig. 9.3). The standard deviation of the residual relative to the exponential curve below 2250 m amounts to only 1.2 ‰. The value of the upwelling velocity w derived from the fit amounts to $2.7\ 10^{-7}$ m/s or 8.5 m/year. It is easy to derive from Eq. (9.13) that where this fit is a useful approximation of the δ^{14}C profile, w can be estimated as the vertical distance between two different levels divided by the radiocarbon age difference between the two different levels. If we apply this to the ages near 2250 and 5250 m, indicated in Fig. 9.3, we obtain a vertical velocity of $2.9\ 10^{-7}$ m/s, a value of the same order of magnitude as the velocity value used for the fit with the function of Eq. (9.11) given by the full line in Fig. 9.3.

For other ocean basins the interpretation of radiocarbon profiles is sometimes less straightforward than in the North Pacific. In the northwestern Atlantic at the latitude of Newfoundland (~43°N) nearly all water is relatively young (Fig. 9.4, upper left panel). This region is close to the main sources of NADW, the Iceland–Scotland and Denmark Strait overflow and the Labrador Sea deep convection. The $\delta^{14}C$ values here also are high because of the influence of bomb radiocarbon produced in nuclear test explosions on the deepwater properties. This caused the DSOW near 60°N to have a positive $\delta^{14}C$ (+2 ‰, from WOCE section A01) in the 1990s. In the north-western Atlantic the core of LSW is generally found at ~1500 m, while the saline NADW core is located near 3000 m. The uncorrected radiocarbon age for these water mass cores at 43°N are, respectively, 180 and 480 year, shorter than the expected reservoir effect of the surface waters. This confirms a measurable influence of bomb-derived ^{14}C in the deep North Atlantic. With remnants of old AABW in the bottom layer, although diluted by mixing with overlying NADW with its load of bomb radiocarbon in the North Atlantic, this leads to a monotonously upward decrease of ^{14}C age. There is no possibility here to apply either Eqs. (9.11) or (9.13), since bomb radiocarbon is a transient tracer.

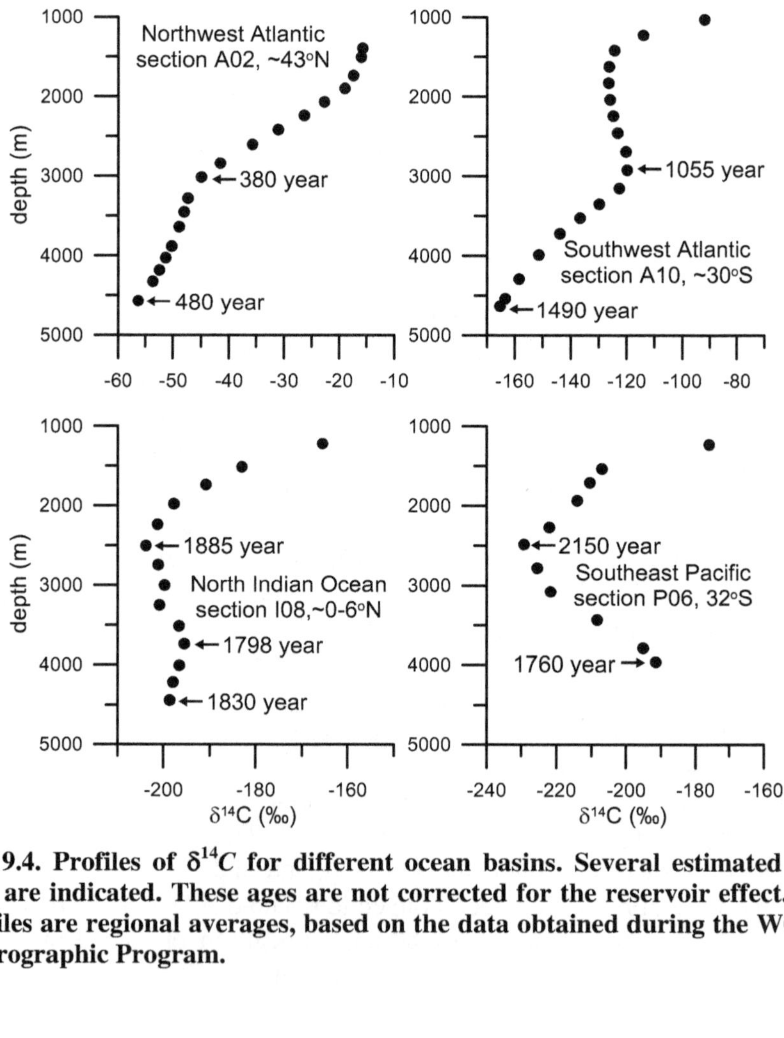

Fig. 9.4. Profiles of $\delta^{14}C$ for different ocean basins. Several estimated $\delta^{14}C$ ages are indicated. These ages are not corrected for the reservoir effect. The profiles are regional averages, based on the data obtained during the WOCE Hydrographic Program.

In the Brazil Basin in the southwestern Atlantic (Fig. 9.4, upper right panel) the presence of relatively young NADW near 3000 dbar causes a δ^{14}C maximum in between the aged waters of southern origin above and below this northern water type. Such an intermediate maximum cannot result from Eq. (9.10); additional horizontal advection or diffusion should be added. The near-bottom AABW has a radiocarbon age of 1490 years, only nearly 100 years older than the typical age of the deep and bottom water in the Antarctic Circumpolar Current.

In the northern Indian Ocean (Fig. 9.4, lower left panel) the youngest deepwater is found at approximately 3750 m (1798 years), probably because the direct flushing of the Indian Ocean at deeper levels is constrained by the topography. When we approximate the $\delta^{14}C$ profile between 2500 m and 3750 m with Eq. (9.13) we arrive at a vertical velocity $w \approx 5\ 10^{-7}$ m/s. This number probably overestimates the upwelling velocity, since both at 2500 and 3700 m the curvature of the $\delta^{14}C$ profile is locally extreme, so that mixing leads to a decrease of the apparent age difference between both levels. The $w = 5\ 10^{-7}$ m/s estimate therefore should be considered as an upper limit of the possible values for w.

In the southeastern Pacific Ocean (Fig. 9.4, lower right panel) we find a δ^{14}C stratification similar to the North Pacific, with a $\delta^{14}C$ minimum near 2500 m. Approximating the profile between 2500 and 4000 dbar with the simplified Eq. (9.13) results in a vertical velocity $w = 1.1\ 10^{-7}$ m/s.

The typical length scale L of the advective–diffusive balance in the semienclosed deep basins is of the order of 1000 m, while a typical value of w, estimated with a least-squares fit of Eq. (9.13) with the data, is of the order of $2\ 10^{-7}$ m/s. Combining these results we come to a characteristic value for the diffusion coefficient for the abyssal ocean below 1000 m of an order of $K \approx 2\ 10^{-4}$ m^2/s, about 2000 times the molecular diffusivity for heat and 200,000 times the molecular diffusivity for salt. This result shows that turbulent diffusion, enhancing the value of K, is required to explain the observed deep stratification of the global ocean. Since the turbulent diffusion constantly brings heat downward from the warmer (sub)tropical waters, the deepwater in the ocean eventually will warm up if this diffusion is not countered by the continuous supply of cold deep and bottom water by the THC. With a typical ocean depth of $D \approx 4000$ m and $K \approx 2\ 10^{-4}$ m^2/s the characteristic timescale of the deep warming of the ocean after a possible stagnation of the deepwater supply for the THC would be $D^2/K \approx 2500$ year. The value of the effective K on a basin scale of $2\ 10^{-4}$ m^2/s is an order of magnitude larger than the observed value for the pelagic ocean far away from bottom and side boundaries. There experiments with microstructure sensors, analysis of thermohaline finestructure, and tracer release experiments all converge to a pelagic K value, K_{PE}, of the order of $1\ 10^{-5}\ m^2/s$. The enhanced turbulence levels over steep slopes and irregular topography appear to have a strong influence on the stratification of whole ocean basins, even in the wide Pacific Ocean that covers nearly half the globe.

The $\delta^{14}C$ profiles, presented in Figs. 9.3 and 9.4, show a monotonic increase of the radiocarbon age at the levels where NADW or its successor water types in the Indian and Pacific Ocean are found (2500–3500 m). Corrected for a reservoir effect of 400 years for the newly formed NADW in the North Atlantic, the radiocarbon age at NADW/LCDW levels increases to 655 year in the South Atlantic (Brazil Basin), 1000 years in the Antarctic Circumpolar Current, 1485 years in the northern Indian Ocean, 1750 year in the South Pacific, to reach 1940 years in the North Pacific. Apparently the flushing time of the deepwater in the ocean by NADW is of the order of 2000 year.

9.3. Tracers with zeroth order sources and sinks: oxygen, and nutrients

The dynamics of the concentrations of dissolved oxygen and nutrients can be approximated as a zeroth order process where the source or sink term does not depend on the concentration itself but is locally constant. To describe the stratification of these tracers, following Eq. (4.1), a concentration-independent source (positive) or sink (negative) term Q has to be added to Eq. (9.5)

$$K\frac{d^2C}{dz^2} - w\frac{dC}{dz} + Q = 0 \quad . \tag{9.14}$$

For simplicity K, w, and Q are assumed to be constant in the depth interval where we apply this model. By introduction of the advective–diffusive length scale L, defined above, this equation can be rewritten as

$$\frac{d^2C}{dz^2} - \frac{1}{L}\frac{dC}{dz} + \frac{Q/w}{L} = 0 \quad . \tag{9.15}$$

It is easy to show that a linear combination of the advective–diffusive solution, Eq. (9.7), and the linear profile of the advection–source balance forms the solution of Eq. (9.15),

$$C = A \cdot e^{z/L} + \frac{Q}{w} z + B \quad , \tag{9.16}$$

where the integration constants A and B have to be determined from the boundary conditions.

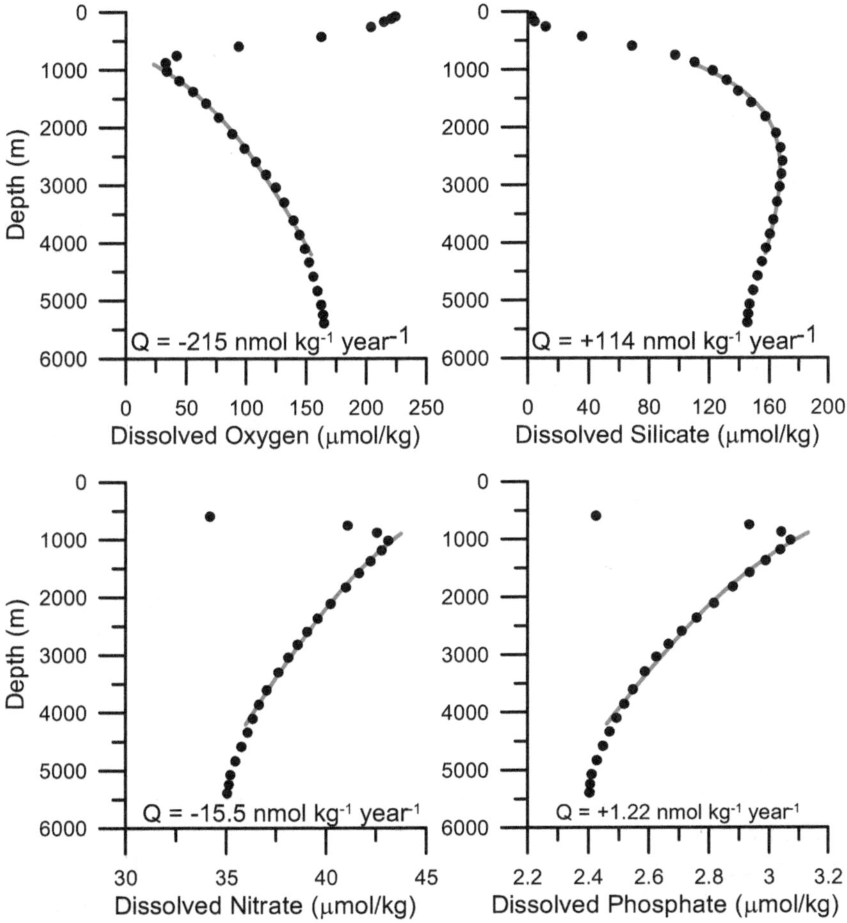

Fig. 9.5. The mean vertical distribution of the concentrations of dissolved oxygen and nutrients in the northeastern Pacific Ocean. The data used for these profiles are from WOCE section P16 between 25 and 30°N. The full lines show least-squares fits of Eq. (9.16) between 1000 and 4000 m depth.

For the profiles of dissolved oxygen and the nutrients silicate, nitrate and phosphate Eq. (9.16) gives an accurate least-squares fit to the mean tracer profiles from the northeastern Pacific Ocean in the depth interval

1000–4000 m, assuming that $L = 885$ m as was derived from the density profile (Fig. 9.5). The standard deviation of the remaining residuals in the 1000–4000 m depth interval is of the order of 1.5% of the range of these tracers in that interval. The minimum in the O_2 profile near 900 m and the maxima in the nitrate and phosphate profiles near 1000 m cannot be simulated accurately with the model of Eq. (9.15). Probably the lateral influx of PAIW complicates the situation at these shallow levels, while our assumption that K, w, and Q are constant also may fail here. This already was shown for the mean profiles of density and radiocarbon, discussed above. However, the deeper silicate maximum near 2700 m is simulated quite adequately by Eq. (9.16).

From the regression coefficient of the linear term in Eq. (9.16), Q/w, the source/sink term Q can be estimated if w is known. When we assume that $w = 2.7\ 10^{-7}$ m/s, as obtained from the model fit of Eq. (9.13) to the $\delta^{14}C$ profile, we obtain source terms $Q = -215\pm18$, $+114\pm10$, 15.5 ± 0.6, and 1.22 ± 0.07 nmol kg^{-1}year^{-1} for, respectively, oxygen, silicate, nitrate, and phosphate. When oxygen is consumed by mineralization of organic material, either particulate or dissolved, nitrate and phosphate are produced in approximately constant stochiometric ratios: the Redfield ratios. The stochiometric ratios of Q for phosphate, nitrate, and oxygen, derived from the model fit of Eq. (9.16) to the data, appear to be

$$Q(P){:}Q(N){:}Q(O_2) = 1{:}12.6{:}-175 \quad . \tag{9.17}$$

This agrees closely with the mean Redfield ratios for the Pacific Ocean between 1000 and 4000 m, derived independently by Anderson and Sarmiento (1994) who have found for P, N, and O_2

$$Q(P){:}Q(N){:}Q(O_2) = 1{:}13{:}-170 \quad . \tag{9.18}$$

The ratios of our estimates of Q from Eq. (9.16) appear to be mutually consistent and consistent with existing biogeochemical knowledge on the Redfield ratios for P, N and O_2. The ratio $Q(P){:}Q(Si) = 1{:}93.4$ is an empiric stochiometric ratio not based on a fixed ratio of these elements in organic material but based on common effects of remineralization and dissolution during aging of a water mass.

9.4. Energy requirements for turbulent mixing

The maintenance of an upward turbulent mass flux (downward heat flux) in the upwelling branch of the THC requires an input of mechanical energy: first because work has to be performed against gravity, second because the turbulent transport has a limited efficiency. Most of the turbulent energy (~80%) is dissipated to heat, leaving only a reduced portion to drive the turbulent mass flux. The turbulent kinetic energy budget for stationary turbulence in a stratified fluid, in the absence of turbulent self-diffusion, can be written as

$$(\Pi - \varepsilon) = -gK\frac{\mathrm{d}\rho}{\mathrm{d}z} \quad , \tag{9.19}$$

with Π being the shear production of turbulent kinetic energy, ε the viscous dissipation, g the gravitational acceleration, and K the turbulent diffusion coefficient that maintains the turbulent mass flux. Experiments have shown that $(\Pi-\varepsilon)/\Pi$ is approximately a constant λ, being about 0.2. Therefore, the mechanical power density in W/m^3, required to maintain the upward turbulent mass flux in the earth's gravity field equals

$$\Pi = -\lambda^{-1}gK\left.\frac{\mathrm{d}\rho}{\mathrm{d}z}\right|_{ad} = -\lambda^{-1}gK\frac{\mathrm{d}\rho_n}{\mathrm{d}z} \quad . \tag{9.20}$$

If K is approximately constant over a large depth interval in the ocean, the power density Π will decrease exponentially with depth, since $\mathrm{d}\rho_n/\mathrm{d}z$ decreases exponentially with depth. Over a layer with a thickness $\Delta z = z_2 - z_1$ the mean power density in becomes

$$\overline{\Pi} = \frac{1}{\Delta z}\int_{z_1}^{z_2}\Pi\,\mathrm{d}z = \lambda^{-1}gK\frac{\Delta\rho_n}{\Delta z} = \lambda^{-1}gwL\frac{\Delta\rho_n}{\Delta z} \quad , \tag{9.21}$$

where $\Delta\rho_n$ is the difference in neutral density between z_1 and z_2, or the density difference corrected for adiabatic compression effects. The typical value for $\Delta\rho$ between 500 and 3000 m depth in the world ocean between 40°S and 40°N, derived from meridional WOCE sections, is about 1.2 kg/m^3. With a characteristic upwelling velocity of the order of 10^{-7} m/s, $L \approx 1000$ m, and $\lambda = 0.2$, this results in a mean power density to maintain the required turbulent mass flux of 2.4 μW/m^3 or in mass units 2.4 nW/kg. Integrated over the volume of the world ocean where upwelling in the THC occurs, V, the total power input is then

$$\iiint_V \Pi\,\mathrm{d}V = \lambda^{-1}gKA\Delta\rho \quad . \tag{9.22}$$

Here A is the surface of the upwelling ocean (assumed to be 80% of the world ocean or 2.9 10^{14} m^2). With $K \approx wL = 10^{-4}$ m^2/s and $\Delta\rho = 1.2$ kg/m^3 the total global input of turbulent kinetic energy needed to maintain the global upwelling between 3000 and 500 m is $1.7 \cdot 10^{12}$ W or 1.7 TW (terawatt). This value is close to the 2.1 TW estimated by Munk and Wunsch (1998) from a much more elaborate analysis. However, they warned that their result may easily be a factor 2 wrong, and that also applies to our estimate. For the turbulent energy needed to drive the interior mixing in the ocean they proposed that the source of the turbulent energy is essentially wind-driven with a major contribution from the tides. Winds and tides generate near-inertial and internal waves that are well able to transport mechanical energy from their source near the sea surface (wind-driven internal and inertial waves) or the upper parts of continental slopes and mid-ocean ridges and other steep topographic features (internal tides) into the ocean interior. There the shear in breaking internal waves may lead to the generation of turbulent kinetic energy. Dimensional analysis suggests that the energy input from the winds into the ocean is proportional to the third power of the wind speed. Since a considerable part of the diapycnal mixing may depend on the input of energy into the oceans by the wind, a change of wind climate over the oceans will lead to a strong change of the effective diapycnal turbulent mixing coefficient in the deep ocean which in its turn may induce a change in the overturning rate of the global THC.

The kinetic energy $\lambda\Pi$, required to maintain the vertical turbulent mass transport according to Eq. (9.20), first will be converted to potential energy of the stratified water column via the vertical turbulent mass transport in the earth's gravity field. In the subtropical regions where deep upwelling takes place, the resulting level of the sea surface is of the order of 1 m higher than in the polar regions. In its turn the potential energy of the subtropical stratified water column is converted to kinetic energy of the overturning THC, when deepwater flows from the formation regions to the upwelling regions in the cold branch of the THC or in the warm return flow from the upwelling region back to the formation region. In a stationary THC, friction ultimately will convert the kinetic energy of the THC to heat. The heat, generated by the dissipation of this turbulent energy in the ocean, $\lambda\Pi$, will, next to ε, lead to a warming of the ocean water of less than 0.02 °C in 1000 year, and therefore is generally ignored in the ocean heat budget.

Equation (9.22) also can be seen from another viewpoint. An immediate consequence for modeling the THC is that the effective diffusion coefficient K may depend on the basinwide production of turbulent kinetic energy, $\overline{\Pi}$, as well as on the density contrast $\Delta\rho$ between the deepwater and the "surface" water. The density of the surface water will depend on the sea surface boundary conditions for the heat and freshwater exchange with the atmosphere. The density of the deepwater will be close to the density of the deep water mass produced in its source region, which depends on the air–sea exchange in the formation areas as well as on the properties of the near-surface return flow to those areas. If the available production of turbulent energy, the left-hand term of Eq. (9.22), as well as the mixing efficiency, λ, are constant for a certain sea area, then in that region the effective diffusion coefficient K will be inversely proportional with the density difference over the upwelling part of the water column.

Following Nilsson and Walin (2001) and Nilsson et al. (2003) we can make a rough estimate of the influence of the diffusivity on the total overturning circulation. It is a rough but robust estimate, since we will perform a scale analysis of the return flow above the thermocline. We consider the circulation in a single hemispheric ocean basin with surface A and assume that the hydrography in this basin is only characterized by the thermocline depth, H, and by the equator-to-pole surface density difference, $\Delta\rho$. The thermocline depth measures the thickness of the stratified ocean above the nearly homogeneous deepwater.

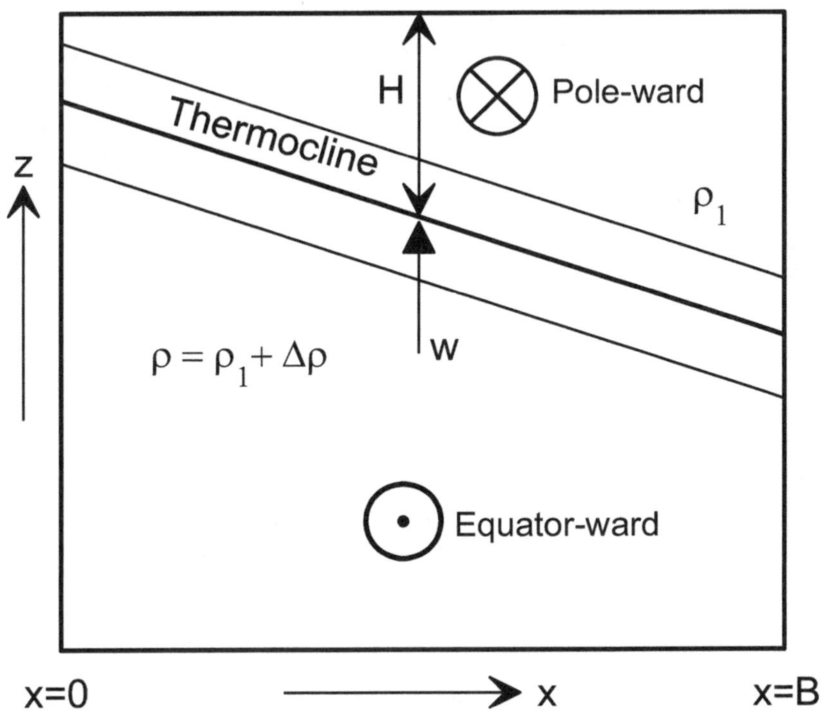

Fig. 9.6. Sketch of a zonal section through the ocean (northern hemisphere), where the geostrophic poleward return flow occurs above the thermocline. The symbols are explained in the text.

For the meridional geostrophic velocity, v, in the return flow above the thermocline, we assume a geostrophic balance, according to

$$\frac{\partial v}{\partial z} = -\frac{g}{f\rho_0}\frac{\partial \rho}{\partial x} \quad , \tag{9.23}$$

where the zonal density gradient is related to the sloping of the thermocline (Fig. 9.6). Further, we assume that the density stratification is described by an advection diffusion balance, similar to Eq. (9.5)

$$w\frac{\partial \rho}{\partial z} = K\frac{\partial^2 \rho}{\partial z^2} \quad . \tag{9.24}$$

Finally we assume continuity of mass transport, making the upwelling intensity, ψ_u, proportional to the poleward return flow in the surface layer, ψ_p.

The determination of the total volume transport in the poleward branch of the THC above the thermocline, ψ_p, requires a vertical integration from the thermocline (= level of no motion) upward to derive the geostrophic velocity and then an integration over the total surface of a zonal section above the thermocline. We assume here that the typical east–west density gradient is proportional to the typical north–south gradient and scales with the density difference $\Delta\rho$ and the width of the ocean, B. With a dimension analysis of Eq. (9.23) this suggests that ψ_p then should be proportional to

$$\psi_p \sim \Delta\rho H^2 \frac{g}{f\rho_0} \quad . \tag{9.25}$$

An independent estimate of the scaling of the total volume transport ψ_u of the upwelling branch at the base of the thermocline can be obtained from the expression for w, derived from a dimension analysis of the advective–diffusive balance of Eq. (9.24). Here we assume that the typical vertical length scale that determines the vertical density gradient and curvature at the base of the thermocline is proportional to the thermocline depth, H. Integration of w over the surface A of the ocean then leads to

$$\psi_u \sim AKH^{-1} \quad . \tag{9.26}$$

The assumption of mass continuity now requires that the intensity of the poleward branch, ψ_p, is proportional to the total upwelling branch, ψ_u. Using this proportionality and omitting the constants A, f, g, and ρ_0 then leads to an expression for the thermocline depth as function of diffusivity and density contrast:

$$H^3 \sim K / \Delta\rho \quad . \tag{9.27}$$

Substitution of Eq. (9.27) into Eq. (9.26), omitting the geographic constant A, results in

$$\psi_u \sim K^{2/3} \Delta\rho^{1/3} \quad . \tag{9.28}$$

The proportionality of the total overturning circulation, given by Eq. (9.28), shows that if the turbulent energy production, Π, is not a limiting factor for the THC, so that K is constant, the intensity of the THC will increase with the $^1/_3$ power of the density contrast between pole and equator. This agrees with the often used intuitive viewpoint that a decrease in the density of the Arctic seas and Polar Ocean due to a decreasing salinity will lead to a decrease of the strength of the THC. In its turn such a decrease will lead to a decrease in the oceanic heat flux to the polar regions, generating a cooling of the local high-latitude climate and possibly inducing a new ice age in Scandinavia (e.g., Vellinga and Wood 2002). The assumption of a constant K is used widely in ocean general circulation models, and such models generally show a decrease in overturning when the density of the newly formed deepwater is reduced by, e.g., admixture of an excess freshwater (e.g., Rahmstorf, 1995). However, if the available turbulent energy production, Π, is a limiting factor, and according to Eq. (9.22) K is proportional to the inverse of the density difference $\Delta\rho$, that dependency will alter Eq. (9.28) to

$$\psi_u \sim \Delta\rho^{-1/3} \quad , \tag{9.29}$$

suggesting that in that case a reduction of the density difference between pole and equator will lead to an increase of the strength of the THC. That increase will bring additional heat to the polar regions, resulting in a warmer polar climate.

The dimensional scaling arguments given above, although very rough, indicate that a very good understanding of oceanic turbulent mixing is essential in order to be able to model the THC for climate prediction purposes. Otherwise even the sign of the change of the THC may be wrong. One has to be aware that the introduction of the turbulent diffusion coefficient, K_v, is an *ad hoc* solution to parameterize the covariance of the turbulent variations of density and velocity, including or not including a variety of processes like real shear-driven turbulent eddies, breaking internal waves, double-diffusive convection, convective plumes, and convective chimneys.

10. Energetics of the THC

10.1. Some thermodynamics

The THC can be schematized in a simplified way as a vertically overturning circulation cell (Fig. 10.1) with a localized downwelling region. There the surface heat loss and/or evaporation to the atmosphere increases the surface density. Global upwelling occurs in the larger part of the ocean, whereby the surface density of the upwelled water is reduced by heating and/or precipitation from the atmosphere. The dense water mass, brought to large depths by the downwelling, spreads to other latitudes (and ocean basins) in the deep branch of the THC. The required return flow to the downwelling region takes place in the near-surface branch of the THC, where the water density increases downstream.

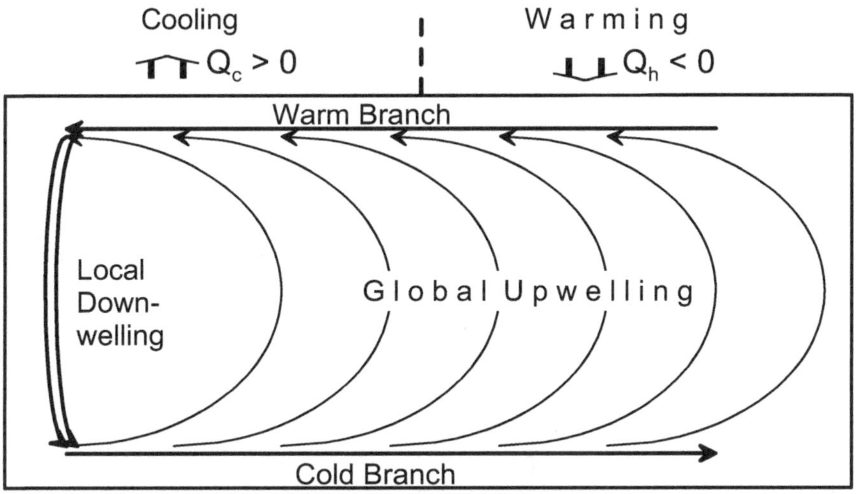

Fig. 10.1. Schematic sketch of a single-hemisphere overturning cell with surface heating in the tropics and subtropics and surface cooling in the subpolar and polar regions.

In the present climate the downwelling regions are found at high latitudes (arctic and antarctic) where surface cooling ($Q_c > 0$) is the main cause of the surface density increase. This causes the deep branch to be a "cold branch", while the near-surface return flow from lower latitudes takes place in the "warm branch" of the THC, since surface warming ($Q_h < 0$) heats the cold upwelled water. The surface flux of freshwater (evaporation and precipitation) also determines the density at the sea surface by affecting the salinity. However, the surface heat flux is presently the dominating component in the density boundary conditions of the THC, as will be shown below. One can pose the question whether and how the surface heat (density) forcing is able to drive the THC. That is: is the ocean a heat engine that obtains the required energy from the differential heat flux at the sea surface or is an additional source of mechanical energy required to drive the THC?

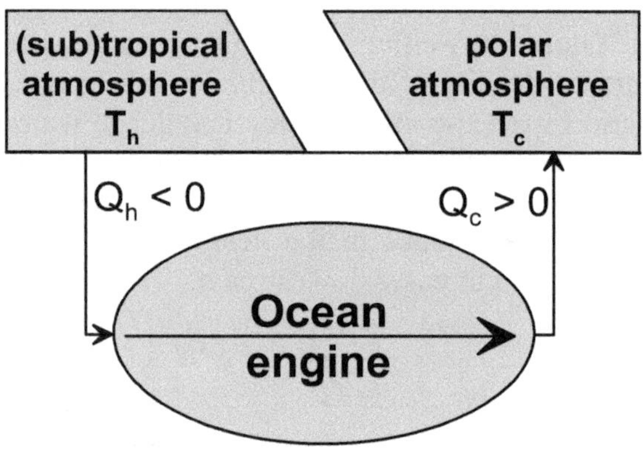

Fig. 10.2. Sketch of the ocean as a heat engine. The ocean receives an amount of heat $-Q_h$ at a temperature T_h from the (sub)tropical atmosphere and releases Q_c to the atmosphere at a temperature T_c ($T_h > T_c$).

The ocean can be considered to be a heat engine that is subject to the laws of thermodynamics, just like a steam engine is. Then the THC will depend on the heat flux and freshwater flux through the system to supply the necessary energy to keep the ocean engine running (Huang 1999). Let us pursue this idea and assume that the THC is driven as a stationary heat engine, receiving an amount of heat per time unit, $-Q_h$, from the warm (sub)tropical atmospheric heat reservoir at an (*absolute!*) temperature T_h (Fig. 10.2). The heat transport Q_h is by definition positive when the ocean releases heat to the atmosphere. After circulating through the THC engine, an amount of heat Q_c is released to the cold polar atmospheric heat reservoir at a temperature T_c. The THC engine is assumed to be stationary. In that case the first law of thermodynamics (energy conservation) states that the amount of work, W, performed by the engine to overcome dissipative processes equals the difference between the heat input into the THC and the heat release from the THC. Following Colin de Verdière (1993) with a slightly different notation, this conservation of energy is written as

$$W = -Q_h - Q_c \quad . \tag{10.1}$$

The second law of thermodynamics states that the total entropy of the atmosphere–THC system has to increase. This entropy increase is related to the resulting decrease of the temperature contrast between the cold polar and warm (sub)tropical atmospheric heat reservoirs due to the meridional heat transport via the THC engine. Since we assume that the THC engine is stationary and that the contribution of the ocean to the total entropy of the ocean–atmosphere system therefore does not change, the total rate of change of entropy, ΔS_{total}, equals

$$\Delta S_{total} = \frac{Q_h}{T_h} + \frac{Q_c}{T_c} \quad . \tag{10.2}$$

Substitution of Eq. (10.1) into Eq. (10.2) then gives

$$W = Q_c\left(\frac{T_h}{T_c} - 1\right) - T_h \Delta S_{total} \quad . \tag{10.3}$$

In order to overcome frictional losses, W must be positive for the heat engine. In the limit of $W = 0$ the THC engine has become ineffective, and then Eq. (10.3) describes the entropy increase due to the heat exchange $Q_c = -Q_h$ between the warm and cold atmospheric heat reservoirs by diffusive transport without any interference of a dissipative THC engine. If we assume that the oceanic processes described by this simple thermodynamic model are reversible, so that the THC has become a Carnot engine and $\Delta S_{total} \approx 0$, the Carnot efficiency of the THC, η, described by the ratio of work performed by the engine and the heat transport from the ocean into the polar atmosphere becomes

$$\eta_{Carnot} = \frac{W}{Q_c} = \left(\frac{T_h}{T_c} - 1\right) \quad . \tag{10.4}$$

Equation (10.4) shows that, as required for every heat engine, the oceanic heat engine can perform work only if the input of heat into the ocean occurs at a higher temperature than the return of the heat to the polar atmosphere.

Observational evidence from satellite data shows that globally on average the net heat loss to the atmosphere occurs at a sea surface temperature of $T_c = 15.7°C$, while the heat gain takes place at $T_h = 27.0°C$. For the Atlantic Ocean alone these temperatures are 15.9 and 26.0°C, respectively. According to Eq. (10.4) the maximum possible efficiency of our THC engine therefore is 3 to 4%. The actual efficiency of the THC will be lower because of turbulent dissipation and in general because the climate is a not a reversible process. But anyway, $T_h > T_c$ so that Eq. (10.3) in principle allows the THC to be driven as a heat engine by the heat exchange with the atmosphere.

10.2. Heat exchange with the atmosphere and heat fluxes

Three German institutes, the University of Hamburg, the Max Planck Institute for Meteorology in Hamburg, and the German Meteorological Institute in Bonn, have processed satellite data to produce a data set containing a climatology of atmosphere–ocean interaction named HOAPS (Hamburg Ocean Atmosphere Parameters and Fluxes from Satellite Data). The HOAPS data set contains the sea surface temperature, the flux densities of latent and sensible heat, and the net long-wave radiation for the years 1987–1997 (www.hoaps.zmaw.de). For the determination of the net heat budget of the sea surface also the net input of shortwave radiation from the sun is required. These data can be obtained from the reanalysis program of the US National Centers for Environmental Prediction and the National Center for Atmospheric Research of the US (NCEP–NCAR, (www.cdc.noaa.gov/cdc/reanalysis/reanalysis.shtml).

The annual mean sea surface temperature (*SST*) (Fig. 10.3) shows the usual distribution with high temperatures of over 26°C in the equatorial zones and low temperatures at high latitudes. The warm belts are the heating zones for the returning upper branch of the THC. Of particular importance for the THC is the asymmetry between the North Pacific Ocean and the North Atlantic Ocean. At high latitudes between about 60 and 80°N in the Atlantic relatively high temperatures are found with an annual mean of 6°C near the Norwegian North Cape. This allows strong cooling and preconditioning of saline water in the Norwegian Atlantic Current well before this water enters the Barents Sea and Greenland Sea where deepwater is formed in winter.

Moreover, the ice-covered Polar Ocean north of 80°N has a direct deep connection with the Nordic seas via the Fram Strait between Greenland and Spitsbergen. Through this passage cooled saline Atlantic water can enter the Polar Ocean at subsurface levels below the pack ice. The Pacific Ocean lacks such a deep polar mediterranean sea where strong cooling and water mass conversion can occur. The Bering Sea between Asia and North America in the North Pacific Ocean is a shallow shelf sea that reaches northward to only about 63°N.

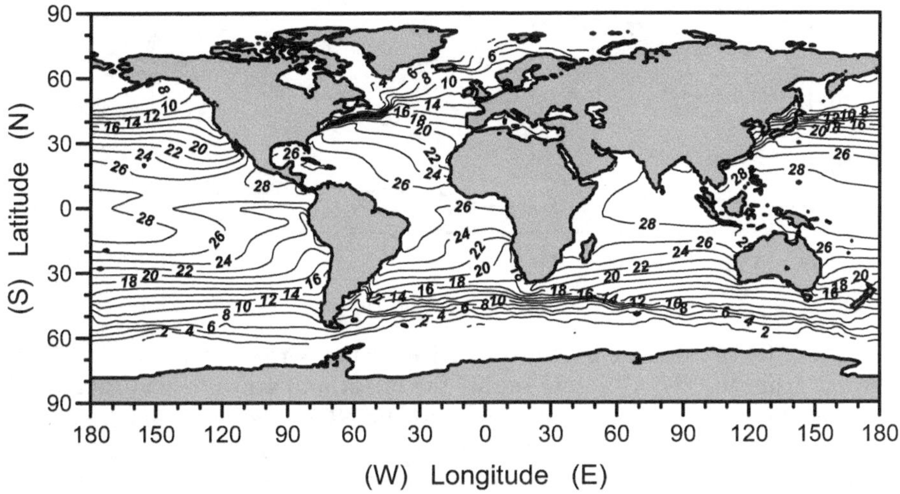

Fig. 10.3. The annual mean sea surface temperature (°C) (*SST*) for the period 1987 to 1997, derived from the HOAPS data set. Source of data: www.hoaps.zmaw.de.

By combining the HOAPS and NCEP–NCAR data, a data set can be constructed of the net sea surface heat flux density. This data set shows a long-term net global heat gain of the oceans of 17.5 W/m^2, probably due to systematic errors in the satellite data and the reanalysis. We have corrected our net heat flux data by subtracting this value from the calculated heat flux density. However, it may be that spatially uniform corrections are not appropriate but should be done locally (Trenberth and Caron 2001). The resulting corrected heat flux density data (Fig. 10.4) anyway can be considered as a reasonable estimate of the large-scale distribution of the heat exchange between the ocean and the atmosphere. They show, as a general trend, strong heating of the ocean near the equator and heat loss at high latitudes. However, the extreme values of mean net heat loss at the sea surface do not coincide with the low temperatures at high latitudes where the deepwater is formed that spreads in the cold branch of the THC, but in regions where the sea surface temperature is relatively high for its latitude. These regions generally are connected with strong poleward boundary currents which form part of the wind-driven gyre circulation in the oceans.

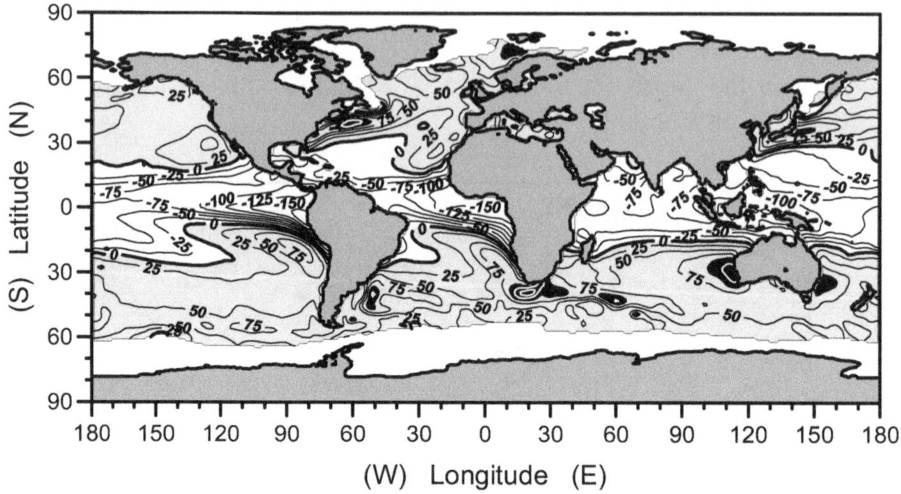

Fig. 10.4. Net sea surface heat flux density for the period 1987–1997 in W/m^2, derived from the HOAPS data set for the latent and sensible heat flux densities and the net long-wave radiation, and the NCEP–NCAR data set for the shortwave radiation. Negative numbers mean heat gain by the ocean (shaded areas). The blackened areas are regions with a net heat loss of over 100 W/m^2. Source of data: www.cdc.noaa.gov/cdc/reanalysis/reanalysis.shtml.

A mean net heat loss of over 100 W/m^2 (blackened areas in Fig. 10.4) is found over the Gulf Stream, the Norwegian Atlantic Current, the Brazil Current, the Agulhas Retroflection and Return Current, the Leeuwin Current west of Australia, and the East Australia Current. The warm branch of the THC, returning to the North Atlantic Ocean, contributes to the Agulhas Current and the Gulf Stream but only in minor amounts compared to the contribution of the wind-driven circulation. Only the strong heat loss between Norway and Spitsbergen is connected for a large part with the return flow, but there the *SST* was already lowered to about 6°C. Apparently strong "precooling" in the warm surface branches of the THC occurs before the returning subtropical waters reach the deep convection areas in the Greenland Sea, the Labrador Sea, and the Weddell Sea.

The change in the sea surface temperature, due to the net air–sea heat flux, will induce a density change because of thermal expansion. A temperature change δT will cause a density change $\delta\rho = -\rho_o\alpha\delta T$, where ρ_o is a reference density and α is the thermal expansion coefficient, defined by Eq. (3.8). The increase in sea surface density due to heat loss to the atmosphere may lead to static instability of the water column and convective mixing, the first steps of newly formed deepwater to the cold branch of the THC.

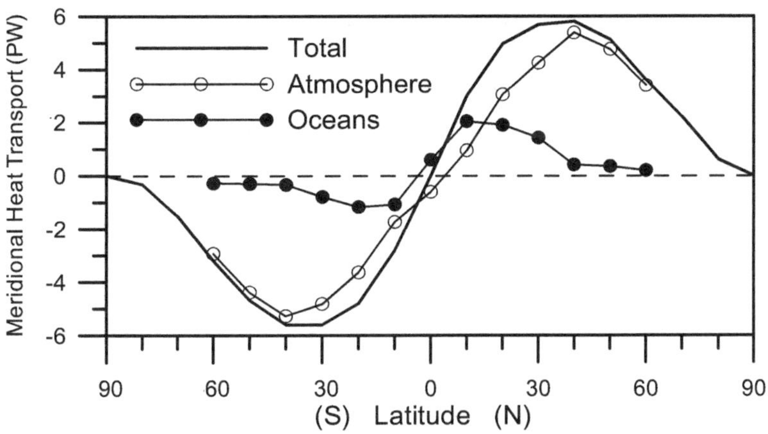

Fig. 10.5. The total meridional heat transport (in PW, 1 PW = 10^{15} W) in the ocean plus atmosphere (thick line) in the atmosphere only (open symbols) and in the ocean only (black symbols) (after Trenberth and Caron 2001). The total heat transport is derived from satellite radiation measurements at the top of the atmosphere and the atmospheric transport from the NCEP–NCAR reanalysis. The oceanic heat transport is the difference between the total and the atmospheric transport.

The meridional heat transport in the ocean can be derived from the meridional integration of the zonally integrated heat flux density from the atmosphere to the ocean. But the presently available estimates of the ocean–atmosphere heat exchange, like the data shown in Fig. 10.4, are not accurate enough to estimate the meridional heat transport in the oceans. Therefore, the meteorologists Trenberth and Caron (2001) have used the satellite measurements of the net radiation transport (short and long wavelengths) at the top of the atmosphere (TOA), from the Earth Radiation Budget Experiment (ERBE) with some corrections. The meridional integration of the zonally integrated radiative flux at the TOA will result in a meridional total heat transport by atmosphere and ocean (the thick line in Fig. 10.5). Hereby it is assumed that, because of geometric reasons, the meridional transport at the poles equals zero. Trenberth and Caron have used the estimates of the atmospheric heat transport produced by NCEP–NCAR and the European Centre for Medium-Range Weather Forecasting (ECMWF) reanalyses of the atmospheric circulation. In Fig. 10.5 the atmospheric heat transport from the NCEP–NCAR reanalysis is shown with open symbols. The difference between the total and the atmospheric heat transport is the transport by the other moving component of the climate system, the ocean (black symbols in Fig. 10.5). The results of the analysis by Trenberth and Caron (2001) clearly show that on a global scale the meridional heat transport by the ocean is mainly important in (sub)tropical band between 40°S and 40°N. At higher latitudes the intense atmospheric eddy activity in the zones with dominantly westerly winds appears to be quite capable to maintain the bulk of the meridional heat transport in the climate system. Note that the oceanic heat transport includes both the transport by the THC and by the wind-driven circulation. Overall the oceanic contribution to the total meridional heat transport is definitely less than suggested by earlier estimates (e.g., Peixoto and Oort 1992).

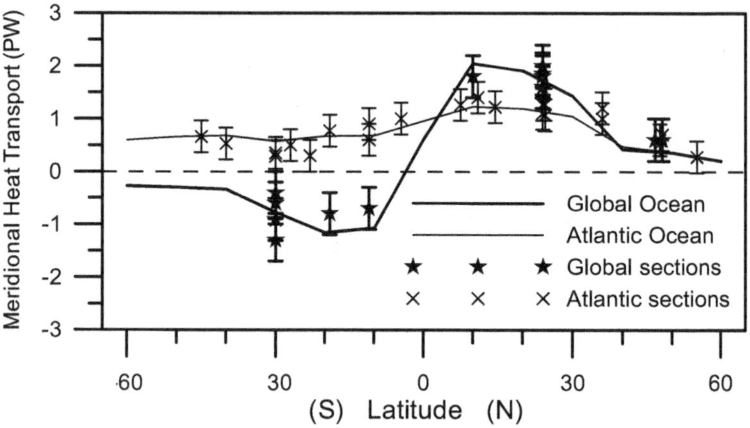

Fig. 10.6. Estimates of the meridional heat transport (PW) in the Global Ocean (thick line) and separately in the Atlantic Ocean (thin line), after Trenberth and Caron (2001). The symbols indicate inverse estimates based on zonal hydrographic sections through the ocean basins.

Trenberth and Caron (2001) also have estimated the total meridional heat transport in the Atlantic Ocean (thin line in Fig. 10.6). Their results confirm that the meridional heat transport is northward in the whole Atlantic Ocean, as can be expected from an overturning circulation system with southward flow of deep cold water and a northward return flow of warmer water in the thermocline. Apparently in the Atlantic Ocean the THC dominates the meridional heat transport over the heat transport in the wind-driven gyres. North of 40°N the oceanic meridional heat transport appears to be solely sustained by the Atlantic circulation which, at those latitudes, is mainly an overturning circulation.

For comparison we have plotted inverse estimates of the Atlantic meridional heat transports by Macdonald and Wunsch (1996), Ganachaud and

Wunsch (2000), and other estimates collected by Trenberth and Caron (2001) (crosses in Fig. 10.6). For the heat transport in the Atlantic Ocean most inverse estimates are within the standard error from the estimate based on the ERBE and NCEP–NCAR data. Also the global inverse oceanic heat transport estimates (stars in Fig. 10.6) appear to confirm the estimates by Trenberth and Caron (2001) for the northern hemisphere. The inverse estimates of the total oceanic heat transport in the southern hemisphere show more scatter. This probably is due to the uncertainty of the effects of the Indonesian through-flow on the heat budgets of the Pacific and Indian oceans. But the mutual agreement of the order of magnitude between the inverse estimates and the estimates by Trenberth and Caron (2001) confirms that both types of estimates are reasonably reliable. The resulting limited contribution of the oceanic heat transport to the total meridional transport casts some doubts on the overwhelming importance for the global climate at high latitudes, attributed to the THC by many oceanographers.

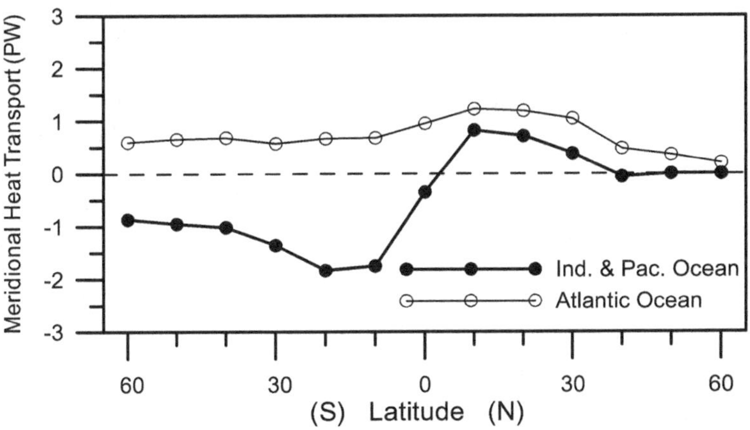

Fig. 10.7. Estimates of the meridional heat transport (PW) in the Pacific and Indian oceans combined (black symbols) and separately in the Atlantic Ocean (open symbols) (after Trenberth and Caron 2001).

When the meridional oceanic heat transport in the Atlantic Ocean is subtracted from the total oceanic meridional heat transport one obtains the meridional oceanic heat transport for the Indian and Pacific oceans combined (Fig. 10.7). At the boundary with the Southern Ocean near 40°S that oceanic heat transport is directed southward, because of the inflow of cold NADW/LCDW and AABW into the Indian and Pacific oceans and a return flow of warmer NIDW and PDW. The negative heat transport nearly doubles in magnitude from 40 to 20°S, because of the additional contribution of the wind-driven gyres in the southern Indian and Pacific oceans. The cross-equatorial transport in the Pacific and Indian oceans is negative but small, in correspondence with a small northward cross-equatorial transport of deep and bottom water and a warmer return flow. In the northern hemisphere the structure of the meridional heat transport, with a maximum near 10°N, seems to be determined mainly by the wind-driven gyre circulation in the Pacific Ocean. A similar enhancement of the meridional heat transport due to the wind-driven gyre can be discerned in the North Atlantic Ocean, also with a maximum near 10°N. North of 40°N the meridional heat transport in the Pacific Ocean is close to zero.

10.3. The influence of the hydrological cycle

The freshwater flux due to the difference between evaporation and precipitation influences the sea surface salinity and will change the density by haline contraction. In this case a salinity change δS will cause a density change $\delta\rho = \beta\rho_0\delta S$ where β is the haline contraction coefficient defined by Eq. (3.7). Since evaporation, *E*, and precipitation, *P*, influence the surface salinity, they also influence the static stability and the density of newly formed deepwater. In order to assess the importance of these freshwater fluxes relative to the surface heat flux, we introduce here virtual salinity flux density F_S which relates the net freshwater flux density at the sea surface, *E–P*, directly to the surface boundary condition for salinity changes. F_S is defined as

$$F_S = -S \cdot (E - P) \quad . \tag{10.5}$$

In this equation E and P are given as velocities (mm/day, m/s, or equivalent). The use of Eq. (10.5) is useful to deal with the development of the salinity field, while it ignores the effects of evaporation and precipitation on the total volume budget of the ocean. This type of parameterization of the freshwater budget also is called the closed lid approximation.

The horizontal distribution of *E–P* from the HOAPS data set (Fig. 10.8) shows that in the subtropics the evaporation E dominates, while at higher latitudes the precipitation is the largest of the two terms. Therefore, the density effect of the precipitation excess in the cooling regions counteracts the density effect of the heat budget. In the tropics the precipitation excess as well as the net heat flux contribute to lower sea surface densities.

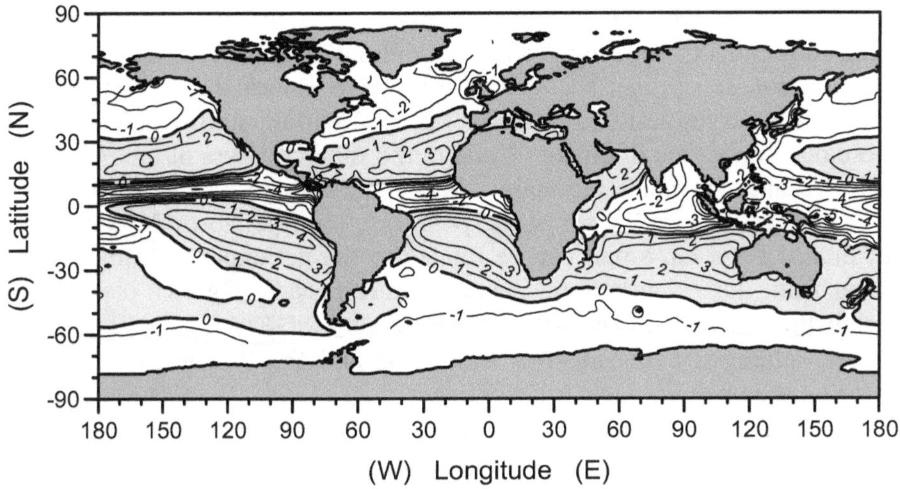

Fig. 10.8. The net surface freshwater flux density *E–P* in mm/day, derived from the estimated evaporation *E* and precipitation *P* in the HOAPS data set. Positive numbers mean an evaporation excess (shaded area). Source of data: www.hoaps.zmaw.de.

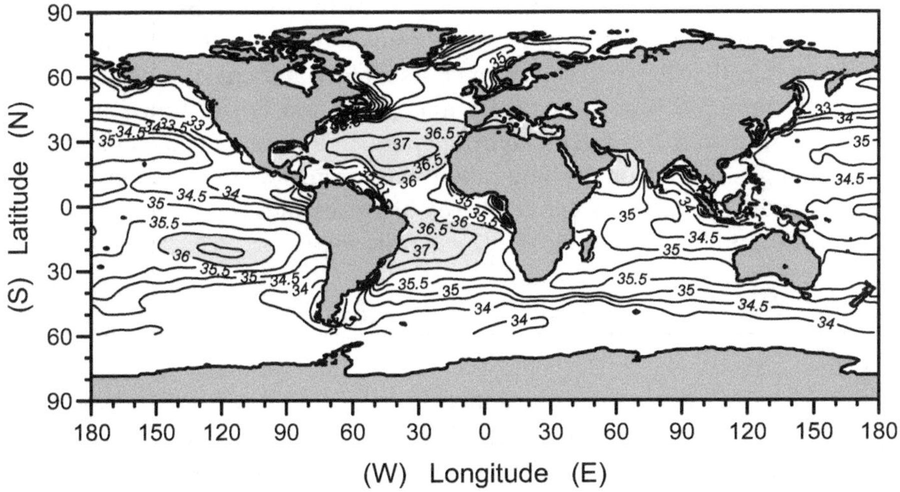

Fig. 10.9. The annual mean sea surface salinity from the World Ocean Database 2001; salinity values above 36.0 have been shaded. Source of data: http://www.nodc.noaa.gov.

The annual mean sea surface salinity (*SSS*) distribution (Fig. 10.9), shows features of importance for the present state of the THC. It reveals a high correlation with the annual mean (*E–P*) distribution. High-salinity regions coincide with the subtropical maxima in the evaporation excess (shaded regions in Fig. 10.9), while at high latitudes and in the tropics the precipitation excess is associated with low *SSS* values. The *SSS* in the Atlantic Ocean is highest of all oceans. Especially in the northern hemisphere these differences are dramatic. While in the subarctic gyre of the North Pacific Ocean *SSS* values are mainly between 32.5 and 33.5, *SSS* values of over 35.0 extend well north of the polar circle in the North Atlantic. This difference for a large part can be related to the net export of atmospheric water from the Atlantic atmosphere to other oceanic regions, particularly to the Pacific Ocean. The net mean difference of $E–P$ from the Atlantic Ocean, averaged between 40°S and 60°N, and the Pacific mean for the same latitude belt amounts to 17 cm/year.

Because of the geographic distribution of $(E–P)$, as shown in Fig. 10.8, the circulation in the atmosphere is required to transport water vapor from the regions with an evaporation excess to regions with a precipitations excess. The oceanic circulation, including the THC, will close the loop, bringing freshwater back from the precipitation-dominated regions to the subtropics. Wijffels et al. (1992) have used the data on precipitation, evaporation, and river runoff published by Baumgartner and Reichel (1975) to estimate the meridional transport of freshwater by the ocean circulation. In their discussion freshwater transport applies to that part of the seawater flux that is pure water. Since oceanic waters have a typical salinity of 35, only 3.5% of the oceanic mass transport is salt transport, while the remaining 96.5% is freshwater transport. In an oceanic basin the meridional changes in the zonally integrated mass transport of seawater are then related to the surface and lateral fluxes of freshwater, E, P, and the runoff R, according to Wijffels at al. (1992) as follows,

$$\frac{\mathrm{d}}{\mathrm{d}y}\iint\limits_{\text{section}}\rho v\,\mathrm{d}x\mathrm{d}z = \int\limits_{width} F(x,y)\mathrm{d}x \quad , \tag{10.6}$$

where x, y, and z are the zonal, meridional, and vertical coordinates, respectively, v is the meridional velocity, and ρ is the density. The function F is the mass flux density of freshwater due to precipitation, evaporation, and river runoff from the land. That is: $F(x,y) = \rho_0(P–E+R)$, where the effects of the river runoff R are attributed to the hole width of the zonal section. The parameter ρ_0 is here the freshwater density. We also can write a similar equation for the mass transport of salt. Since the salt flux by river outflow and across the sea surface can be ignored for the timescales of interest, that salt mass transport will be divergence free, or

$$\frac{\mathrm{d}}{\mathrm{d}y}\iint\limits_{section}\frac{\rho v S}{1000}\mathrm{d}x\mathrm{d}y=0 \quad . \tag{10.7}$$

By subtraction Eq. (10.7) from Eq. (10.6), we get the freshwater mass balance

$$\frac{\mathrm{d}}{\mathrm{d}y}\iint\limits_{section}\rho v(1-\frac{S}{1000})\mathrm{d}x\mathrm{d}y=\int\limits_{width}\rho_0(P-E+R)\mathrm{d}x \quad . \tag{10.8}$$

In order to ascribe the freshwater transport to different aspects of the oceanic circulation we will use the section-averaged salinity $\overline{S}$ and velocity $\overline{v}$ as well as the respective deviations from this mean, S' and v'. Neglecting the covariance of density variations with S' and v' the net meridional freshwater transport then becomes

$$\frac{\mathrm{d}}{\mathrm{d}y}\iint\limits_{section}\rho\left(\overline{v}-\frac{\overline{v}\overline{S}}{1000}-\frac{\overline{v'S'}}{1000}\right)\mathrm{d}x\mathrm{d}y \quad . \tag{10.9}$$

In Eq. (10.9) the first two terms in the integral represent the freshwater transport due to the mean oceanic mass transport across the section. There is no *a priori* reason to expect that $\overline{v}$ equals zero, and it is known that, e.g., in the Bering Strait $\overline{v}$ supports a northward volume transport of ~0.8 Sv, while $\overline{v}$ in the Indonesian through-flow maintains a volume transport of the order of 10 Sv between the Pacific and Indian oceans. Ignoring turbulent transport the covariance term $\rho\,\overline{v'S'}$ in Eq. (10.9) represents the freshwater transport due to the spatial structure of the oceanic circulation, caused by the overturning THC and the wind-driven gyre circulation.

Using the estimates of P, E, and R by Baumgartner and Reichel (1975) and assuming the mass transport through Bering Strait to be 0.8 10^9 kg/s, Wijffels et al. (1992) derived the meridional freshwater transport through the different oceans by integration of Eq. (10.8). In the Atlantic Ocean the freshwater transport was directed to the south at all latitudes (× in Fig. 10.10) and decreased in magnitude southward because of the evaporation excess of the Atlantic Ocean. Since the large Indonesian through-flow would dwarf all other transport estimates, it was ignored for the transport estimates in the Pacific and Indian oceans (+ and ★ in Fig. 10.10). Near the equator the freshwater transport through the Pacific was close to zero, increasing northward. Apparently the freshwater transport through the Bering Strait can be maintained completely by the precipitation and runoff excess in the North Pacific Ocean. The resulting global oceanic meridional freshwater transport (● in Fig. 10.10) compares well in spatial structure and order of magnitude with the independent estimate from Peixoto and Oort (1983), shown as a thick line. To obtain this oceanic transport it is assumed that the contributions of rivers and glaciers to the meridional freshwater transport can be neglected.

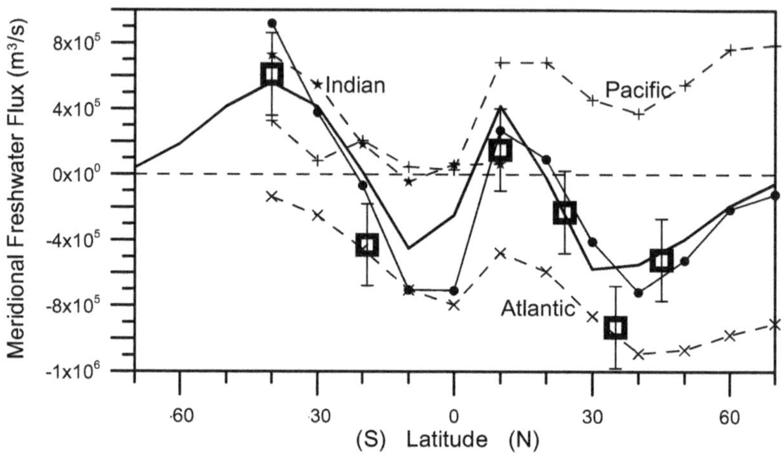

Fig. 10.10. Estimates by Wijffels et al. (1992) of the oceanic meridional freshwater transport through the Atlantic Ocean (×), the Pacific Ocean (+), and the Indian Ocean (★). For the latter two estimates the Indonesian throughflow was ignored. Addition of the estimates for the three oceans resulted in a global oceanic freshwater transport (●). The thick line shows the independent global estimate by Peixoto and Oort (1983). Direct estimates of the total meridional oceanic freshwater transport from hydrographic sections (◘) were summarized by Wijffels (2001).

The few available direct estimates of the meridional freshwater transport, obtained from hydrographic sections by means of inverse modeling techniques, were summarized by Wijffels (2001) and compare reasonably well with the other estimates in (◘ in Fig. 10.10). Apparently these direct estimates, although not perfect, have a useful accuracy.

The meridional freshwater transport sets some constraints on the description of the THC. At 40°S the freshwater flux in the Atlantic Ocean is close to zero. Apparently no large freshwater input from the return flow of the THC from the Southern Ocean is required to balance the strong evaporation in the Atlantic Ocean, the transport of water from the Pacific Ocean through the Bering Strait seems to suffice. The relatively high sum the of the freshwater transport from the Southern Ocean into the Indian and Pacific oceans at 40°S is probably maintained by the inflow of AAIW into the wind-driven southern hemisphere subtropical gyres (Wijffels 2001).

This is a $\overline{v'S'}$ contribution due to the lateral structure of the circulation and not due to overturning in the global THC.

10.4. The density boundary conditions

The combined effects of the freshwater budget and the heat budget at the sea surface can be analyzed more quantitatively with a virtual mass flux density M at the sea surface, defined as

$$M = \beta \cdot \rho_0 \cdot F_S - \alpha \cdot Q_{net} / c_p \quad , \tag{10.10}$$

where the parameter c_p stands for the specific heat of seawater. The mass flux density M according to Eq. (10.10) is an approximation of the surface boundary condition required to model the evolution of the density field in the ocean. It only ignores any effects of F_S and Q_{net} on the volume budget of the ocean.

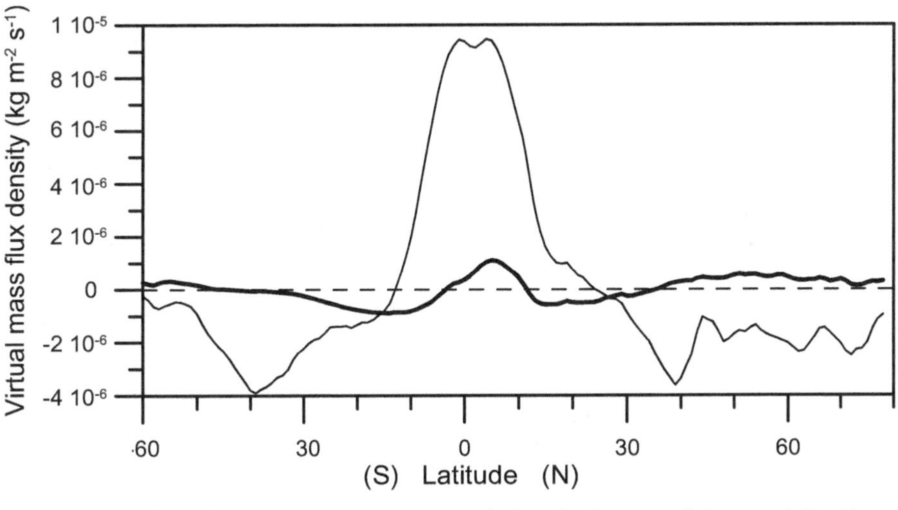

Fig. 10.11. The zonal mean values in the Atlantic Ocean of the contributions $\beta\rho_o F_S$ (thick line) and $-\alpha Q_{net}/c_p$ (thin line) to the virtual mass flux density M, plotted versus latitude.

Based on the HOAPS satellite data and the NCEP–NCAR reanalysis we have estimated the zonal mean value of the right-hand terms of Eq. (10.10) in the Atlantic Ocean in order to assess their relative importance (Fig. 10.11). It appears that the effect of the freshwater budget is, on a global scale, an order of magnitude smaller than the effect of the heat budget.

A high surface salinity increases the surface density and therefore generally favors convective mixing to ventilate the deepwater. The low *SSS* in the North Pacific effectively prevents deep convection that does occur in the more saline North Atlantic. Similar to the effects of the freshwater and heat fluxes on the virtual mass flux density we can quantify the contributions of the *T* and *S* on the sea surface density. First we approximate the equation of state by a linearization relative to a reference situation with reference values S_0, T_0, and $\rho_0 = \rho(S_0,T_0)$ as

$$\rho = \rho_0\{1+\beta(S-S_0)-\alpha(T-T_0)\} \text{ or } \delta\rho = \rho_0(\beta\delta S - \alpha\delta T) \quad , \qquad (10.11)$$

where $S_0 = 35.0$ and $T_0 = 0$ °C are used.

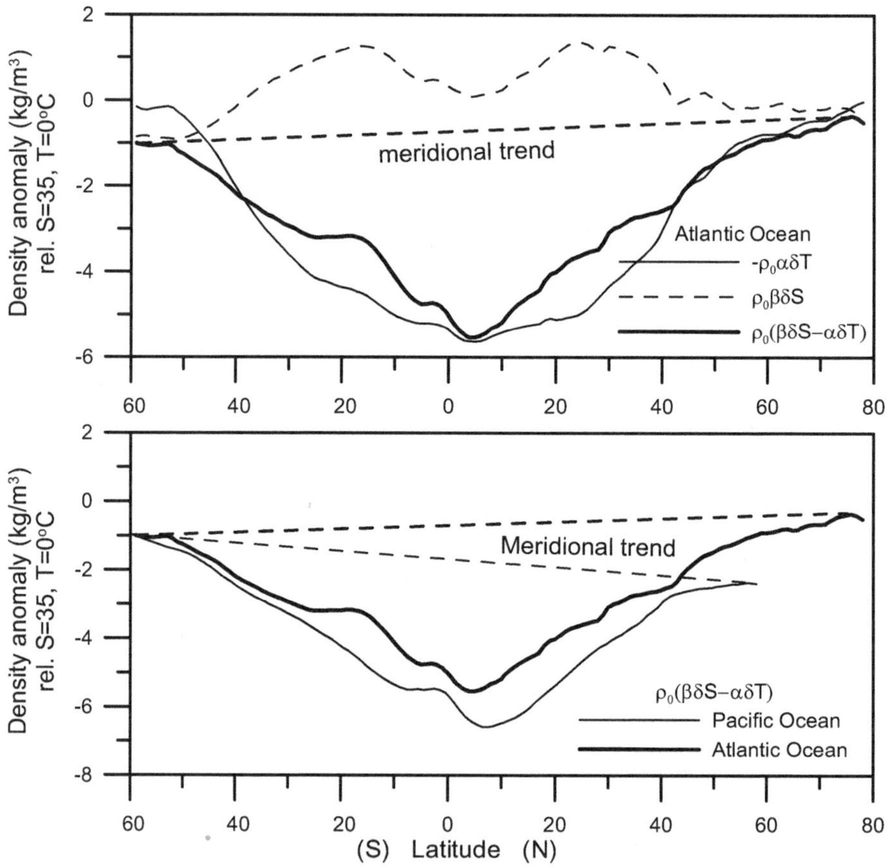

Fig. 10.12. Meridional distribution of the zonally averaged values of $-\rho_0\alpha\delta T$, $\rho_0\beta\delta S$, and $\delta\rho = \rho_0(\beta\delta S - \alpha\delta T)$ in the Atlantic Ocean (upper panel) and of the zonally averaged values of $\delta\rho$ for the Atlantic and Pacific oceans (lower panel).

The zonally averaged values of the salinity and temperature contributions to the meridional variation of the sea surface density anomaly $\delta\rho$ for the Atlantic Ocean (Fig. 10.12, upper panel) show that the magnitude of the salinity contribution to the right-hand side of Eq. (10.11) is about one third of the contribution of the temperature to the meridional density variation. The overall shape of the meridional variation of the sea surface density (low density at the equator, high at high latitudes) is caused mainly by the meridional variation in *SST*. However, the large-scale meridional density trend increasing from antarctic to arctic surface water mainly is related to the *SSS* difference of about 0.5 between the ice edge in the Southern Ocean and the cold waters at the latitude of Spitsbergen in the northern Atlantic Ocean. This trend in *SSS* is connected with the salinification of water in the warm return flow in the upper branch of the THC through the evaporation-dominated Atlantic Ocean. Temperature differences at the sea surface between the cold waters of the arctic and antarctic seas hardly matter for the large-scale density trend, since the high-latitude *SST* will never come below the freezing point of nearly −1.9°C.

Comparison of the meridional distribution of $\delta\rho$ in the Atlantic Ocean and in the Pacific Ocean shows that between ~60°S and ~60°N (the latitude of the Bering Sea in the North Pacific) this large-scale density trend due to north–south salinity differences is reversed in the Pacific Ocean (Fig. 10.12, lower panel). The present state of the THC with the main deepwater formation in the North Atlantic and the absence of deepwater formation in the North Pacific is connected with the opposite large-scale meridional surface salinity trends in these oceans.

Note that in this discussion we did not reckon with the local effects of river runoff, iceberg release from coastal glaciers and subsequent melting, and the salinity effects of the formation and melting of sea ice. The Polar Ocean receives large amounts of freshwater from Eurasian and North American rivers. Regular seasonal ice formation and melting are dominant processes in the arctic Greenland Sea and antarctic Weddell Sea.

10.5. The THC engine and Sandström's theorem

If the THC engine in principle can be driven by differential heat exchange with the atmosphere as discussed above in relation to Eq. (10.3) and similarly by freshwater fluxes, does it actually operate as such? In an engineering analogue, the boiler and condenser of the steam engine provide the right circumstances (heat input, temperature difference), but are the steam tubes, valves, pistons, etc., in good order to let the engine provide any work? One of the earliest examples that something may be wrong with an overturning circulation, merely driven by differential heat fluxes at the sea surface, is found in a publication by Sandström (1908), who carried out laboratory experiments with seawater in a tank. The essence of his experimental results is that a sustained stationary thermal circulation cannot exist when the heat source and the heat sink are situated at the same horizontal level, e.g., the sea surface. Only when the heat source is placed deeper than the heat sink is potential energy generated, which in its turn can be converted into the kinetic energy of the overturning circulation at the levels above the heat source. In a later paper Sandström (1916) states that this effect is a consequence of the Carnot cycle in the earth's gravity field. In order to envisage this, let us assume that the stationary motion of the THC occurs along closed streamlines with a velocity vector **u**, and that for simplicity friction can be approximated linearly by –*r***u**. The momentum equation for a stationary motion on a rotating earth is then

$$0 = -\frac{1}{\rho}\nabla P - f\mathbf{k}\times\mathbf{v} - \nabla\phi - r\mathbf{u} \quad , \tag{10.12}$$

where $f\mathbf{k}$ is twice the local vertical vector component of the earth rotation and ϕ is the gravitational potential gz. Integration of Eq. (10.12) along a steady-state closed streamline, **s**, then gives

$$0 = -\oint\frac{1}{\rho}\nabla P d\mathbf{s} - \oint r\mathbf{u}d\mathbf{s} = -\oint V dP - R \quad , \tag{10.13}$$

or, given that the amount of work to overcome frictional dissipation R is positive along a streamline,

$$-\oint V dP = R > 0 \quad , \tag{10.14}$$

where $V(T,S,P)$ is here the specific volume, $1/\rho$. The effects of the Coriolis acceleration $-f\mathbf{k}\times\mathbf{v}$ will disappear in the integral of Eq. (10.12), since this acceleration is everywhere perpendicular to **ds**. Similarly, the gravitational potential disappears from the integral, since the gradient of a potential is free of rotation. In order to be able to maintain a steady THC in a dissipative ocean, the left-hand side of the resulting Eq. (10.14) always has to be positive, since R is positive.

The left-hand side of Eq. (10.14) also can be written as a closed integral of P in V space. That leads to the equation known as Sandström's theorem (Colin de Verdière 1993),

$$\oint P dV = R > 0 \quad . \tag{10.15}$$

According to Eq. (10.15) the work of pressure forces must be such that expansion ($dV > 0$), or decrease of density, occurs at high pressures and contraction ($dV < 0$), or increase of density, occurs at low pressure. If the water properties do not change along a streamline, the left-hand side of Eq. (10.15) will equal zero and no overturning circulation can be maintained. Specific forcing boundary conditions for temperature and salinity, which influence ρ and V, are required to maintain such a density-driven circulation.

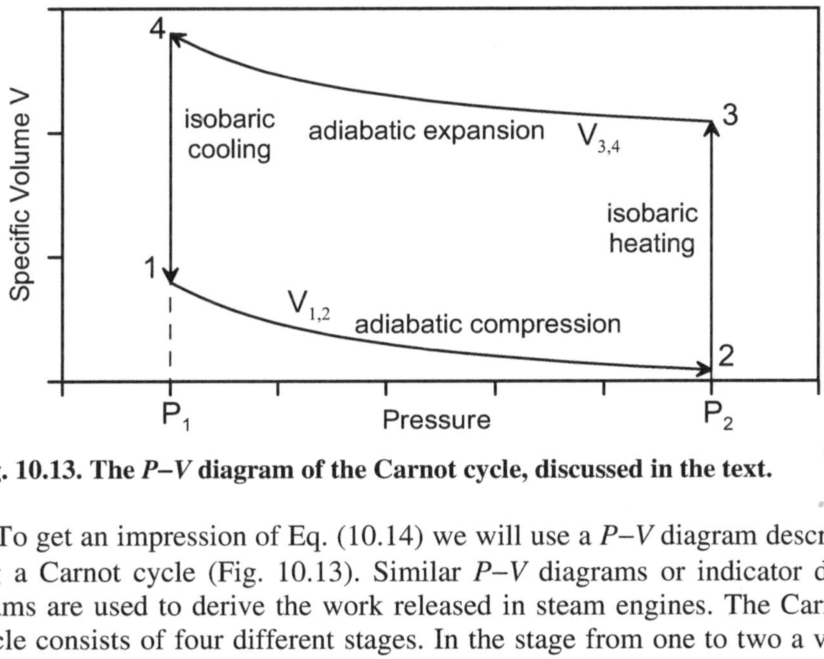

Fig. 10.13. The *P–V* diagram of the Carnot cycle, discussed in the text.

To get an impression of Eq. (10.14) we will use a *P–V* diagram describing a Carnot cycle (Fig. 10.13). Similar *P–V* diagrams or indicator diagrams are used to derive the work released in steam engines. The Carnot cycle consists of four different stages. In the stage from one to two a volume of fluid with specific volume V_1 at pressure P_1 is compressed adiabatically to a pressure P_2 and obtains a specific volume V_2. Then in the stage from two to three, isobaric expansion (heating) to a specific volume V_3 occurs. From three to four the fluid volume expands adiabatically to a specific volume V_4 at pressure P_1. Finally in the last stage isobaric compression (cooling) brings the volume back to its initial specific volume V_1. The left-hand side of Eq. (10.14) then will obtain a value

$$-\oint V dP = \int_{P_1}^{P_2} (V_{3,4} - V_{1,2}) dP \quad , \tag{10.16}$$

where $V_{1,2}$ and $V_{3,4}$ are the pressure-dependent specific volumes during stage$_{1-2}$ and stage$_{3-4}$, respectively. The work produced in the cycle is positive, since $V_{3,4} > V_{1,2}$, and a stationary process is possible. Note that if we exchange the locations of isobaric heating and cooling, the cycle in Fig. 10.13 will change from anticlockwise to clockwise, and the value of Eq. (10.14) will become negative. In that case additional input of mechanical energy is required to force the steady-state process: a heat pump.

Contrary to the Carnot cycle in Fig. 10.13, the cooling and heating in the ocean both occur at the same low-pressure level near the sea surface. In that case the stationary oceanic heat engine cannot generate work, since the curves for the adiabatic compression stage and the adiabatic expansion stage will overlap, and the integral (10.14) will have a value of zero. This effect was already shown in the laboratory experiments by Sandström (1908), but it is still ignored in many popular accounts of the THC.

What are the properties of the overturning circulation of the real ocean in the indicator diagram? Is there enough energy produced to overcome friction in the system? In Chapter 3 it was shown that in the ocean the density or specific volume mainly changes because of pressure changes. In a P–V diagram of the meridional overturning THC that effect will dominate all other changes of V. Therefore, we will use the specific volume anomaly δ instead of V. This parameter is defined as

$$\delta(T,S,P) = V(T,S,P) - V(0,35,P) \quad . \tag{10.17}$$

It is easy to derive that the closed integral of V in Eq. (10.14) will equal the closed integral of δ), which implies:

$$-\oint \delta dP = R > 0 \quad . \tag{10.18}$$

In the P–δ diagram of the circulation of NADW in the ocean (Fig. 10.14) it is assumed that the source of NADW (the cooling) is in the surface layers of the Nordic seas from where it descends into the Atlantic Ocean through the Denmark Strait. In this simplified example the Nordic seas are considered as the heat sink (density source) for the cooling stage. The overflow across the sill in the Denmark Strait near 500 dbar is the main source of NADW after leaving the cooling area. Between the sill in the Denmark Strait and Cape Farewell at the southern tip of Greenland this NADW descends over the continental slope of Greenland to a pressure level of 3000 dbar. From there NADW is assumed to flow at 3000 dbar through the Atlantic Ocean to the abyssal basins in the South Atlantic, Indian, and Pacific oceans. From those abyssal basins the NADW ascends to

500 dbar by global upwelling, and from there it is assumed to return isobarically at 500 dbar to the cooling region in the Nordic seas.

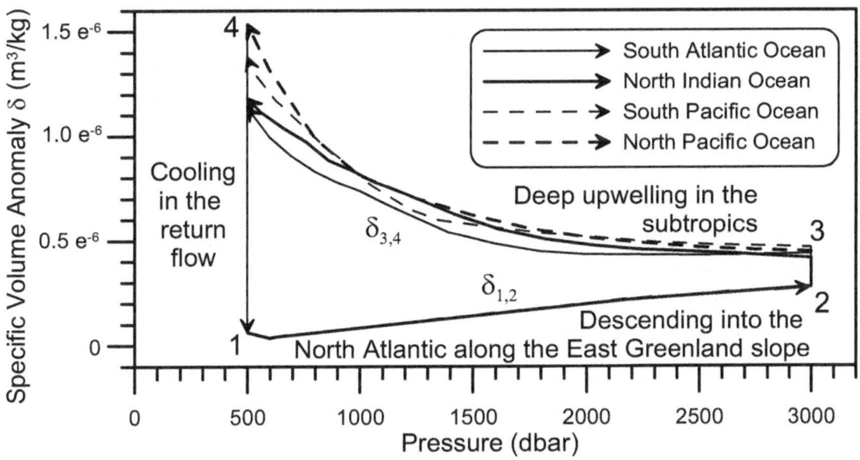

Fig. 10.14. The *P*–δ diagram of the cycling of NADW through the THC, with formation of NADW by cooling in the Nordic seas and upwelling in several abyssal basins. The deep cold branch of the THC is assumed to be at 3000 dbar and the warm return branch at 500 dbar.

We have determined the specific volume of the descending overflow water (stage$_{1-2}$) from a series of North Atlantic hydrographic sections of the WOCE Hydrographic Program. For the upwelling regions (stage$_{3-4}$) in the South Atlantic, North Indian, and South and North Pacific oceans, profiles of the specific volume anomaly were obtained from characteristic hydrographic stations in those basins also occupied during the WOCE Hydrographic Program. The circulation sense in the P–δ diagram (Fig. 10.14) appears to be in the right direction to generate work required to overcome friction in a steady cycle (anticlockwise). Stage$_{3-4}$ does not differ strongly between the different ocean basins. We have estimated the mean value of the work R produced in one THC cycle from Eq. (10.14) for the four upwelling regions in Fig. 10.14 to be about 11.5 J/kg. This seems to be a lot of energy, comparable to the kinetic energy of water moving at a speed of 4.8 m/s. However, it takes hundreds of years for a water parcel to complete the THC cycle, reducing the power density of the THC to a few nW/kg.

When we consider Fig. 10.14 in more detail, a few striking differences with the Carnot cycle from Fig. 10.13 are noted. First, the compression or contraction stage from 1 to 2 is not adiabatic. If it were adiabatic, δ would have been nearly constant, while along the East Greenland slope δ increases significantly due to mixing with surrounding warmer water. This causes an increase of the potential temperature in the NADW core with 1.9°C between the sill in the Denmark Strait at 66.9°N and Cape Farewell at 59.6°N. Second, the upwelling in the abyssal basins in $stage_{3-4}$ is even less adiabatic than the descent into the North Atlantic Ocean. On average the potential temperature in the abyssal basins increases with 7.1°C between 3000 and 500 dbar. However, the direct heating of the upper branch of the THC by air–sea interaction and the absorption of shortwave radiation is limited to a thin near-surface layer, and does not reach 500 to 3000 dbar. There appears to be a virtual heat source at subsurface levels in the upwelling region. The only way heat from the sea surface can reach such depths is by turbulent diffusion which mimics a subsurface heat source. In Chapter 9 it was shown already that in order to maintain turbulent diffusive transport of heat and mass an input of turbulent kinetic energy is required, which is converted to potential energy. That potential energy is released to drive the overturning THC according to Eq. (10.18). Similar to the Carnot cycle of Fig. 10.13 some additional isobaric heating occurs between 2 and 3 at 3000 dbar due to turbulent mixing. On average the potential temperature increases with 0.9°C while the salinity decreases with 0.2 at 3000 dbar. Also at this deep pressure level turbulent mixing with the surrounding water is the main agent for the increase in specific volume in this stage, which also contributes to the value of Eq. (10.18).

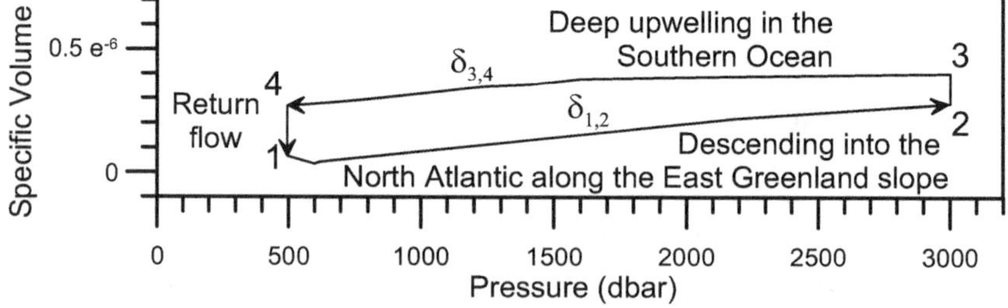

Fig. 10.15. The *P*–δ diagram for the overturning of NADW in the southern South Atlantic Ocean. The deep cold branch of the THC is assumed to be at 3000 dbar, the return branch at 500 dbar. The data for the upwelling stage were obtained from the WOCE hydrographic section A23 and represent the ascending maximum salinity core in the Atlantic sector of the Southern Ocean.

In Chapter 8 the wind-driven upwelling of NADW/CDW in the Southern Ocean was discussed. But the wind forcing itself does not supply the energy needed to overcome frictional effects in the whole THC loop and to bring the NADW core near the sea surface. Also for this overturning loop that power has to be supplied according Eq. (10.18). The stratification in the Southern Ocean is such that it allows the production of the power to overcome frictional effects, reflected by the anticlockwise path in the P–δ diagram (Fig. 10.15). The energy produced along this Southern Ocean path of the THC amounts to 5.0 J/kg, which also here has to be supplied by turbulent mixing.

Summarizing, it can be stated that the THC is dominantly determined by the meridional distribution of the air–sea heat exchange and the *SST*, which allows the operation of the THC as a rather inefficient heat engine according to Eq. (10.4). The large-scale structure of the net freshwater exchange with the atmosphere and the *SSS* are responsible for the north–south asymmetry of the THC in the Atlantic Ocean as well as for the asymmetry between the Atlantic and the Pacific oceans. The requirements of Sandström's theorem, Eq. (10.15), show that additionally turbulent mixing is required to mimic a deep heat source. Without turbulent mixing a stationary overturning THC is not possible. The resulting THC supports, together with the wind-driven near-surface circulation, the meridional oceanic transports of heat and freshwater which form essential parts of the global climate system. In the Atlantic Ocean the THC is dominant for the northward heat transport in both the southern and northern hemispheres, due to the large-scale north–south salinity differences. These differences are maintained by the net atmospheric transport of water vapor from the Atlantic Ocean to the Pacific Ocean.

11. Simple models, boundary conditions, and feedbacks

11.1. Models and boundary conditions

The Atlantic THC is a self-sustained nonlinear phenomenon, prone to instability and multiple equilibria. At present formation of deepwater in the Atlantic THC seems to be thermally driven, with a source of cold deepwater in the high-latitude Atlantic Ocean. However, variations of salinity in the deepwater source region may modify the THC. If, for example, the northward transport of saline subtropical waters decreases by ice berg melt, increased precipitation, and river runoff, the density of high-latitude waters may be reduced so that these waters are no longer able to sink according to Eq. (10.13). Then the THC with a near-polar deepwater source will terminate and the now more stagnant subtropical surface water may increase in salinity and density due to ongoing evaporation, until it reaches a density that allows deep convective mixing and subtropical downwelling. Then the circulation direction of the THC is reversed, and a salinity-driven THC will emerge. In order to study the occurrence of these different equilibria for the THC and the transition from one equilibrium to another, simple models for the THC are used with simplified boundary conditions that are assumed to present the essentials of the feedback processes in the THC.

Models of the THC and of the oceanic circulation in general consist of a set of equations that explicitly describe the flow dynamics, the equation of state, and the budgets of heat and salt (or freshwater), as well as initial and boundary conditions. Similar to the use of turbulent diffusion and viscosity coefficients, any choice of specific boundary conditions is equivalent to an implicitly conceptual model of that part of the system that is not modeled explicitly, including the interactions across the boundary (J.S. Marotzke, unpublished lecture notes, 2001). An example is the boundary condition of the homogeneous upwelling velocity, used for the derivation of the Stommel–Arons model in Chapter 8. Simplified models of the THC are used to evaluate the existence of one or more stable solutions given a specific forcing prescribed by the boundary conditions and the specific dynamics prescribed by the model equations and model geometry. Another question, studied by means of mathematical models of the THC, is the existence of unstable, variable solutions of the model and the dependence of the variability of the THC on the boundary conditions, the model dynamics, and the model geometry.

One can start the study of the dependence of the THC on the forcing by the boundary conditions for heat and freshwater by asking a few questions:

- Is there only one possible solution for the THC, given a certain set of boundary conditions, or are multiple equilibrium solutions possible, some of which have a contrary overturning sense?
- Do we expect nonstationary oceanic changes when the atmospheric forcing is stationary?
- If so, can we determine how they are caused by ocean–atmosphere feedbacks?

Given the nonlinearity of the equations that describe the fluid dynamics of the ocean, it can be expected that under certain (most?) circumstances the solutions of THC models will show some internal variability. But even then, is the variability dominant or is the solution statistically stabile?

In this chapter we will introduce boundary conditions that represent the atmospheric forcing of the temperature and salinity fields of the THC, as well as some simple models. These are chosen to illustrate certain feedbacks of the THC. It will be simple models, not full-grown ocean general circulation models (OGCMs) or coupled ocean–atmosphere models. Such complex models perform probably better for the simulation of the real ocean dynamics, but because of their complexity they generally fail to give insight into the main processes that determine the behavior of the THC.

The forcing of the THC is expected to be determined by the boundary conditions that enforce the temperature and salinity fields in the ocean and the feedbacks between the ocean state and these boundary conditions. The net vertical upward heat flux density at the sea surface, Q_{net}, represents the exchange of thermal energy between ocean and atmosphere. It can be divided into contributions of shortwave radiation from the sun, long-wave infrared thermal radiation, the direct turbulent exchange of sensible heat, and the heat required for the evaporation of sea water, referred to as the latent heat flux. These different factors are generally expressed in so-called bulk formulae (Gill 1982) which depend on atmospheric and oceanic parameters like wind speed, air and sea surface temperature, cloudiness etc. Similarly, the evaporation density E can be derived via a bulk formula from atmospheric and oceanic parameters. The precipitation density P over the ocean is the least well known of the contributions to the freshwater flux density at the sea surface. Precipitation over the ocean is difficult to measure and to model.

In the early days of ocean modeling boundary conditions for the heat and freshwater fluxes were derived from climatologies like those shown in Figs 10.4 and 10.8. This often led to very unrealistic thermocline structures (J.S. Marotzke, unpublished lecture notes, 2001). Another, easier to control approach for these boundary conditions is to prescribe the sea surface temperature and salinity from climatologies as shown in Figs. 10.3 and 10.9. The accompanying heat and freshwater fluxes are then derived from the model results.

11.2. Random boundary conditions

Hasselmann (1976) and Frankignoul and Hasselmann (1977) presented a simple model, where the climatic state of the ocean is determined by a random atmospheric forcing, without any feedback where the climatic state of the ocean in turn influences the forcing. Such a simple system appears to generate a complicated solution with somewhat counter-intuitive properties. As an example we assume to have a large homogenized ocean basin, like the deep Labrador Sea, where the salinity change only depends on a small but randomly varying virtual salt flux. The random contribution to the forcing may result from the varying severity of successive winters, the occurrence of specific strong cold storms in winter, melting sea ice that is transported to the basin by occasional favorable winds, precipitation from storm depressions which will or will not pass over the ocean basin, etc. A climatic parameter, e.g., the salinity of the basin at a certain time step (i+1), S_{i+1} depends on the salinity at the previous time step (i), S_i, and the stochastic virtual salinity flux density for that time step, F_{Si} according to

$$S_{i+1} = S_i + \frac{F_{Si}}{D} \quad , \tag{11.1}$$

where D is the depth of the basin. We have applied a near-Gaussian random "virtual salt flux" F_s/D with a standard deviation 0.01 (Fig. 11.1) and an initial value $S_0 = 35.0$. For 60 000 successive time steps the development of the salinity, calculated with Eq. (11.1), was followed during two separate model runs. Since during each run the random values of F_{Si}/D were determined independently, the resulting time series of S_i differed for both experiments (Fig. 11.2). In both experiments the resulting parameter S showed large variations, of the order of 1, with a characteristic timescale of several thousand time steps. Moreover, the development of S during experiment 1 showed a long-term trend of 2 over 60,000 time steps, while during experiment 2 the final salinity S was close to its initial value, 35.0.

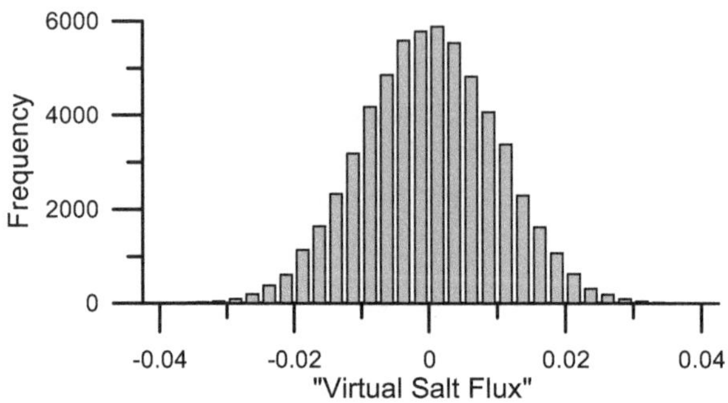

Fig. 11.1. The statistical distribution of the virtual salt flux F_{Si}/D, used in the first experiment with the random forcing of the salinity in and ocean basin with stochastic salinity forcing.

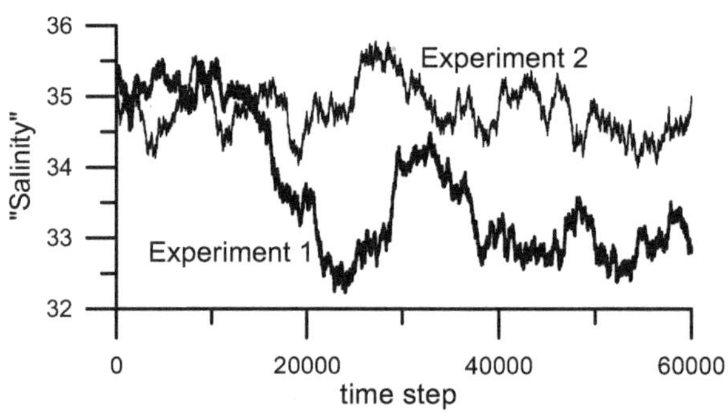

Fig. 11.2. Two independent time series of the salinity S_i for 60,000 successive time steps, calculated according to Eq. (11.1). The thick line shows the results of experiment 1, the thin line of experiment 2. In both experiments the standard deviation of the random near-Gaussian salt change F_{Si}/D was 0.01 and the initial value of S was 35.0.

Although this result seems counterintuitive, the strong low-frequency variability of S resulting from a random white noise forcing can be understood from the spectral behavior of a signal and its time derivative. When a signal S_i has a certain power spectrum $Sp(\omega)$, depending on the frequency ω, the spectrum of the derivative of S will have a spectrum of the form $\omega^2 Sp(\omega)$. The time derivative in our experiment is F_{Si}/D and has the character of white noise. Then $\omega^2 Sp(\omega)$ is constant, with a value C_S. The resulting $Sp(\omega)$ will have the form $C_S\omega^{-2}$. That is a so-called red spectrum, characteristic for many geophysical signals. Such a red spectrum has large values at low frequencies, indicative of large variations on longer time-scales.

The character of S, as shown in Fig. 11.2, hardly can be discriminated from many real oceanic time series, e.g., the salinity in the Rockall Channel or the volume transport through Florida Straits (Wunsch 1992), apparently signals which are determined by many small forcing distributions. This has consequences for the observation of climate change. Many oceanographers have been attracted by the idea of observing the THC at certain "choke points" and "key processes." These words turn up as a shibboleth in many research proposals. The idea behind choke points and key processes is that at certain locations or ocean currents, e.g., in a DWBC, Drake Passage, Denmark Strait, or the Agulhas Retroflection, the volume transport and the thermohaline properties filter out the effects of many high-frequency uncorrelated contributions, while the remaining signal represents the "true climate variability" of the THC (Wunsch 1992). In the British Rapid Climate Change program it is assumed that by monitoring the overturning circulation of the THC near 24°N we can determine the true climatic variability of the THC and link this to specific forcing variations at high latitudes. Given the long-term variability that may result from completely stochastic forcing without a strong feedback mechanism, as illustrated in Fig. 11.2, it may be that long-term time series of oceanographic parameters at choke points or of key processes will show instead the variability of the ocean without any possibility to find specific forcing events which can be linked directly to the variability. Fortunately, several feedback mechanisms exist, which may dampen the long-term variations of the THC resulting from stochastic forcing. The assumptions (or beliefs and hope) behind many efforts to model the THC are that the feedbacks of the forcing and the internal dynamics of the THC are more important than the stochastic part of the forcing or the stochastic behavior of the ocean.

11.3. Boundary conditions for temperature and salinity with feedback

Prescription of constant or stochastic boundary conditions from existing climatologies, either as sea surface temperature and salinity or heat and freshwater fluxes, does not allow the study of the effects of atmospheric feedbacks on the THC. This section is focused on the influence of varying surface ocean conditions on a formulation of the heat and freshwater fluxes which may reveal feedbacks that can cause multiple equilibria and/or variability of the THC.

The heat flux density at the sea surface contains several terms that depend on the *SST*, e.g., the flux density of latent and sensible heat, the outgoing infrared radiation, and the albedo or reflection of shortwave radiation from the sun when sea ice is involved. In order to assess the effects of changing *SST* on the heat flux, one can write the heat flux density as a truncated Taylor series around the state where there is no temperature contrast between atmosphere and ocean (Haney 1971). Hereby we assume that the air temperature and humidity over the ocean is determined by the large-scale atmospheric circulation. The infrared radiation and the exchange of sensible and latent heat depend on the *SST*, which therefore allow an approximation of the net surface heat flux as

$$Q(T_O) = Q(T_A) + \left.\frac{\partial Q}{\partial T_O}\right|_{T_A} \cdot (T_O - T_A) \quad , \tag{11.2}$$

where Q is the heat flux density, T_O is the *SST* and T_A is the surface air temperature. The derivative $\partial Q/\partial T_O$ at a temperature T_A represents the effects of a varying ocean temperature on the turbulent transports of sensible and latent heat fluxes and on the infrared radiation from the sea surface. All three will increase with increasing *SST*, and therefore $\partial Q/\partial T_O$ will be positive. Haney (1971) derived from data that the typical value of $\partial Q/\partial T_O$ amounted to ~35 $Wm^{-2}{}^\circ C^{-1}$, with only a slight change with latitude.

The terms in the right-hand side of Eq. (11.2) can be rearranged to

$$Q(T_O) = \left.\frac{\partial Q}{\partial T_O}\right|_{T_A} \cdot \left(T_O - \left[T_A - \frac{Q(T_A)}{\partial Q / \partial T_O|_{T_A}} \right] \right) = \rho_0 C_P w^* \left(T_O - T_A^*\right) \quad . \tag{11.3}$$

In this equation T_A^* is an apparent equilibrium temperature to which the SST ultimately will be restored. The parameter w^* has the dimension of a velocity and can be interpreted as a heat flux transport velocity. The temperature development of a basin or mixed layer with depth D subjected to a forcing given by Eq. (11.3) can be described as

$$D\frac{\mathrm{d}T_O}{\mathrm{d}t} = D\frac{\mathrm{d}\left(T_O - T_A^*\right)}{\mathrm{d}t} = -w^*\left(T_O - T_A^*\right) \quad . \tag{11.4}$$

The resulting $T_O - T_A^*$ will decrease exponentially with an e-folding time τ equal to

$$\tau = \frac{D}{w*} = \frac{\rho_0 C_P D}{\partial Q / \partial T_O|_{T_A}} \quad . \tag{11.5}$$

With $C_P \approx 4000\ \text{Jkg}^{-1}{}^{\circ}\text{C}^{-1}$, $\rho_0 \approx 1000\ \text{kg/m}^3$, and $D = 50$ m, the e-folding time τ equals about 2 months. This agrees reasonably well with the observed delay between the time of maximum insolation and the time of maximum *SST*. The flux transport velocity w^* will depend to a certain degree on the prevailing wind speed over the ocean, and therefore will vary with latitude. Haney (1971) found from observations that the latitudinal mean value of w^* varies only about 20% around its global mean value.

The apparent equilibrium temperature T_A^* is neither an atmospheric nor an ocean temperature. When T_O has a value equal to T_A^*, the net heat flux density equals zero, which implies that the latent heat flux is balanced by the sensible heat flux and the short- and long-wave radiation. In that case T_A^* can better be interpreted as a bulb temperature for a wet bulb thermometer without radiative protection.

The boundary condition (11.3) assumes that while the sea surface air temperature T_A determines the sea–air heat flux density, and ultimately the *SST*, T_A itself is not influenced by the air–sea heat exchange. This implies that the atmosphere drives the ocean, and atmospheric processes other than air–sea interaction, e.g., the mean winds and atmospheric eddies, mainly determine T_A. Therefore, Eq. (11.3) cannot be used for coupled atmosphere–ocean climate models where there is a feedback from T_O to T_A. In such models the heat flux has to be modeled more explicitly, e.g., by also introducing an atmospheric model to determine T_A or T_A^*. But for ocean-only models Eq. (11.4) is a very useful boundary condition to assess the influence of *SST* variations on the atmospheric forcing.

In order to model the salinity field, a boundary condition in the form of a virtual salt flux similar to Eq. (10.5) is required, which connects the salinity with the net freshwater flux (E–P). The evaporation E is a function of *SST*, surface air humidity, surface air temperature, and wind speed. The *SST* is involved in two ways: it determines the saturated specific humidity in equilibrium with the ocean and it influences the stability of the lower atmosphere, and thereby the bulk exchange coefficient for water vapor. The precipitation P is a complicated function of atmospheric processes and the atmospheric state and is in first order assumed to be independent of the sea surface properties. Van der Schrier and Maas (1998) have proposed a truncated Taylor expansion of E around a reference temperature T_A (our notation) similar to Eq. (11.3) for the heat transport. This gives

$$E(T_O) = E(T_A) + \left.\frac{\partial E}{\partial T_O}\right|_{T_A} \cdot (T_O - T_A) \quad . \tag{11.6}$$

From Eq. (11.6) the following expression for (E–P) can be derived:

$$\left(E(T_O)-P\right)=\left.\frac{\partial E}{\partial T_O}\right|_{T_A}\cdot\left(T_O-\left[T_A-\frac{E(T_A)-P}{\partial E/\partial T_O|_{T_A}}\right]\right)=\frac{w_E^*}{\Delta T_{ref}}\left(T_O-T_E^*\right). \quad (11.7)$$

In this equation the term $\partial E/\partial T_O$ is positive. The parameter w_E* has the dimension of a velocity, and ΔT_{ref} is some reference constant with the dimension of a temperature interval. From Eq. (11.7) the virtual salt flux of Eq. (10.5) can be derived equal to

$$F_S=-S\cdot(E-P)=w_E^*S_O\frac{\left(T_E^*-T_O\right)}{\Delta T_{ref}} \quad . \quad (11.8)$$

According to Eq. (11.6) an increase in T_O will lead to an increase in evaporation which in its turn, following Eq. (11.8), results in a decrease in the upward or an increase in the downward virtual salinity flux, depending on the value of T_E* relative to T_O. The influence of S_O is very limited, of the order of 10% or less, given the limited range of the *SSS* in the world ocean. Boundary condition (11.8) approaches the reference boundary condition $S_O(E(T_A)–P)$ if T_O approaches T_A, while at the equilibrium temperature T_E^* the precipitation precisely balances the evaporation leading to a zero virtual salt flux density.

Boundary condition (11.8) does not have a restoring character like Eq. (11.4). This makes it *a priori* difficult for models of the THC to reach an equilibrium circulation. More popular but devoid of any physical basis is the restoring boundary condition for salinity of the type

$$F_S=w_S^*\left(S_O-S_A^*\right) \quad , \quad (11.9)$$

where w_S^* is a virtual salt flux transport velocity and S_A^* is an equilibrium salinity to be reached when the ocean is in balance with the equilibrium atmosphere in the absence of ocean circulation. A boundary condition in this restoring form was used already by Stommel in his 1961 paper on a box model of the THC. Similar to boundary condition (11.3) for the temperature, an exponential decrease of the deviation ($S_O–S_A^*$) will follow with an e-folding period $\tau_S = D/w_S^*$ when a basin or mixed layer with depth D is subject to Eq. (11.9).

Boundary condition (11.9) mainly assumes that the influence of the interaction with the atmosphere has a tendency to force the surface salinity S_O back to the equilibrium salinity S_A^*. Contrary to the restoring boundary condition for the temperature field which influences the heat flux density, the *SSS* does not influence the evaporation or precipitation. Therefore, it is expected that the drift toward equilibrium will be slower for the salinity field than for the temperature field with Eq. (11.4). Surface salinity anomalies in the North Atlantic Ocean are known to have survived for several years (Belkin 2004). This can be realized with the additional assumption that the flux transport velocity for salt, w_S^*, is definitely smaller than the flux velocity for heat, w^*.

In the distinguished limit case of $w_S^* \rightarrow 0$ and $S_A^* \rightarrow \infty$ with $w_S^* S_A^*$ = constant, the virtual salt flux F_S becomes constant, independent of S_O. The combination of the *SST*-dependent Eq. (11.4) for the temperature field with a prescribed virtual salt flux boundary condition for the salinity field is known as "mixed boundary conditions" that often are used for numerical experiments with THC models.

11.4. A consequence of *SST*-dependent evaporation

The *SST*-dependent salinity boundary condition in the form (11.8) may explain the large difference in *SSS* between the North Atlantic and the North Pacific Ocean. In the subtropical and moderate latitudes between 20 and 58°N the mean *SST* in the North Atlantic is on average about 2.8°C higher than in the North Pacific Ocean in the same latitude range. Warren (1983) already speculated that the lower *SST* in the North Pacific leads to smaller evaporation and lower *SSS* values in that ocean basin. Assuming that T_E^*, which is mainly determined by the atmosphere, does not differ strongly between both oceans, the net freshwater flux (E–P) over the North Pacific described by Eq. (11.7) will be smaller than over the North Atlantic Ocean. According to Eq. (11.8) this leads to a larger upward or smaller downward virtual salinity flux. The result of this boundary condition will be a lower *SSS* over the North Pacific Ocean compared with the North Atlantic. Warren (1983) attributed the *SST* difference between both ocean basins to the strong cross-equatorial northward heat flux in the Atlantic Ocean compared to the Pacific Ocean (see Fig. 10.7). This difference in the equatorial heat flux is a result of the existing mode of the thermohaline circulation with deepwater formation in the North Atlantic, not in the North Pacific. The deepwater formation in the North Atlantic Ocean requires a (relatively warm) northward return flow in the near-surface layers. The positive correlation between the difference of the zonally averaged evaporation between the Atlantic and Pacific oceans and the difference in zonally mean *SST* between these oceans between 59°S and 59°N (Fig. 11.3) agrees with Eq. (11.6). These differences are significantly correlated, suggesting that on an even larger scale Eq. (11.8) indeed determines the difference of *SSS* between the Pacific and Atlantic oceans. This effect forms a feedback loop in the THC, since it leads to a more saline North Atlantic resulting in more deepwater formation in that ocean basin.

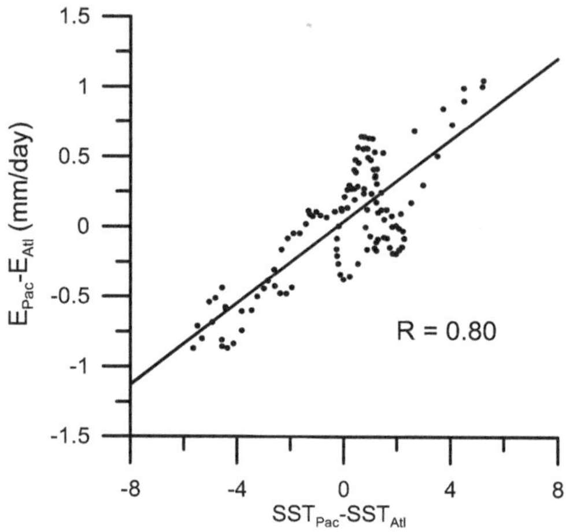

Fig. 11.3. The difference between the zonally averaged evaporation in the Pacific Ocean and the Atlantic Ocean, plotted versus the difference in zonally average *SST* between both oceans. The data points represent zonal averages per degree between 59°S and 59°N. The straight line shows the linear regression between both parameters. The correlation coefficient R of this regression amounts to 0.80. The data have been derived from the HOAPS data set.

This result seems to contradict the results from the use of boundary condition (11.9). Assuming a zonally constant S_A^*, such a restoring boundary condition will lead to a smaller upward or larger downward virtual salinity flux in the North Pacific Ocean, and therefore to a decrease in the salinity contrast between both oceans. Boundary condition (11.8) maintains the salinity contrast independent of the *SSS* in both ocean basins.

The example given here suggests that understanding of the boundary conditions, presented in the previous section, can be very useful not only in a verbal argument like the one given in this section but also in simple analytical models of the THC. The mathematical structure of the solutions of such models may give insight into the main feedback processes which determine the structure of the THC and its stability.

11.5. Consequences of restoring boundary conditions

The feedback of restoring boundary conditions in first order will dampen the variability of oceanic parameters which results from random forcing, as illustrated in Fig. 11.2. Such a damped system with restoring feedback can be described as

$$S_{i+1} = S_i + \frac{F_{Si}}{D} - \frac{\left(S_i - S_A^*\right)}{N} \quad , \tag{11.10}$$

where F_{Si} is the stochastic forcing boundary condition for S and N is equal to the restoring timescale D/w_S^*, expressed in number of time steps.

The resulting time series with stochastic forcing and damping feedback according to Eq. (11.10) shows less variability when it is compared with a time series calculated without damping, according to Eq. (11.1) when a similar stochastic forcing term F_S is used (Fig. 11.4). The strength of the feedback determines the amount of variability in S in the resulting time series. Especially the long-term oscillations with a large amplitude, which are so prominently present in experiment 1, are dampened by the restoring feedback used in experiment 3. A restoring timescale of $N = 365$ time steps was used. Increase of N diminishes the damping effect.

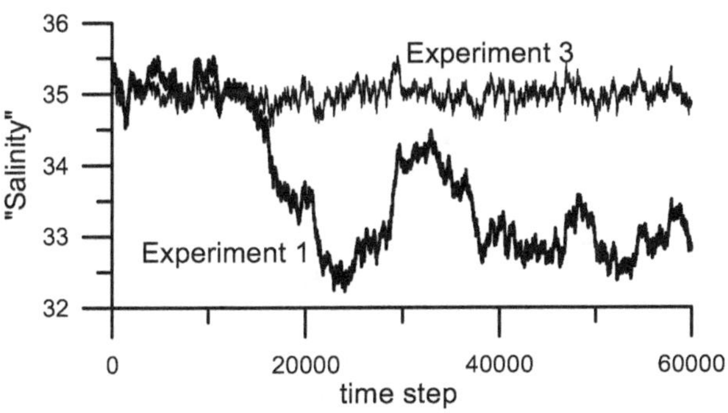

Fig. 11.4. Two time series of the salinity S_i for 60,000 successive time steps. The thick line represents experiment 1, also shown in Fig. 11.2, calculated according to Eq. (11.1). The thin line shows the results of experiment 3, where the development of S was calculated with the damped-stochastic boundary conditions of Eq. (11.10) with $N = 365$ and $S_A^* = 35.0$. In both experiments an identical random near-Gaussian salt change F_{Si}/D was used with standard deviation of 0.01, while for both series the initial value of S was 35.0.

The difference between the restoring surface flux transport velocity for the surface heat flux density Q according to Eq. (11.4) and the surface flux transport velocity for the virtual salinity flux F_S according to Eq. (11.9) may lead to a peculiar behavior of the surface density, which may or may not favor deep convection. When we assume that the near-surface return flow from subtropical to arctic latitudes can be described as a channel flow with a depth D (e.g., thermocline depth) and flow has a velocity u, the restoring boundary conditions for heat and salt will lead to the following equations:

$$\frac{\mathrm{d}\left(T - T_A^*\right)}{\mathrm{d}x} = -\frac{w^*}{u}\frac{\left(T - T_A^*\right)}{D} \quad , \tag{11.11}$$

$$\frac{\mathrm{d}\left(S - S_A^*\right)}{\mathrm{d}x} = -\frac{w_S^*}{u}\frac{\left(S - S_A^*\right)}{D} \quad . \tag{11.12}$$

Further the equation of state is linearized as a truncated Taylor expansion relative to T_A^*, S_A^*, as

$$\delta\rho = \rho(S,T) - \rho_A^* = \rho_A^*\left[\beta\left(S - S_A^*\right) - \alpha\left(T - T_A^*\right)\right] \quad , \tag{11.13}$$

with ρ_A^* being the ocean density at the equilibrium temperature T_A^* and salinity S_A^*, while α and β represent, respectively, the thermal expansion coefficient and haline contraction coefficient. The resulting temperature and salinity anomalies then develop exponentially as

$$\left(T(x) - T_A^*\right) = \left(T(0) - T_A^*\right) \cdot e^{-x/L^*} \quad \text{and} \tag{11.14}$$

$$\left(S(x) - S_A^*\right) = \left(S(0) - S_A^*\right) \cdot e^{-x/L_S^*} \tag{11.15}$$

with the decay lengths L^* and L_S^* being, respectively, the e-folding distance Du/w^* for the development of the temperature field and the e-folding distance Du/w_S^* for the salinity field. Generally L^* is smaller than L_S^* because of the physical coupling between heat flux density and *SST*. If the initial temperature anomaly has the same sign as the salinity anomaly, while the density effect of the initial temperature anomaly is larger than the density effect of the salinity anomaly, the density anomaly will change sign downstream (Fig. 11.5). This is caused by the fact that the dominant initial temperature-controlled density anomaly decays faster than the salinity-controlled density anomaly. At some downstream distance these effects will balance each other, while further downstream the slower decaying positive salinity anomaly will be determining the density.

In the near-surface return flow of the THC in the North Atlantic Ocean the fast decay of a temperature anomaly compared to a salinity anomaly will make advected salinity anomalies more important for the sea surface density in those parts of the North Atlantic where deepwater is convectively formed. Anomalies of the surface salinity in these regions are assumed to modify the strength of the winter convection, the depth reached by convective mixing, and the density of the newly formed deepwater. These effects are feedback mechanisms that may be of importance for variations in the strength of the THC. Note here that the decay lengths L^* and L_S^* both depend linearly on the advection velocity u.

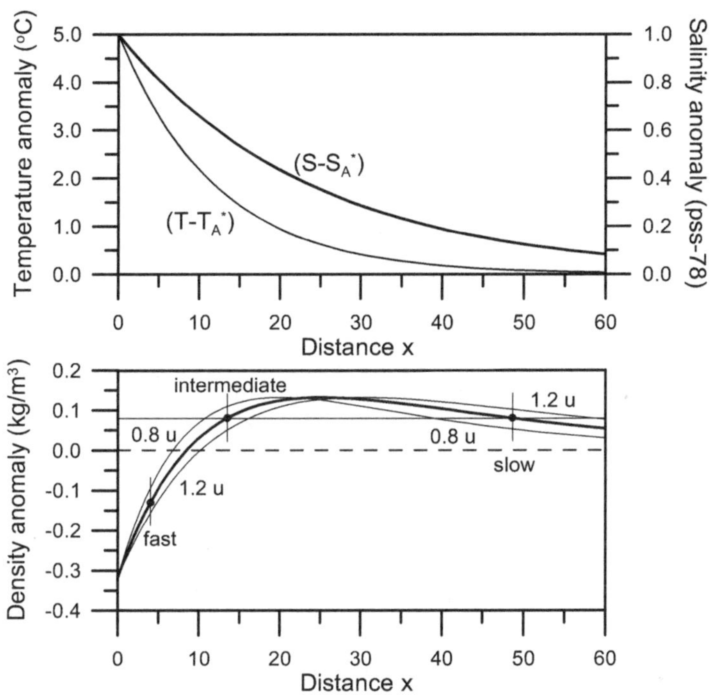

Fig. 11.5. Downstream development of the temperature for the basic experiment (thin curve in the upper panel) and salinity (thick curve in the upper panel), according to Eqs. (11.14) and (11.15), with $L^* = 12$ and $L_S^* = 24$. The lower panel shows the accompanying change of the density anomaly (thick curve) and the solutions with a velocity, u, 20% higher or lower than in the basic experiment (thin curves). Three characteristic regions of the solution for relatively fast, intermediate, and slow velocities are indicated with black dots.

One can divide the solution of the density development in Fig. (11.5) in three subregions (black dots) that can represent the deepwater formation area. If the flow is fast (large u) or if the deepwater formation area is on relatively short distance, the temperature-dependent density anomaly will dominate, with the same sign as for $x = 0$ and decreasing downstream. With intermediate velocities the salinity anomaly will dominate until a maximum in the density anomaly is reached. With slow velocities (low u) the density anomaly decreases again downstream. This behavior of the density due to different response times of the temperature and the salinity field invokes a feedback that may influence the stability of the THC. In many models of the THC the water mass formation and overturning rate are assumed to increase with increasing density of the water brought in the return flow to the formation area. A positive (or less negative) density anomaly in the formation region will invoke an increase in the overturning rate, and therefore an increase in the velocity u. In both the fast and intermediate regimes, a positive perturbation of u (the 1.2 u curve in the lower panel of Fig. 11.5) causes a decrease in the density anomaly, which leads to a decrease in the overturning rate restoring the solution in the direction of the undisturbed state (thick curve in the lower panel of Fig. 11.5). With an initially negative perturbation (the 0.8 u curve) this feedback also leads to a restoration to the undisturbed state in the fast and intermediate regimes. But if the flow is slow (or the formation region far away), a positive perturbation velocity of the return flow, u, will lead to a positive density anomaly that will strengthen the already existing velocity anomaly. The same occurs for an initially negative perturbation of u in the slow-flow regime. Here the feedback due to differences in restoring time for the temperature and velocity fields leads to an amplification of the velocity perturbation.

11.6. The single-hemispheric Stommel box model

The present THC is mainly thermally driven, with cold NADW formed in the Arctic seas descending at high latitudes and warm, saline water flowing from equatorial latitudes to the northern North Atlantic Ocean. However, it is in principle also possible that the evaporation excess in the subtropics produces surface water with such a high salinity that deepwater formation takes place at low latitudes. Then warm but saline high-density subtropical deepwater spreads poleward where, via upwelling, it can enter an equatorward cold return branch of the THC.

Stommel (1961) devised a simple (single-hemispheric) two-box model of the THC in order to explore the existence of stable responses to a constant atmospheric forcing with either a temperature-driven or a salinity-driven THC. The model consists of two boxes (Fig. 11.6). Stommel explicitly described these boxes as "well mixed." That implies that diapycnal mixing is not a limiting factor for the model. The high-latitude box is characterized by a low temperature T_1 and salinity S_1, while the low latitude box is characterized by high T_2 and S_2. The boxes are connected with two ducts, one representing the near-surface branch of the THC and the other representing the abyssal flow. The poleward flow through the near-surface duct amounts to Ψ, equal to the equatorward volume flux in the abyssal duct. The high-latitude box loses a total of Q (positive) heat to the atmosphere and receives a net amount of freshwater by precipitation phere and receives a net amount of freshwater by precipitation minus evaporation, equivalent to a positive virtual salt flux F_S. The boundary conditions ensure that $T_1 < T_2$, and $S_1 < S_2$. Because of conservation of heat and freshwater (zero atmospheric storage is assumed) the heat and virtual salinity exchanges with the atmosphere in both boxes have the same magnitude but opposite signs.

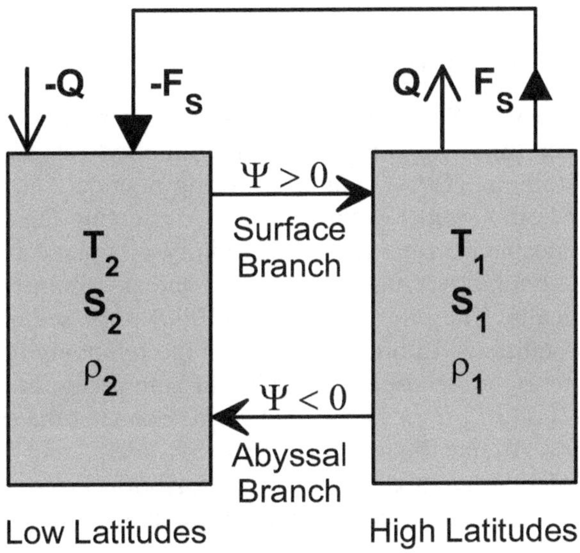

Fig. 11.6. Single-hemispheric box model of the THC with two well-mixed boxes, following the scheme introduced by Stommel (1961). The left box represents the low latitudes of the ocean and the right box the high latitudes. If the abyssal branch transports water from the high to the low latitudes, its volume transport Ψ is defined as positive. Because of conservation of heat and salt the total heat flux, *Q*, and virtual salinity flux, F_S, at high latitudes have the same magnitude but opposite signs to the heat and virtual salt transport from the low-latitude ocean toward the atmosphere.

The prescription of the dynamics of the two-box system, the boundary conditions, and the equation of state will determine the possible solutions. Stommel (1961) proposed for the abyssal dynamics a capillary linear resistance and assumed the surface level in both boxes to be nearly equal, while the abyssal pressure gradient is determined by the density difference between the boxes. That condition is described by

$$\Psi = \frac{k}{\rho_0}(\rho_1 - \rho_2) \quad , \tag{11.16}$$

where k is a dimensional linear friction coefficient and ρ_0 is a reference density. The surface branch has no dynamics of its own but follows the continuity of the overturning circulation because of mass conservation. If Ψ is positive, the near-surface flow is a warm return flow like in the present-day North Atlantic Ocean. When Ψ is negative, relatively saline water with high density, formed at low latitudes, is transported to high latitudes by the abyssal circulation. With a linear equation of state for the water in the model c.f. Eq. (11.13), Eq. (11.16) becomes

$$\Psi = k\left[\alpha\left(T_2 - T_1\right) - \beta\left(S_2 - S_1\right)\right] \quad . \tag{11.17}$$

If both salinities S_1, S_2, as well as the temperatures T_1, T_2, are prescribed as boundary conditions, a single stationary solution results, according to Eq. (11.17). Stommel (1961) applied restoring boundary conditions (11.3) and (11.9) for both Q and F_S, with $w_S^* < w^*$, reflecting the slower adaptation to the atmospheric forcing of the salinity compared to temperature. However, that boundary condition restricts the possibilities to solve the system analytically. Therefore, Marotzke (1990) proposed a simplified set of boundary conditions. In order to simulate the relatively fast response of the temperature to the atmospheric forcing in both boxes, he prescribed the temperatures T_1 and T_2, $T_2 > T_1$. The slower response of the salinity field to the atmospheric forcing then can be modeled adequately by prescribing constant virtual salinity fluxes F_S and $-F_S$ for both boxes. This boundary condition is in principle even more physical than a restoring boundary condition for salinity. The prescription of the temperature removes the necessity to formulate a separate conservation equation for the temperature in both boxes. For the salinity we assume that the flow Ψ brings water with the salinity from its box of origin to the other box. With a unit volume for both boxes the salinity equations become

$$\frac{\mathrm{d}S_1}{\mathrm{d}t} = -F_S + |\Psi|\left(S_2 - S_1\right) \quad , \tag{11.18a}$$

$$\frac{\mathrm{d}S_2}{\mathrm{d}t} = F_S - |\Psi|\left(S_2 - S_1\right) \quad . \tag{11.18b}$$

The sum of Eqs. (11.18a) and (11.18b) equals zero, indicating that the mean salinity stays fixed. The modulus of Ψ is used in both salinity equations because that volume is exchanged between both boxes anyway, with a poleward flow in either the near-surface duct or in the abyssal duct.

The equations can be simplified further by introducing $\Delta T = T_2 - T_1 > 0$ and $\Delta S = S_2 - S_1 > 0$. By subtracting Eq. (11.18a) from Eq. (11.18b) the evolution of the salinity difference ΔS is described as

$$\frac{d\Delta S}{dt} = 2F_S - 2|\Psi|\Delta S \quad . \tag{11.19}$$

In this equation the latitudinal variation of the virtual salinity flux leads to the generation of a salinity difference between the boxes, which is damped by a restoring term due to the THC, $-2|\Psi|\Delta S$. Combining Eq. (11.19) with the dynamic Eq. (11.17) ultimately leads to a single equation for the development of ΔS

$$\frac{d\Delta S}{dt} = 2F_S - 2k|\alpha\Delta T - \beta\Delta S|\Delta S \quad . \tag{11.20}$$

Equation (11.20) is nonlinear and allows in principle multiple stationary solutions. We can derive stationary solutions, which occur when the right-hand side of Eq. (11.20) equals zero. Because of the presence of the modulus in this equation we have to discriminate between the case where the argument of the modulus is positive and the case where the argument is negative. First we take the situation of a temperature-driven THC, which is the situation with $\Psi > 0$, $\alpha\Delta T > \beta\Delta S$. In that case the stationary situation is described by

$$\begin{gathered} F_S - (\alpha\Delta T - \beta\Delta S)\cdot\Delta S = 0 \quad \text{or} \\ (\beta\Delta S)^2 - (\alpha\Delta T)\cdot(\beta\Delta S) + \beta F_S / k = 0 \quad . \end{gathered} \tag{11.21}$$

This quadratic equation has in principle two roots, $\beta\Delta S_1$ and $\beta\Delta S_2$, given by

$$(\beta\Delta S)_{1,2} = (\alpha\Delta T)\cdot\left[\frac{1}{2} \pm \sqrt{\frac{1}{4} - \frac{\beta F_S}{k(\alpha\Delta T)^2}}\right] \quad . \tag{11.22}$$

The requirement that the roots are real implies that the right-hand term under the square root is smaller than one quarter. If that condition is fulfilled, the square root is always less than ½. Then the two roots of $\beta\Delta S$ represent two possible stationary THC circulations for a single forcing. The (+) branch implies a large salinity difference between box 2 and box 1. Then the salinity difference compensates the density difference due to the difference in temperature for a large part, and the overturning rate of the THC, expressed as Ψ, will be small. The smaller (–) solution for ΔS results in a larger density gradient and therefore in a strong THC. If the salinity forcing βF_S exceeds $\frac{1}{4}k(\alpha\Delta T)^2$, no real solution of Eq. (11.21) is possible, and therefore no stationary thermally driven THC can exist.

If the THC is driven by the salinity excess in the low-latitude box, that is, if $\Psi < 0$ because $\beta\Delta S > \alpha\Delta T$, the stationary THC described by Eq. (11.20) can be written as

$$F_S + (\alpha\Delta T - \beta\Delta S)\cdot \Delta S = 0 \quad \text{or}$$
$$(\beta\Delta S)^2 - (\alpha\Delta T)\cdot(\beta\Delta S) - \beta F_S / k = 0 \quad , \tag{11.23}$$

which has a single root, $\beta\Delta S_3$,

$$(\beta\Delta S)_3 = (\alpha\Delta T)\cdot\left[\frac{1}{2} + \sqrt{\frac{1}{4} + \frac{\beta F_S}{k(\alpha\Delta T)^2}}\right] \quad . \tag{11.24}$$

The possible root $\beta\Delta S_4$ of Eq. (11.23) with a minus sign before the square root is not a solution of our problem. Then both terms under the square root are positive; that root implies that $\beta\Delta S_4$ is negative, contradicting our assumption that $\beta\Delta S > \alpha\Delta T$.

From the form of Eqs. (11.22) and (11.24) it becomes clear that we can write the stationary solutions as a dimensionless salinity gradients $G^* = \beta\Delta S/\alpha\Delta T$ which is a function of a dimensionless forcing $F^* = \beta F_S/\mathrm{k}(\alpha\Delta T)^2$. Then the dimensionless overturning rate $\Psi^* = \Psi/k\alpha\Delta T = (1-G^*)$. From the solutions (11.22) and (11.24) for G we then get the dimensionless solutions for Ψ^*

$$\Psi^*_{1,2} = (1 - G^*_{1,2}) = \frac{1}{2} \mp \sqrt{\frac{1}{4} - F^*} \quad , \tag{11.25a}$$

$$\Psi^*_3 = (1 - G^*_3) = \frac{1}{2} - \sqrt{\frac{1}{4} + F^*} \quad . \tag{11.25b}$$

Without any salinity forcing Ψ^* will equal 1. With a dimensionless time $t^* = 2k\alpha\Delta T t$ Eq. (11.20) can be written as

$$\frac{\mathrm{d}G^*}{\mathrm{d}t^*} = F^* - |1 - G^*|G^* \quad , \tag{11.26}$$

which, given the dynamic Eq. (11.17), results in an equation for the derivative of Ψ^*:

$$\frac{\mathrm{d}\Psi^*}{\mathrm{d}t^*} = -F^* + |\Psi^*|(1 - \Psi^*) \quad . \tag{11.27}$$

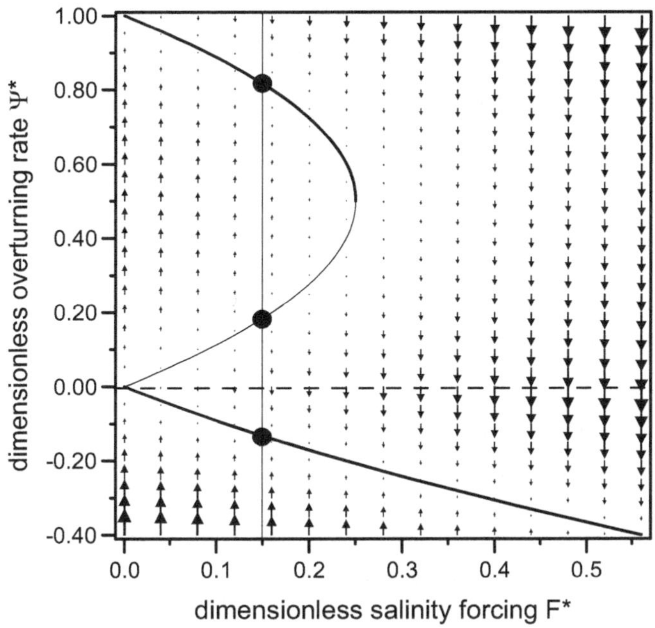

Fig. 11.7. The solution of Eq. (11.25) in the dimensionless salinity forcing–overturning rate response space. The full line shows the possible stationary responses. The black dots illustrate the three possible dimensionless salinity gradients Ψ^* for a dimensionless salinity forcing F^* smaller than 0.25. The arrows represent the dimensionless time derivative $d\Psi^*/dt^*$ given by Eq. 11.27).

Figure 11.7 represents a part of the dimensionless forcing response space, where the stationary solutions are represented by a full line. Solutions with $\Psi^* > 0$ represent the thermally driven THC according to Eq. (11.25a). Solutions with $\Psi^* < 0$ represent the salinity-driven THC, according to Eq. (11.25b). The stationary thermally driven solution itself is divided into two parts. The upper part (thick line) corresponds with a fast thermally driven circulation with low salinity contrast between the boxes. The lower half of the upper parabola (thin line) corresponds with the slow thermally driven mode, with large salinity differences between the boxes. These large salinity gradients compensate the density gradient due to the temperature difference ΔT for over 50%. It appears that for a dimensionless salinity forcing F^*, smaller than 0.25, three possible stationary responses Ψ^* exist, shown by the black dots in Fig. 11.7.

If, with a given dimensionless salinity forcing F^*, the actual model is not in a stationary mode the time derivative $d\Psi^*/dt^*$ can be calculated according to Eq. (11.27). The arrows in Fig. 11.7 represent that time derivative. Close to the stationary solutions the time derivative is very small. For $F^* < 0.25$, $d\Psi^*/dt^*$ is directed to the upper branch of the upper parabola, the fast thermally driven solution, for all initial positions between the zero forcing line and that parabola. For initial positions below the thermally driven upper parabola as well as for all initial positions with $F^* > 0.25$, the solution is attracted toward the negative, salinity-driven stationary solution. For all initial positions close to the slow, lower branch of the thermally driven circulation (thin line), $d\Psi^*/dt^*$ is away from that branch. That slow stationary circulation, with a low density difference between both boxes, therefore is unstable. Apparently the stationary state of the THC in Stommel's box model that will be reached from an arbitrary initial condition will depend on the position of that initial state of the THC in the forcing–overturning space, shown in Fig. 11.7.

If the circulation in Stommel's two-box model of the THC is in the fast thermally driven stationary mode (upper thick line, Fig. 11.7) and the salinity forcing F^* slowly increases, the response will follow that fast thermal branch closely. That will go on until a forcing $F^* = 0.25$ with $\Psi^* = 0.5$ is reached. For a larger forcing no thermally driven stationary solution is possible, and at all Ψ^* levels $d\Psi^*/dt^*$ will force the solution to the salinity-driven circulation (lowest thick line, Fig. 11.7). In the theory of dynamic systems, such a point is called a saddle point bifurcation. In order to return the circulation of the THC back to the thermally driven mode, it is not enough to bring F^* back to a value below 0.25, since the salinity-driven stationary mode is stable, with for small deviations of Ψ^* values of $d\Psi^*/dt^*$ which restore Ψ^* to the stationary salinity-driven mode. This behavior also is found for more sophisticated numerical models of the THC. Only if F^* is reduced considerably (in our model to zero) can the thermally driven mode be reached again. In the present climate system F^* is small enough to prevent low latitude, evaporation-driven downwelling of deepwater.

11.7. The interhemispheric Rooth box model

Stommel's box model, discussed in the previous section, offers a fundamental explanation of multiple steady states in a single-hemisphere THC. However, its applicability in simulating the behavior of the global THC, or even only that of the whole Atlantic Ocean or Pacific Ocean, is of course limited, since it does not contain a cold branch from the cold northern North Atlantic to the even colder Southern Ocean or from the Southern Ocean to the North Pacific Ocean. In order to study the interaction of thermal and hydrological (saline) forcing of the THC on a global scale,

Rooth (1982) developed a simple interhemispheric box model for the NADW branch of the THC (Fig. 11.8).

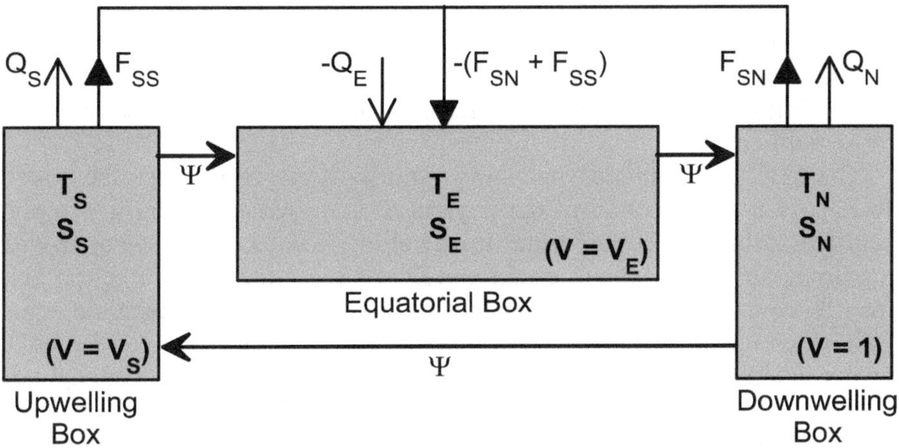

Fig. 11.8. Scheme for the interhemispheric box model, introduced by Rooth (1982). Two "polar" boxes exist (volumes 1 and V_S), where surface cooling (Q_N and $Q_S > 0$) and an upward virtual salt flux (F_{SN} and F_{SS}) due to a precipitation excess maintain a relatively low temperature and salinity. At the levels in and above the thermocline these boxes are connected with a subtropical and tropical box (volume V_E) that extends across the equator in both hemispheres. This "equatorial" box is warmed by air–sea interaction ($Q_E < 0$). Because of salt conservation in the ocean a virtual mass flux equal to $-(F_{SN} + F_{SS})$ due to a net evaporation excess tends to increase the equatorial salinity. The deep branch of the THC transports water between both polar boxes with an overturning rate Ψ.

The discrimination between NADW and AABW is ignored in this model. A single source of deep and bottom water is assumed instead. In principle the forcing of the temperature and salinity fields by air–sea interaction in the model may be asymmetric. The boxes are assumed to be well mixed and are characterized by a mean salinity and temperature. Contrary to Stommel's box model, no upwelling directly into the equatorial box is allowed. The northern downwelling box of the model is the present NADW formation area (Greenland Sea, Polar Ocean, and Labrador Sea), while the southern upwelling box may be considered to begin in the region south of the latitude of the southern tip of Africa (~35°S), where NADW is introduced into the circumpolar current. For the basic dynamics of Rooth's model it is not important whether the NADW rises there directly to close the loop or whether it follows a more complex route through the Indian and Pacific oceans, governed by the dynamics outside the scope of the box model (Rahmstorf 1996). The volume of the downwelling north polar box is taken as unit volume, while the cross-equatorial box has a volume V_E and the volume of the southern, downwelling box is V_S. While Rooth (1982) mainly focused on the stability with symmetric salinity and temperature forcing, his model was further elaborated by Rahmstorf (1996) and Scott et al. (1999), allowing asymmetric forcing of the temperature and salinity field. Much of what follows in this section is derived from those publications.

The THC in this model maintains a transport rate Ψ between the boxes. In Fig. 11.8 the direction of the abyssal Ψ is drawn according to the present-day Atlantic THC and the downwelling box can be associated with the arctic seas. But under certain conditions the circulation direction also may be contrary to the present Atlantic THC ($\Psi < 0$). The dynamic equation which relates the abyssal Ψ to the density difference between the polar boxes is similar to the dynamic equation used in the single-hemispheric Stommel model. With the use of a linear equation of state the dynamics are described by

$$\Psi = k\left[\alpha\left(T_S - T_N\right) - \beta\left(S_S - S_N\right)\right] \quad . \tag{11.28}$$

Note that the value of the hydraulic constant k in Eq. (11.28) will differ strongly from the value of k in Eq. (11.17), since the typical density differences between both high-latitude boxes are much smaller than the pole to equator density contrast. The shallow return flow has no dynamics of its own but is governed by volume conservation without storage terms in any of the boxes.

The equations that describe the development of the temperature of each box when, as in the present THC, $\Psi \geq 0$ are (Scott et al. 1999)

$$\frac{dT_N}{dt} = -Q_N + \Psi\left(T_E - T_N\right) \quad , \tag{11.29a}$$

$$\frac{dT_E}{dt} = -\frac{Q_E}{V_E} + \frac{\Psi\left(T_S - T_E\right)}{V_E} \quad , \tag{11.29b}$$

$$\frac{dT_S}{dt} = -\frac{Q_S}{V_S} + \frac{\Psi\left(T_N - T_S\right)}{V_S} \quad , \tag{11.29c}$$

where Q_N and Q_S are the integrated heat losses to the atmosphere of the downwelling and the upwelling polar box (both > 0), while $-Q_E$ is the heat gain in the cross-equatorial box ($Q_E < 0$). These fluxes can be parameterized by, e.g., the restoring boundary conditions as formulated in Eq. (11.3) or by prescribed temperatures.

Since the total salt content is conserved, while changes in the box volumes due to evaporation and precipitation are ignored (closed lid approximation), the sum of the virtual salinity flux F_S for the three boxes will be zero. This condition is maintained by the atmospheric branch of the hydrological cycle. That implies that for each box F_S is determined by the combination of evaporation, precipitation, and river runoff into that box approximated by $F_S = -S_0(E-P+R)$, where S_0 is a reference salinity (~35) and the net amount of water in the ocean does not change. That will result in the following equations for the salinity of the boxes:

$$\frac{\mathrm{d}S_N}{\mathrm{d}t} = -F_{SN} + \Psi(S_E - S_N) \quad , \tag{11.30a}$$

$$\frac{\mathrm{d}S_E}{\mathrm{d}t} = \frac{(F_{SN} + F_{SS})}{V_E} + \frac{\Psi(S_S - S_E)}{V_E} \quad , \tag{11.30b}$$

$$\frac{\mathrm{d}S_S}{\mathrm{d}t} = -\frac{F_{SS}}{V_S} + \frac{\Psi(S_N - S_S)}{V_S} \quad , \tag{11.30c}$$

where F_{SN} and F_{SS} are the virtual salinity fluxes for, respectively, the northern and southern polar box (both > 0). Stationary solutions (indicated with an overbar) have zero time derivatives in Eqs. (11.29) and (11.30). That also implies that the volumes of the different boxes do not determine the stationary solution. Substitution of the stationary forms of Eqs. (11.29c) and (11.30c) into Eq. (11.28) then results in an equation for the stationary overturning $\overline{\Psi}$:

$$\overline{\Psi}^2 = k(-\alpha Q_S + \beta F_{SS}) \quad . \tag{11.31}$$

This implies that the stationary overturning rate in this model is proportional to the square root of the net virtual mass flux due to air–sea interaction over the upwelling box only. It is easy to derive from Eqs. (11.29) and (11.30) that in a stationary ocean the total virtual mass flux equals zero, so that the downward virtual mass flux over the upwelling box equals the total upward virtual mass flux over the equatorial plus the downwelling box. Thereby the density difference that ensures a positive value for Eq. (10.14) is maintained, allowing overturning circulation according to Sandström's theorem. That is implicitly determined by the condition that each of the boxes is completely mixed; there is no energy limitation to the mixing.

Generally the surface heat fluxes Q are not prescribed but depend on the overturning rate Ψ as with restoring boundary conditions or prescribed temperatures. According to Eq. (11.29c), Eq. (11.31) also can be written as

$$\overline{\Psi} = k\left[-\alpha\left(\overline{T}_N - \overline{T}_S\right) + \beta\frac{F_{SS}}{\overline{\Psi}}\right] \quad . \tag{11.32}$$

This quadratic equation of the stationary overturning rate $\overline{\Psi}$ has as solution (Rahmstorf 1996)

$$\overline{\Psi} = -\frac{\alpha k\left(\overline{T}_N - \overline{T}_S\right)}{2} \pm \frac{1}{2}\sqrt{k^2\alpha^2\left(\overline{T}_N - \overline{T}_S\right)^2 + 4k\beta F_{SS}} \quad , \tag{11.33}$$

which depends on the virtual salinity flux over the upwelling box, as well as on the temperature difference between both polar boxes. The temperature equations (11.29 a to c) that determine $\overline{T}_N$ and $\overline{T}_S$ depend on $\overline{\Psi}$ and the thermal boundary conditions, but do not contain any salinity term. The equilibrium overturning rate $\overline{\Psi}$ will depend on F_{SS}, and therefore on the freshwater forcing in the upwelling box only (Scott et al. 1999). The variable F_{SS} in Eq. (11.33) is $F_{SS} = (F_{SS}+F_{SN})-F_{SN} = -F_{SE}-F_{SN}$, where F_{SE} is the virtual salinity flux over the equatorial box. According to Eq. (11.33) the mutual distribution of F_{SE} and F_{SN} over the equatorial and downwelling boxes does not influence the resulting $\overline{\Psi}$.

Mutual changes in F_{SN} and F_{SE} may affect the transient response of the model, especially when $V_E >> 1$, which causes a slow response of the equatorial box. A rapid increase of F_{SN} will dilute the water in the northern downwelling box, leading to a temporary reduction $\overline{\Psi}$. With an advective timescale of centuries the model will adjust the salinities in the boxes until the original stationary equilibrium transport, which is governed by F_{SS}, is restored again. In extreme cases like a massive freshwater influx, e.g., from a melting Greenland ice cap, a transition to a new equilibrium solution with inversed overturning direction may be triggered. But such an event falls outside of the scope of this model which ignores water storage outside the ocean, e.g., in ice caps.

The major control parameter for the value of $\overline{\Psi}$, next to $(\overline{T}_N - \overline{T}_S)$, is the virtual salinity flux in the upwelling box, F_{SS}, which equals $-(F_{SE} + F_{SN})$, the net virtual salinity flux over the return flow in the cross-equatorial box and the downwelling box. This allows us to classify the overturning circulation in three dynamic regimes (Fig. 11.9). If $\Delta T = \overline{T}_N - \overline{T}_S > 0$, then the deep temperature gradient counteracts the positive overturning rate $\overline{\Psi}$. Then only the (+) solution gives a positive $\overline{\Psi}$ and the normalized overturning rate $\overline{\Psi}^*$, derived from Eq. (11.33), equals

$$\overline{\Psi}^* = \frac{\Psi}{k\alpha\Delta T} = -\frac{1}{2} + \sqrt{\frac{1}{4} + \frac{\beta F_{SS}}{k\alpha^2\Delta T^2}} = -\frac{1}{2} + \sqrt{\frac{1}{4} - \frac{\beta(F_{SE} + F_{SN})}{k\alpha^2\Delta T^2}} = -\frac{1}{2} + \sqrt{\frac{1}{4} - F^*} \quad (11.34a)$$

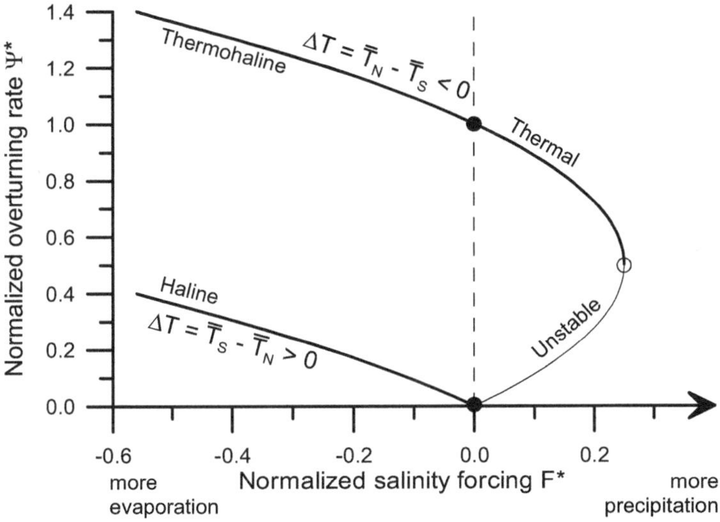

Fig. 11.9. The stationary solution of Eq. (11.34) in the dimensionless overturning rate salinity forcing response space. The normalized forcing *F reflects the net virtual salinity flux for the cross-equatorial and the downwelling box due to evaporation and precipitation. The full lines are associated with the possible stationary responses. The thin line shows the unconditionally unstable part of the stationary solution and the open symbol indicates the critical forcing.**

If, on the contrary, the deep temperature gradient supports the positive overturning rate, that if $\Delta T = T_S - T_N > 0$, the positive normalized overturning rate equals

$$\overline{\Psi}^* = \frac{\Psi}{k\alpha\Delta T} = \frac{1}{2} \pm \sqrt{\frac{1}{4} + \frac{\beta F_{SS}}{k\alpha^2\Delta T^2}} = +\frac{1}{2} \pm \sqrt{\frac{1}{4} - \frac{\beta(F_{SE} + F_{SN})}{k\alpha^2\Delta T^2}} = +\frac{1}{2} \pm \sqrt{\frac{1}{4} - F^*} \quad (11.34b)$$

Rahmstorf (1996) identified in the solutions (11.34a, b) three different regimes. In case the temperature gradients counteract the positive $\overline{\Psi}$, the overturning according to Eq. (11.34a) is completely driven by the precipitation excess over the upwelling box, which equals the evaporation excess over the cross-equatorial and the downwelling box (the haline circulation in Fig. 11.9). But when $\overline{T}_N < \overline{T}_S$ the deep temperature gradient supports the driving of the THC. When there is a net evaporation excess over the return flow and the downwelling box, the resulting overturning THC is driven by both the temperature and salinity gradient, the thermohaline regime in Fig. 11.9. A peculiar regime is the purely thermally driven flow when $\overline{T}_N < \overline{T}_S$ drives the THC, counteracted by the precipitation excess $F^* > 0$ over the return flow and the downwelling box. The temperature condition reflects an asymmetric forcing. This thermal branch in Fig. 11.9 ends in a so-called bifurcation at a critical normalized virtual salinity flux $F^* = 0.25$. With a larger F^*, a stationary positive overturning rate is no longer possible, and the whole box system will switch to another circulation mode with time-dependent behavior. According to the meridional distribution of potential temperature and salinity in the Atlantic Ocean, shown in Figs. 4.2 and 10.12 (upper panel), we are at present in the haline regime of Fig. 11.9. The deep temperature gradient counteracts the overturning circulation ($T_S < T_N$). The connected relatively high salinity of the NADW in the deep branch of the THC is due to the evaporation excess of the Atlantic Ocean. In this regime an increase in the salinity forcing F^*, that is an increase in the total precipitation over the equatorial and northern Atlantic Ocean, will slowly decrease the overturning rate Ψ. Note that this haline regime in Rooth's model refers to the salinity difference between the two high-latitude boxes, contrary to the previous section on Stommel's box model, where the haline regime refers to downwelling of warm but extremely saline water at low latitudes.

In the Pacific Ocean the deep horizontal temperature and salinity gradients (Fig. 5.13) indicate that in that ocean the deep meridional gradients of both temperature and salinity agree with a deep circulation of the thermohaline regime in Fig. 11.9. Especially the precipitation excess in the relatively cold North Pacific Ocean will decrease the salinity there, while the relatively low surface temperatures in that ocean basin also will reduce the heat loss to the atmosphere.

The same solution structure as we discussed so far of course also exists for a THC with a negative overturning rate $\overline{\Psi}$, but then the downwelling will be found in the South Atlantic Ocean and the upwelling in the North Atlantic box. To derive the appropriate temperature and salinity equations similar to Eqs. (11.29) and (11.30) only the advective terms have to be adapted. This allows two stationary circulations with pole-to-pole flow in either direction even for a completely symmetrical box model (i.e. $\overline{T}_N = \overline{T}_S$ and $F_{SN} = F_{SS}$).

Another type of cross-equatorial box model for the global THC was devised by Welander (1986) who combined two Stommel box models back to back, where the equatorial boxes where combined into a single cross-equatorial fully mixed box. He found nine steady state solutions, of which four where stable. One of these stationary solutions corresponds to the northern sinking, southern upwelling solution. Studies with much more sophisticated three-dimensional ocean general circulation models (OGCMs) by, e.g., Tziperman (1997) and Wang et al. (1999), showed that although the global deepwater formation is relatively insensitive to the distribution of the virtual salinity flux via freshwater forcing, the formation rate of NADW strongly depended on the freshwater forcing at high latitudes. Especially Wang et al. (1999) observed in their OGCM that similar to the simple box model of Rooth the production rate of NADW in the northern North Atlantic Ocean increased when the salinity in the South Atlantic Ocean decreased through a larger atmospheric poleward moisture flux in the southern hemisphere. But Tziperman (1997) also mentioned inherently unstable behavior of the THC in his model. That subject is discussed in the next section.

11.8. The stability of Rooth's model

The stationary solution for Rooth's model appears quite simple. However, Scott et al. (1999) found that in a more complicated version of Rooth's box model with asymmetric forcing the northern downwelling solution was not always found, and slight initial perturbations of the salinity and temperature field could lead to oscillating as well as inherently unstable solutions. Here we will study the stability of the model by means of a so-called linear perturbation expansion around the steady state. For simplicity we will assume that the relatively fast response of the temperature field to atmospheric forcing is mimicked by prescribing the temperature in each box and that the atmospheric temperature forcing is symmetric, $\overline{T}_N = \overline{T}_S$, so that temperature variations can be ignored. Then the temperature does not influence the overturning rate and we can skip Eq. (11.29).

We assume that both the salinity in each box and the overturning rate can be written as a sum of the stationary simulation (with overbar) and a (small) perturbation (S'_N, S'_E, S'_S, Ψ'). From Eqs. (11.28) and (11.30) the stationary solutions in this simplified model are derived as

$$\overline{\Psi} = k\beta(\overline{S}_N - \overline{S}_S) \quad , \tag{11.35a}$$

$$\overline{\Psi}\left(\overline{S}_E - \overline{S}_N\right) = F_{SN} \quad , \tag{11.35b}$$

$$\overline{\Psi}\left(\overline{S}_S - \overline{S}_E\right) = -\left(F_{SN} - F_{SS}\right) \quad \text{and} \tag{11.35c}$$

$$\overline{\Psi}\left(\overline{S}_N - \overline{S}_S\right) = F_{SS} \quad . \tag{11.35d}$$

Substitution of Eq. (11.35d) into Eq. (11.35a) gives

$$\overline{\Psi} = \sqrt{k\beta F_{SS}} \quad , \tag{11.36a}$$

the simplified form of Eq. (11.33). Substitution of Eq. (11.36a) into Eq. (11.35b to d) will give expressions for a salinity gradient in terms of virtual salinity fluxes. If we substitute Eq. (11.36a) into Eq. (11.35d) we get an expression for the north–south salinity difference:

$$\left(\overline{S}_N - \overline{S}_S\right) = \sqrt{\frac{F_{SS}}{k\beta}} \quad . \tag{11.36b}$$

Substitution of Eq. (11.36a) into Eq. (11.35b) gives

$$\left(\overline{S}_E - \overline{S}_N\right) = \frac{F_{SN}}{\sqrt{k\beta F_{SS}}} \quad , \tag{11.36c}$$

while the salinity difference between the equatorial and southern boxes can be derived as

$$\begin{aligned} &\left(\overline{S}_E - \overline{S}_S\right) = \left(\overline{S}_E - \overline{S}_N\right) + \left(\overline{S}_N - \overline{S}_S\right) = \\ &\frac{F_{SN}}{\sqrt{k\beta F_{SS}}} + \sqrt{\frac{F_{SS}}{k\beta}} = \frac{F_{SN} + F_{SS}}{\sqrt{k\beta F_{SS}}} \quad . \end{aligned} \tag{11.36d}$$

A parameter that will be used in the stability analysis of the perturbed model is the asymmetry of the freshwater or virtual salinity forcing, R_F, which is defined as the ratio of Eqs. (11.36c) and (11.36b):

$$\frac{\left(\overline{S}_E - \overline{S}_N\right)}{\left(\overline{S}_N - \overline{S}_S\right)} = \frac{F_{SN}/\sqrt{k\beta F_{SS}}}{\sqrt{F_{SS}/k\beta}} = \frac{F_{SN}}{F_{SS}} \equiv R_F \quad . \tag{11.37}$$

The forcing of the perturbed model with prescribed virtual salt fluxes is assumed to be constant. That implies that the weighted mean salinity remains constant or that, because of conservation of salt,

$$S'_E = -\frac{1}{V_E} S'_N - \frac{V_S}{V_E} S'_S \quad . \tag{11.38}$$

The dynamic equation of the perturbed system derived from Eq.(11.28) is

$$\Psi' = k\beta\left(S'_N - S'_S\right) \quad . \tag{11.39a}$$

The derivative of S'_N derived from Eq. (11.30a) is

$$\begin{aligned} \frac{\mathrm{d}S'_N}{\mathrm{d}t} &= \Psi'\left(\overline{S}_E - \overline{S}_N\right) + \overline{\Psi}\left(S'_E - S'_N\right) = \\ & k\beta\left(S'_N - S'_S\right)\frac{F_{SN}}{\sqrt{k\beta F_{SS}}} + \sqrt{k\beta F_{SS}}\left(S'_E - S'_N\right) \end{aligned} \quad . \tag{11.39b}$$

Using the expression for S'_E in Eqs. (11.38) and (11.36a) this becomes

$$\frac{\mathrm{d}S'_N}{\mathrm{d}t} = \overline{\Psi}\left[\left(R_F - \frac{1}{V_E} - 1\right)S'_N - \left(R_F + \frac{V_S}{V_E}\right)S'_S\right] \quad . \tag{11.39c}$$

The equation for S'_S derived from Eq. (11.30c) is

$$\begin{aligned} \frac{\mathrm{d}S'_S}{\mathrm{d}t} &= \overline{\Psi}\left(S'_N - S'_S\right) + \Psi'\left(\overline{S}_N - \overline{S}_S\right) = \\ & \sqrt{k\beta F_{SS}}\left(S'_N - S'_S\right) + k\beta\left(S'_N - S'_S\right)\sqrt{\frac{F_{SS}}{k\beta}} = 2\overline{\Psi}\left(S'_N - S'_S\right) \end{aligned} \quad . \tag{11.39d}$$

The derivatives of S'_N and S'_S are now written as functions of S'_N and S'_S. In vector notation Eqs. (11.39c) and (11.39d) then become

$$\frac{\mathrm{d}}{\mathrm{d}t}\begin{pmatrix} S'_S \\ S'_N \end{pmatrix} = \mathbf{A}\begin{pmatrix} S'_S \\ S'_N \end{pmatrix} \quad , \tag{11.40a}$$

with the 2×2 matrix $\mathbf{A}$ written as

$$\mathbf{A} = \overline{\Psi}\begin{pmatrix} -2 & 2 \\ \left(-R_F - \frac{V_S}{V_E}\right) & \left(R_F - \frac{1}{V_E} - 1\right) \end{pmatrix} \quad . \tag{11.40b}$$

In the perturbation analysis we assume that the salinity vector from Eq. (11.40a) has the form $S' \sim e^{\overline{\Psi}\lambda t}$, where the imaginary part of λ, $\mathrm{Im}(\lambda)$, represents an oscillatory solution of the perturbation and the real part of λ, $\mathrm{Re}(\lambda)$, represents the exponentially changing part of the perturbation. If $\mathrm{Re}(\lambda)$ is positive, any small perturbation will grow exponentially, and the THC in Rooth's box model will be unstable.

and the THC in Rooth's box model will be unstable. Solutions for parameter λ equal the eigenvalues of $\mathbf{A}$ and are obtained from

$$(\lambda+2)\cdot\left(\lambda+1+\frac{1}{V_E}-R_F\right)+2\left(\frac{V_S}{V_E}+R_F\right)=0 \quad \text{or}$$
$$\lambda^2-\lambda\left(R_F-\frac{1}{V_E}-3\right)+2\left(\frac{1+V_S}{V_E}+1\right)=0 \qquad . \qquad (11.41)$$

which results in

$$\lambda_{1,2}=\frac{1}{2}\left(R_F-\frac{1}{V_E}-3\right)\pm\sqrt{\frac{1}{4}\left(R_F-\frac{1}{V_E}-3\right)^2-2\left(\frac{1+V_S}{V_E}+1\right)} \quad . \qquad (11.42a)$$

For a simple box model with equal volumes for the three boxes, Eq. (11.42a) reduces to

$$\lambda_{1,2}=\frac{1}{2}(R_F-4)\pm\sqrt{\frac{1}{4}(R_F-4)^2-6} \quad . \qquad (11.42b)$$

With a symmetrical freshwater forcing ($R_F = 1$), the first term in the right-hand side of Eq. (11.42b) will be −1.5 and the square root will always have an imaginary value (~1.94i); all perturbations will have a strongly damped oscillatory character. In that case the stationary solution with NADW formation in the North Atlantic Ocean will be stable. The formation area of NADW (the downwelling box) has a much smaller surface than the Southern Ocean where upwelling takes place, while the net freshwater flux density and therefore the virtual salinity flux density in both areas do not differ strongly. Therefore it can be expected that the freshwater forcing is asymmetric, with $R_F > 1$. For an R_F above 4, the THC in the box model with equal volumes and NADW formation in the northern box, small perturbations will initially grow exponentially. Then the THC is unconditionally unstable for small perturbations and the only stable solution will be the one with a southern source of deepwater. For such a reversed circulation the appropriate $R_F = F_{SS}/F_{SN} < ¼$. The second term in Eq. (11.42b) will become real if $R_F >$ ~8.9. For lower values Ψ' also will have an oscillatory character, while for larger values the second term will be real, allowing an even larger exponential growth for the (+) solution of Eq. (11.42b).

The feedback that leads to Eq. (11.42b) can be illustrated by the derivative of Ψ'', derived from Eq. (11.39a)

$$\frac{d\Psi'}{dt} = k\beta\left(\frac{dS'_N}{dt} - \frac{dS'_S}{dt}\right) = \\ k\beta\left[\Psi'\left(\overline{S}_E - \overline{S}_N\right) + \overline{\Psi}\left(S'_E - S'_N\right) - \Psi'\left(\overline{S}_N - \overline{S}_S\right) - \overline{\Psi}\left(S'_N - S'_S\right)\right] \quad . \tag{11.43a}$$

With Eqs. (11.37) and (11.38) for equal box volumes Eq. (11.43a) can be rewritten as

$$\frac{d\Psi'}{dt} = -3k\beta\overline{\Psi}S'_N + k\beta\Psi'\left(\overline{S}_N - \overline{S}_S\right)\cdot\left(R_F - 1\right) \tag{11.43b}$$

The first term of the right-hand side of Eq. (11.43b) represents a mean flow feedback. If Ψ' is positive, S'_N is larger than S'_S. Then the mean flow will decrease the salinity difference between the downwelling and upwelling box according to Eq. (11.39c), and that will lead to a decrease of Ψ'. The mean flow eliminates salinity anomalies and therefore is stabilizing Ψ'. The second term in Eq. (11.43b) is an anomalous flow feedback. An anomaly of Ψ' will cause an increase of S'_N proportional to $\Psi'\left(\overline{S}_E - \overline{S}_N\right)$ and an increase of S'_S proportional to $\Psi'\left(\overline{S}_N - \overline{S}_S\right)$. According to Eq. (11.37) these terms have a ratio R_F. With $R_F > 1$, as with northern downwelling of NADW, a positive Ψ' anomaly therefore will lead to an increase of the salinity difference between the downwelling and the upwelling box, and therefore to an increase of Ψ'. If $R_F > 4$, the stabilizing mean flow feedback is overcome by the destabilizing anomalous flow feedback and the steady-state solution of the box model has become unstable.

With unequal volumes for the boxes in the model the stability condition, Eq. (11.42a), alters slightly relative to the simplified equal volume form of Eq. (11.42b), although the main structure remains. In the limit case where the downwelling box is very small relative to the cross-equatorial box, $V_E >> 1$, instability of small perturbations already starts with an asymmetric freshwater forcing with $R_F > 3$. An increase of the volume of the southern upwelling box will extend the domain of asymmetric forcing where perturbations will have an oscillatory character. Although it is questionable whether one can apply box models any further than to exemplify certain feedbacks in the THC, one may guess in which stability domain the present THC is located. Since the net (P–E) at high latitudes in the northern and southern hemisphere does not differ strongly [P–$E \approx$ O(1 mm/day), Fig. 10.8], it mainly will be the differences in surface area of the boxes which determine R_F. It is in a certain way arbitrary where one locates the boundaries between the boxes, but an estimate of R_F close to 3 or 4 seems to be of the right order of magnitude. That suggests that the overturning of NADW at present may be close to an unstable situation, for which oscillatory perturbations of the THC can be expected. Such oscillating perturbations may be a cause of decadal to centennial climate change as well as to a sudden climate change (Scott et al. 1999). In the unstable situation any small perturbation initially will grow exponentially and be the cause of a reversal of the THC.

11.9. Two-dimensional meridional models of the THC

In Chapter 10 we have shown that the THC can be maintained only if turbulent mixing is able to enforce a density distribution in the ocean that meets the requirement of Sandström's theorem. In the box models from the previous sections the turbulent mixing is modeled in a very crude way. In each box the turbulent diffusion coefficient is infinite, maintaining a completely mixed density, while no mixing occurs in the ducts that connect the different boxes. In order to study the influence of mixing, while keeping the model simple and the required computational effort limited, two-dimensional numerical models of the THC have been developed (Marotzke et al. 1988; Wright and Stocker 1991). In these models the circulation is projected onto a vertical meridional plane by zonally averaging the three-dimensional equations. The original three-dimensional dynamic equations are near-geostrophic with friction, with horizontal and vertical viscosity coefficients A_h and A_v, ignoring all inertial terms, and the hydrostatic balance:

$$-fv = -\frac{1}{\rho_0}\frac{\partial P}{\partial x} + A_h\left(\frac{\partial^2 u}{\partial x^2} + \frac{\partial^2 u}{\partial y^2}\right) + A_v\frac{\partial^2 u}{\partial z^2} \quad , \tag{11.44a}$$

$$fu = -\frac{1}{\rho_0}\frac{\partial P}{\partial y} + A_h\left(\frac{\partial^2 v}{\partial x^2} + \frac{\partial^2 v}{\partial y^2}\right) + A_v\frac{\partial^2 v}{\partial z^2} \quad \text{, and} \tag{11.44b}$$

$$\frac{\partial P}{\partial z} = -\rho g \quad . \tag{11.44c}$$

Lateral friction will be ignored in the following discussion ($A_h = 0$) because it is an ill-understood process and may lead to negative viscosity coefficients near the western boundary currents like the Gulf Stream (Starr 1968). The continuity equation, Eq. (3.30), is repeated here:

$$\frac{\partial u}{\partial x} + \frac{\partial v}{\partial y} + \frac{\partial w}{\partial z} = 0 \quad . \tag{11.44d}$$

The equations for the development of the temperature and salinity fields contain both horizontal and vertical turbulent diffusion:

$$\frac{\partial T}{\partial t}+\frac{\partial uT}{\partial x}+\frac{\partial vT}{\partial y}+\frac{\partial wT}{\partial z}=K_h\left(\frac{\partial^2 T}{\partial x^2}+\frac{\partial^2 T}{\partial y^2}\right)+K_v\frac{\partial^2 T}{\partial z^2} \quad , \qquad (11.44e)$$

$$\frac{\partial S}{\partial t}+\frac{\partial uS}{\partial x}+\frac{\partial vS}{\partial y}+\frac{\partial wS}{\partial z}=K_h\left(\frac{\partial^2 S}{\partial x^2}+\frac{\partial^2 S}{\partial y^2}\right)+K_v\frac{\partial^2 S}{\partial z^2} \quad . \qquad (11.44f)$$

The complete equation of state has to be used to allow the effects of a temperature-dependent thermal expansion coefficient. Here, however, we will use the linearized form of the EOS,

$$\rho=\rho_0\left(1-\alpha\left(T-T_0\right)+\left(S-S_0\right)\right) \quad . \qquad (11.44g)$$

Zonally averaged parameters are indicated with an overbar, while the local deviation from the zonal average is indicated with a prime. The zonal average of a parameter a is defined as

$$\overline{a}(y,z)=\frac{1}{B}\int_{x_W}^{x_E} a(x,y,z)dx \quad , \qquad (11.45)$$

where x_W and x_E are the zonal positions of, respectively, the western and eastern boundary of the ocean, which are at a zonal distance B from each other. It is assumed, as boundary condition, that the horizontal velocity perpendicular to the zonal boundaries of the ocean equals zero, as is the turbulent diffusion of heat and salt perpendicular to that boundary. The zonal average of the dynamic Eqs. (11.44a, b, and c), neglecting A_h, is given by

$$-f\overline{v}=-\frac{P_E-P_W}{\rho_0 B}+A_v\frac{\partial^2\overline{u}}{\partial z^2} \quad , \qquad (11.46a)$$

$$f\overline{u}=-\frac{1}{\rho_0}\frac{\partial\overline{P}}{\partial y}+A_v\frac{\partial^2\overline{v}}{\partial z^2} \quad , \qquad (11.46b)$$

$$\frac{\partial\overline{P}}{\partial z}=-\overline{\rho}g \quad . \qquad (11.46c)$$

In Eq. (11.46a) a term with the pressure difference between the eastern and western boundary of the ocean, $(P_E - P_W)/\rho_0 B$ emerges, which is not expressed in zonally averaged parameters. Some closure method has to be devised to parameterize this term, as will be discussed below. Zonal integration of the continuity equation Eq. (11.44d) gives

$$\frac{\partial B\overline{v}}{\partial y}+\frac{\partial B\overline{w}}{\partial z}=0 \quad , \tag{11.46d}$$

where the convergence of the meridional transport density, $B\overline{v}$, is balanced by the divergence of the vertical upwelling transport density. The equations for heat and salt become, after zonal integration,

$$\frac{\partial \overline{T}}{\partial t}+\frac{1}{B}\left[\frac{\partial B\overline{v}\overline{T}}{\partial y}+\frac{\partial B\overline{v'T'}}{\partial y}+\frac{\partial B\overline{w}\overline{T}}{\partial z}+\frac{\partial \overline{w'T'}}{\partial z}\right]= \frac{K_h}{B}\frac{\partial^2 B\overline{T}}{\partial y^2}+\frac{K_v}{B}\frac{\partial^2 B\overline{T}}{\partial z^2} \quad , \tag{11.46e}$$

$$\frac{\partial \overline{S}}{\partial t}+\frac{1}{B}\left[\frac{\partial B\overline{v}\overline{S}}{\partial y}+\frac{\partial B\overline{v'S'}}{\partial y}+\frac{\partial B\overline{w}\overline{S}}{\partial z}+\frac{\partial \overline{w'S'}}{\partial z}\right]= \frac{K_h}{B}\frac{\partial^2 B\overline{S}}{\partial y^2}+\frac{K_v}{B}\frac{\partial^2 B\overline{S}}{\partial z^2} \quad . \tag{11.46f}$$

In these equations the Reynolds terms with the covariance of the deviations from the zonal averages cannot be expressed in terms of these average parameters. For these Reynolds terms a closure method is required, although it may represent a first-order process like the heat transport by the wind-driven gyre circulation. Finally, the linearized equation of state can be written as

$$\rho=\rho_0\left(1-\alpha\left(\overline{T}-T_0\right)+\left(\overline{S}-S_0\right)\right) \quad . \tag{11.46g}$$

For the closure of the dynamic equations Marotzke et al. (1988) propose replacing Eqs. (11.46a and b) by a balance between the zonally mean meridional pressure gradient and the turbulent friction term from Eq. (11.46a), thereby ignoring all effects of the Coriolis force on the dynamics of the THC, which are dominant in the Stommel–Arons model:

$$A_v\frac{\partial^2 \overline{v}}{\partial z^2}=\frac{1}{\rho_0}\frac{\partial \overline{P}}{\partial y} \quad , \tag{11.47}$$

where A_v may have a value that differs from the turbulent viscosity used in Eqs. (11.44a and b). Equation (11.47) corresponds with the intuitive notion that the meridionally overturning THC is driven by the pressure gradient connected with the meridional density gradient, resulting from density gradients according to Eq. (11.46c). However, there are no physical arguments why Eq. (11.47) should be valid for the overturning THC on a rotating earth. Wright and Stocker (1991) have tried to find such physical arguments for the closure of Eqs. (11.46a and b). In their argument it is assumed that in the THC the friction is mainly meridional, caused by the relatively fast DWBC. Therefore, they start to ignore the friction term in Eq. (11.46a). In Eq. (11.46b) they replace the turbulent friction term by a more general Rayleigh friction term $-\mu\overline{v}$. This formulation excludes Ekman layers but still allows a near-geostrophic flow. The resulting dynamic equations then become

$$f\overline{v} = \frac{P_E - P_W}{\rho_0 B} \quad \text{, and} \tag{11.48a}$$

$$f\overline{u} = -\frac{1}{\rho_0}\frac{\partial \overline{P}}{\partial y} - \mu\overline{v} = f\overline{u}_g - \mu\overline{v} \quad , \tag{11.48b}$$

where $\overline{u}_g$ equals the mean zonal geostrophic velocity. Using Eq. (11.48b) we now can write an expression for the mean meridional velocity:

$$\overline{v} = \frac{f}{\mu}\left(\overline{u}_g - \overline{u}\right) \quad . \tag{11.49}$$

Substitution of Eq. (11.49) into Eq. (11.48a) and division by the mean zonal geostrophic velocity then leads to

$$\frac{\left(P_E - P_W\right)}{\rho_0 B} = -\frac{f}{\mu}\left(1 - \frac{\overline{u}}{\overline{u}_g}\right)\frac{1}{\rho_0}\frac{\partial \overline{P}}{\partial y} = -\varepsilon\frac{1}{\rho_0}\frac{\partial \overline{P}}{\partial y} \quad . \tag{11.50}$$

In a very narrow ocean basin $\overline{u}/\overline{u}_g \approx 0$, and then $\varepsilon = f/\mu$ is the ratio of the damping timescale and the inertial timescale. For a realistic ocean basin it is expected, as we have assumed in Chapter 8, that the zonal velocity nearly equals the geostrophic velocity. That will reduce ε to a small value and suggests that the mean zonal pressure gradient, the left-hand side of Eq. (11.50), is only a small fraction of the mean meridional pressure gradient. From Eqs. (11.50) and (11.48a) we obtain an expression for the mean meridional velocity:

$$\overline{v} = -\frac{\varepsilon}{f}\frac{1}{\rho_0}\frac{\partial \overline{P}}{\partial y} \quad , \tag{11.51}$$

which is equivalent to the dynamic equation used for the box models presented above in Sections 11.6 and 11.7. One should be aware that Eq. (11.51) completely depends on the specific formulation of the meridional friction term. Wright and Stocker (1991) note, based on comparison with three-dimensional general circulation models, that although ε is expected to be generally positive (in the northern hemisphere), it still may vary spatially with latitude and depth. Wright et al. (1995) used considerations on vorticity dissipation to come to yet another closure of the dynamic equations in a two-dimensional THC model, in which they also introduced the effects of lateral turbulent friction.

Another closure required for the formulation of a two-dimensional THC model is for the Reynolds terms in Eqs. (11.46e and f). These terms not only represent a transport term due to ocean eddies, but also the heat and salt transport by the near-surface wind-driven ocean gyres and by the abyssal circulation like in the Stommel–Arons model. The meridional heat transport by the wind-driven gyres may affect the sea surface temperature, and thereby the heat flux and evaporation boundary conditions. But since it is assumed that in the (North) Atlantic Ocean the overturning heat transport is dominant relative to heat transport by the gyre circulation, the Reynolds terms for the meridional transport of heat and salt are generally neglected or they are parameterized by the horizontal diffusion terms in Eqs. (11.46e and f). In general the Reynolds terms for the vertical transport are also neglected or incorporated into the vertical turbulent diffusion term. The vertical turbulent diffusion term in Eqs. (11.46e and f) also may be adapted to support the complete vertical mixing when the stratification becomes locally unstable during convective adjustment in the high latitude deepwater formation regions.

The two-dimensional models of the THC in a vertical meridional plane consist of Eqs. (11.47) or (11.51) together with Eqs. (11.46c and d), simplified forms of Eqs. (11.46e and f) without Reynolds terms and Eq. (11.46g). Generally these equations have no analytical solution but are solved by numerical methods. These methods approximate the continuous differential equations by a series of algebraic difference equations on a set of grid points in a computational grid in the meridional plane. This requires also a set of boundary conditions for velocity, temperature, and salinity at the surface, bottom, and meridional boundaries.

In experiments with two-dimensional THC models, with the model as well as the forcing by the heat flux density and virtual salt flux density symmetric to the equator, the symmetric solution was found to be unstable and an asymmetric pole-to-pole THC developed in all cases, similar to the results of Rooth's box model presented above (Marotzke at al. 1988). Numeric experiments by Stocker and Wright (1991) have used two two-dimensional models, representing the Atlantic and Pacific oceans, coupled at their southern side with a short two-dimensional Southern Ocean. In these experiments it appeared that the stability of the stationary solutions depended on the type of boundary condition representing that air–sea interaction, restoring boundary conditions for both heat and salt fluxes or mixed boundary conditions. Also the final equilibrium stage, which resulted after perturbing the model with an anomalous freshwater flux, depended on the type of surface boundary condition. In a study with a two-dimensional THC model Vellinga (1997) concluded that multiple stationary solutions were possible in part of the forcing parameter space. These areas of multiplicity were determined mainly by the choice of parameter values for the parameterization of the convection in the model, while the vertical resolution of the computational grid was found to be of great importance too.

Despite the existence of competing closure schemes for two-dimensional models that ignore the effects of wind forcing and the earth's rotation, these models have been used for the exploration of possible feedbacks in the THC, of the influence of the formulation of the surface boundary conditions, and of the influence of turbulent mixing on the modes and stability of the THC. The results of the computations with such models, superior in physics to the conceptual box models since they allow vertical stratification and mixing, however should not be used for realistic forecasts of the THC, since still too much other essential physics are lacking. The Stommel–Arons model of the abyssal circulation (Chapter 8), which agrees in several aspects with the observed deep circulation, has shown that a specific horizontal structure of the deep flow is required to reconcile a net deep meridional flow in the DWBC with upwelling in the ocean's interior. To approach the circulation in the real three-dimensional ocean on a rotating earth, the use of a three-dimensional ocean general circulation model therefore is required.

11.10. Three-dimensional ocean general circulation models

Ocean general circulation models (OGCMs) are used for more realistic simulations of the full three-dimensional ocean circulation, including the THC and the wind-driven circulation. The development of such models started in the 1960s with the study of the near-equilibrium state of the ocean with wind and thermal forcing of an idealized ocean basin with a three-dimensional OGCM (Bryan and Cox 1968). However, the early OGCMs were seriously limited in their ability to so simulate the details of the global THC. In these early days high-quality data sets of the global wind stress and the flux densities of the heat and freshwater forcing were incomplete and very inaccurate. This made it quite premature to validate the outcome of OGCM simulations with the then-available hydrographic data. In fact, the lingering disparity between the early ocean models and the high-precision, deep ocean data sets, meticulously gathered by individual oceanographic expeditions, may be regarded as a characteristic of the THC studies with OGCMs until the 1990s (Böning and Semtner 2001).

OGCMs are based on the full set of equations for the horizontal velocity components, the vertical balance of forces, the continuity equation, equations for the development of the temperature and salinity field, and the equation of state. Examples of such a set of equations are Eqs. (11.44a to g), but other approximations of the dynamics, e.g., geostrophic flow or complete primitive equations, can be used too. Boundary conditions for the flux densities for heat and freshwater (or the virtual salinity flux density) and the wind stress are prescribed at the sea surface. At the ocean bottom and lateral boundaries zero turbulent fluxes of heat and salt, perpendicular to the bottom, are required. The boundary conditions for the velocity at the bottom require that the flow is parallel to the bottom, while the velocity at the bottom is zero (no-slip condition). The continuous equations of an OGCM are solved by approximation with finite difference schemes on a discrete three-dimensional grid in an ocean basin with a topography, discretized at the horizontal positions of the grid points (e.g., Bryan 1969).

Although the reliability of OGCMs for the accurate simulation of the THC was still questionable, such models were used around 1990 for the study of feedback processes and the existence of stationary solutions with more realism and less *ad hoc* closure schemes than was possible with box models of meridional two-dimensional models. Bryan (1986) used an OGCM to investigate the interaction between the Atlantic-scale THC and the salinity distribution. The basin geometry of the model was a 60°-wide sector of the globe, extending from pole to pole, with a flat bottom at 5 km depth. The horizontal resolution of the computational grid was 3.75° longitude × 4.5° latitude with 12 levels in the vertical. The wind stress τ approximated the observed zonal and hemispheric mean stress. The surface heat flux density was approximated as a restoring boundary condition conforming to Eq. (11.3) where the zonally and hemispheric averaged *SST* was used for the equilibrium temperature T_A. A constant virtual salt flux density F_S was prescribed, derived from earlier single-hemisphere models. The forcing, that is τ, T_A, and F_S, were symmetric relative to the equator. After 1300 years of integration of the model, starting with a symmetric initial state, a near-symmetric overturning circulation was obtained. This solution was close to equilibrium, as measured by deepwater temperatures and heat flux densities. Bryan (1986) concluded that this three-dimensional, equatorially symmetric circulation was stable to very small perturbations, arising from numerical round-off errors, or that the growth rate of any such instability was extremely small. This contrasts with the results from two-dimensional meridional models and box models. By adding an asymmetric salinity perturbation to the near-symmetric solution, while maintaining the symmetric forcing, a stationary pole-to-pole THC developed similar to the pole-to-pole THC in Rooth's box model or in two-dimensional meridional models.

Marotzke and Willebrand (1991) used an OGCM with a highly idealized geometry to investigate multiple equilibria of the global-scale THC. Their model comprised two identical flat-bottomed ocean basins, representing the Atlantic and Pacific oceans, which were connected by a circumpolar channel in the south, representing the Southern Ocean. Similar to Bryan (1986) a prescribed wind stress and mixed boundary conditions for heat and salt were used for both oceans, which were symmetric relative to the equator. Four different stable steady states were found for the same set of boundary conditions. Two of the equilibria showed both oceans in the same state, with formation of deepwater occurring either in both northern or in both southern oceans, respectively. Two additional equilibria existed, in which the THC of the basins differed fundamentally from each other; one ocean formed deepwater a northern high latitudes, while the other ocean had a much weaker circulation, with downwelling motion in the southern hemisphere. The present THC pattern seems to correspond with one of these equilibria. It was demonstrated that a transition from one of the four equilibria to another could be accomplished by relatively small differences in the freshwater fluxes. The preference and sensitivity of these steady states depended critically on the freshwater fluxes applied.

Due to significant improvements in numerical methods, advances in computer capacity, and better data sets for initialization and forcing, OGCMs have become much more realistic over the years. The substantially finer horizontal and vertical resolution in the models (down to ¼° or less, with > 30 layers) allows the numerical diffusion and viscosity coefficients in a model to be chosen that are small enough to avoid fictitious non-physical damping, New methods to represent the vertical coordinate and the bottom topography, like isopycnic coordinates (density as vertical coordinate) or terrain following (or sigma coordinates) are used to represent advective transport and the stepwise representation of the bathymetry. This ultimately has led to a significant expansion in the use of numerical models in the study of the THC, next to the analysis of data from oceanographic observations (Böning and Semtner 2001).

However, it has to be emphasized that there still is a gap between availability of high-resolution OGCMs which even can resolve or at least permit mesoscale ocean eddies and narrow western boundary currents and the type of models required for the study of the role of the THC in climate change. The use of high resolution OGCMs requires heavy investments in computer power. Only a few high-resolution, eddy-permitting OGCMs have been integrated for more than 50 years. Given the timescales involved in the THC and its variability, the hydrographic properties of the abyssal branch of the THC will hardly differ from the initial conditions over such a short period, although the currents may be adjusted to changing external forcing within that integration period. In order to investigate the complete change of the THC to a different near-equilibrium solution, lower-resolution models have to be used where more physical processes are approximated by a parameterization in terms of diffusion and viscosity coefficients.

Another aspect of even the present-day high-resolution OGCMs is that they are still too coarse to resolve important processes like convective water mass formation, which takes place on scales of 1 km (convective plumes) to 100 km (chimneys), while the downwelling due to eddy formation in the rim current of the chimneys also is not resolved. This implies that in OGCMs the effects of a changing surface salinity on the convective formation of deepwater will strongly depend on the way this process is parameterized. The entrainment of lower-density water by plumes of high-density water descending from the arctic and antarctic continental shelves or from the sills in the passages of the Greenland–Scotland Ridge cannot be modeled adequately by the present generation of OGCMs. Also the deep flow through narrow passages in the topography which connect different ocean basins is not completely resolved. Given the anticipated development of computational resources, the OGCMs for climate variability studies with the required long-term integrations will be restricted in the coming years to horizontal resolutions of about 1°. The simulation of important aspects of the "mean" THC depends critically on small-scale processes in the ocean that are sub–grid-scale, even for models with a higher resolution. The model simulations are strongly sensitive to parameterization of such processes. Therefore, the development of improved parameterization schemes is a top priority in climate research. Application of such schemes already has led to significant improvement in realistic model behavior.

12. The THC and different climates

12.1. Climate variability in numerical simulations

The simple models discussed in the previous chapter suggest that the THC and climate may change due to feedback processes. The available oceanographic observations only cover the last 100 years, a period too short to display large changes of the THC. Geologists are able to study past changes in the climate and ocean circulation and numerical modelers are able to simulate such changes with coupled ocean–atmosphere models.

Paleoclimatology is the science that deals with the history of the climate and climate change. Paleoceanography is that part of the paleoclimatology which studies the history of the ocean. Subjects are the development the shape of ocean basins, the temperature distribution in the ocean, shallow and deep currents, and biological productivity. Perhaps the most important result of paleoceanographic research is the reconstruction of past oceanic climatic and environmental conditions for various time intervals and at various levels of precision (Wefer et al. 1999). These fields of science are developing fast. In this chapter some results will be presented that illustrate the developing techniques to reconstruct past climate and the accompanying changes in the THC. During the last 30 years paleoclimatologists and paleoceanographers have documented considerable changes in the hydrography of the THC during the geological past. Although most paleoclimatologists would agree that the past is unlikely to provide true analogues for the future, past climate synopses are considered valuable for confronting the results of modeling experiments or for illustrating global warming (Weaver and Hillaire-Marcel 2004). From the combination of paleoceanographic studies and climate models an improved understanding of the variability of the THC can be expected.

The THC is generally assumed to be an essential factor for the ocean's influence on climate and climate change. The climate is known to vary on a range of timescales, resulting from different key components in the climate system. The THC may respond passively to changed boundary conditions at the sea surface. Via feedback processes as discussed in the previous chapter, the THC also may play an active role in triggering climatic shifts and sustaining specific climatic modes. One of the mechanisms by which this can occur is the increase of precipitation in the North Atlantic region due to a change in the global hydrological cycle. There is observational evidence that during the climatic warming in the second half of the 20th century evaporation in the subtropical oceans increased, while precipitation increased at higher latitudes, changing the meridional salinity gradients (Boyer et al. 2005). At the end of the last ice age melting of ice sheets and sea ice also may have strongly altered the global hydrological cycle on a larger temporal scale.

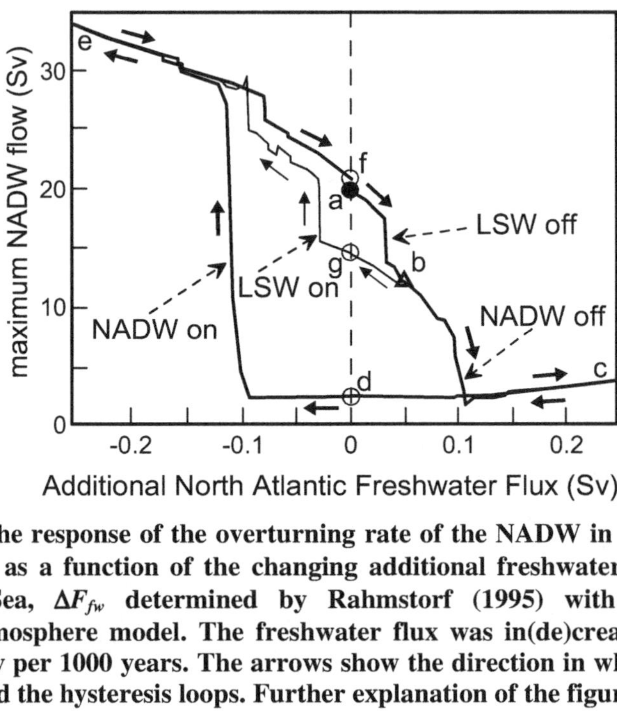

Fig. 12.1. The response of the overturning rate of the NADW in the Atlantic Ocean, Ψ , as a function of the changing additional freshwater flux to the Irminger Sea, ΔF_{fw} determined by Rahmstorf (1995) with a coupled OGCM–atmosphere model. The freshwater flux was in(de)creased linearly with 0.05 Sv per 1000 years. The arrows show the direction in which the system followed the hysteresis loops. Further explanation of the figure is given in the text.

The German climate modeler Stefan Rahmstorf (1995) has carried out interesting numerical experiments with a coarse-resolution OGCM coupled to a simple diffusive atmospheric energy balance model, to determine the near-stationary response of the THC to changes in the hydrological cycle. The resulting overturning rate of the THC, Ψ, defined as the maximum meridional overturning transport anywhere in the Atlantic Ocean, was determined (Fig. 12.1). After a spin-up period of 5000 years, driven with the observed annual mean wind stress and an initial surface flux of heat and freshwater, a steady equilibrium state was reached in this climate model (point a in Fig. 12.1). In this equilibrium state 20 Sv NADW was formed in the North Atlantic Ocean and 18 Sv AABW was formed around Antarctica. From that equilibrium solution an additional precipitation excess was applied south of Greenland (the Irminger Sea), slowly increasing with 0.05 Sv per 1000 years. The additional freshwater at the sea surface in high latitudes increased the vertical stability near the surface and reduced the convective formation of high-density water. Initially Ψ decreased nearly linearly with the additional freshwater flux, ΔF_{fw}. When ΔF_{fw} had reached a value of about 0.035 Sv after 7000 years the formation of LSW collapsed, leading to a sudden reduction in Ψ. With ΔF_{fw} slightly over 0.1 Sv, the formation of NADW collapsed completely. The small remaining value of Ψ was due to the fact that even in the complete absence of NADW formation, the meridional overturning stream function did not become zero or negative everywhere in the Atlantic Ocean.

With a further increase of ΔF_{fw} to 0.25 Sv (point c) the formation rate of NADW stayed zero. When from that point ΔF_{fw} decreased slowly with 0.05 Sv per 1000 years, the NADW formation did not recover directly, even when the freshwater forcing was brought back to the initial value (point d). Only when ΔF_{fw} had reached a negative value below –0.1 Sv the formation of NADW and LSW were turned on suddenly and increased to a value of ~34 Sv for $\Delta F_{fw} = -0.25$ Sv (point e). An increase of ΔF_{fw} from that point onward brought the overturning rate back to a value close to the initial situation for $\Delta F_{fw} = 0$ Sv (point f). The resulting ΔF_{fw}–Ψ curve (thick line in Fig. 12.1) describes the response of Ψ with a hysteresis loop, which has characteristics similar to the hysteresis loops of the box models of the THC by Stommel and Rooth (Figs. 11.7 and 11.9). When the NADW formation broke down, a relatively shallow (~1500 m) single-hemisphere overturning cell was formed in the North Pacific Ocean. When the change of ΔF_{fw} was reversed earlier, at a value of 0.05 Sv (point b), the hysteresis loop (thin line) remained small and more symmetric. Only the LSW formation was shut down, resulting in an overturning rate for NADW of 16 Sv when the freshwater forcing anomaly, ΔF_{fw} was brought back to zero. In this small hysteresis loop the THC switched to the upper branch of this loop when ΔF_{fw} reached –0.05 Sv.

With his model exercises Rahmstorf (1995) has shown that variations in freshwater forcing in the northern North Atlantic may cause fundamental shifts in the mode in which the global THC operates, which are not restored to normal when the freshwater forcing returns back to normal. However, one has to keep in mind that an additional freshwater forcing of

0.1 Sv to the Irminger Sea is equivalent with an additional precipitation of 7 mm/day. This quite high value compared to the actual annual mean value of *P–E* in the Irminger Sea of 1 to 2 mm/day (Fig. 10.8). More climatologic research of the hydrological cycle is required before we can determine whether Rahmstorf's computer simulations used realistic magnitudes of the long-term changes in the freshwater forcing.

Manabe and Stouffer (1999) carried out experiments with a much more advanced global climate model (GCM). They also found evidence for the existence of two distinctive climate modes, with and without NADW formation. However, these climate modes were not recognized in simulations with other modern GCMs (e.g., Schiller et al. 1997; Vellinga and Wood 2002). There also is no agreement among these GCMs about the response of the THC to an altered hydrological cycle due to global warming, although none of these has predicted a THC collapse over the next century (Wood et al. 1999; Latif et al. 2000; Thorpe et al. 2001). The cause of the different outcomes of climate simulations has to be sought in the details of the models, e.g., the horizontal and vertical resolution of the models, the parameterization of air–sea interaction, clouds and precipitation, and diapycnal mixing.

In order to assess the climatic consequences of a shutdown of the Atlantic THC *per se*, Vellinga and Wood (2002) from the British Hadley Center for Climate Prediction and Research in Bracknell have studied the results of a GCM in which the THC was suppressed by a sudden reduction of the salinity of the top 800 m of the North Atlantic Ocean north of 50°N. The $\Delta S = 2$ salinity reduction is equivalent with an addition of a freshwater surface layer with a thickness of 45 m. The GCM that they used was a coupled ocean–atmosphere model with additional sea-ice and land surface schemes. This model maintains a stable surface climate and THC throughout a control run of over 1000 years without the need to correct the air–sea fluxes of heat and freshwater. After the introduction of the surface salinity change the model was allowed to adjust freely to the new salinity field during an integration period of 150 years, and the transient response of the climate model was analyzed. The introduction of the low-salinity surface layer effectively suppressed water mass formation by deep convection in the North Atlantic Ocean, and the overturning of the Atlantic THC collapsed completely. The poleward heat flux due to the THC ceased, causing a reduction of the surface air temperature over much of the northern hemisphere after 20 years (average 1 to 2°C), especially over the northwestern North Atlantic Ocean, where atmospheric cooling well above 5°C was found (Fig. 12.2). For the southern hemisphere, a slight warming was found (average 0.2°C). The strong cooling over the northwest Atlantic, the

Labrador Sea, and the Sea of Okhotsk was caused by increased sea-ice cover that isolated the cold atmosphere from the warmer sea surface and augmented the cooling. The anomaly in sea-ice cover in the Norwegian and Greenland Seas after 20 years was much less (Vellinga et al. 2002). The surface wind field changed, causing enhanced westerlies at the moderate latitudes of the northern hemisphere. The intertropical convergence zone shifted to the south, causing large regional changes in precipitation over South America and Africa. About 120 years after its collapse, the model's THC had largely recovered and most climatic anomalies had disappeared. The reestablishment of the THC was closely linked with the recovery of density in the North Atlantic Ocean, largely determined by the evolution of salinity. Vellinga et al. (2002) identified three important processes for the recovery of the THC. First, the intensified wind-driven subtropical gyre transported salt from the subtropical North Atlantic northward to the mid-latitudes, where the meridional overturning was no longer able to do so. Second, the southward shift of the intertropical convergence zone due to the collapse of the THC created positive salinity anomalies that compensated for the fresher water arriving from the North Atlantic. Finally, the absence of a persistent anomalous sea-ice cover over much of the Norwegian and Greenland Seas allowed intermediate and deepwater formation that helped to start the THC again.

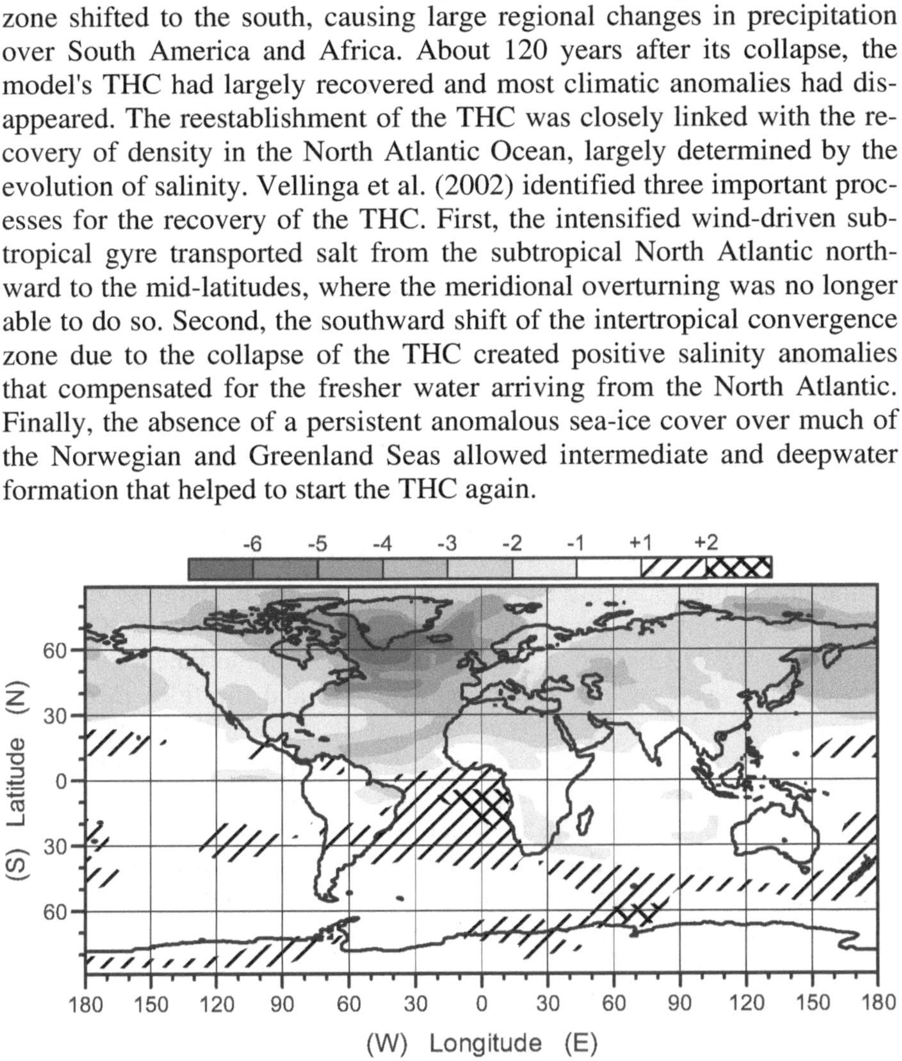

Fig. 12.2. The change in the surface air temperature during 20–30 years after the collapse of the THC in a GCM model simulation. The areas where the air temperature did not change significantly are shown in white. After Vellinga and Wood (2002).

The experiments with GCMs presented here have shown that when the THC is suppressed, the global climate may show a strong response. When the cause of the suppression of the THC is a transient phenomenon, like in the experiment by Vellinga and Wood (2002), the strong response of the climate system may cause a feedback in the hydrological cycle that re-introduces saline water into the Nordic seas which helps to restart the THC. Such a feedback is missing in the atmospheric part of the GCM, used by Rahmstorf (1995), who used a slow, continuous change in the hydrological cycle to force the collapse of the THC instead of a transient one. These GCM studies have revealed the potential importance of the THC in climate change and highlight the need to reduce the uncertainties in climate models with regard to the stability of the THC, including a variety of feedback processes. At present, as stated in the Third Assessment Report of the Intergovernmental Panel on Climate Change (IPCC 2001, p. 474) with regard to the global coupled atmosphere–ocean models, "The very complexity of climate models means that there are severe limits placed on our ability to analyse and understand the model processes, interactions and uncertainties."

One of the less well-understood parts in the OGCMs that are used in climate research is the parameterization of the diapycnal turbulent mixing. Mixing is very crudely represented in climate models, and it is not known whether this deficiency has a significant effect on the model thermohaline circulations (IPCC 2001, p. 487). Since the mixing determines the strength of the upwelling of deep water masses, any change in the deep turbulent regime will result in a change in the overturning rate of the THC, as has been discussed in Chapters 9 and 10. The level of the turbulent kinetic energy in the deep ocean is related to the wind and the tides (Munk and Wunsch 1998). Wunsch (2003) argues that during the last ice age the atmospheric surface temperature gradient was steeper than nowadays. This will have strengthened the zonal wind systems and the winds related to the atmospheric eddy activity. Second, during the maximum of the last ice age the sea level was more than 120 m lower than today. Such a lowered sea level removes most of the strongly dissipative continental shelves from the system, potentially greatly increasing the deep ocean tidal amplitudes. According to Wunsch (2003) it is expected that the increased winds and tides did amplify the mixing rate in the ocean during ice ages, and he anticipates a strengthening of the overturning THC for these cold periods. Better understanding of the deep mixing in the ocean is required to model the THC in past and future climates adequately.

12.2. Paleoclimate changes

Qualitative clues for large climate changes in the past few million years, the ice ages or glacials, can be found in the erosional and depositional traces left behind by large continental ice sheets and glaciers over Europe and North America. More quantitative clues for climate change can be found in the stratigraphic change of so-called "proxies" in oceanic and lake sediments, ice sheets, etc. Proxies in the paleoclimatological sense are measurable descriptors of some climate parameters (element and isotope ratios in sediments and ice sheets, width of tree rings, etc.) which depend on a climatological state in the past and can be regarded as a climate archive. Preferentially, climate proxies have a close relationship to one singular climate variable (the target parameter), e.g., surface air temperature (*SAT*) or *SST*. An important proxy for climate variability is the ratio of the stable oxygen isotopes ^{16}O and ^{18}O. The most common form of oxygen is ^{16}O with eight protons and eight neutrons in its atomic nucleus (~99.8% of all oxygen on earth), while the heavy isotope ^{18}O with 8 protons and ten neutrons is found in 0.2% of the atoms. Water in the atmosphere and ocean can have either isotope of oxygen, $H_2{}^{16}O$ or $H_2{}^{18}O$. Due to the difference in molecular weight of these water molecules, a phase change (evaporation/condensation or freezing/thawing) causes fractionation, which changes the proportion of the heavy oxygen isotope in the remaining water. That proportion of the heavy isotope expressed as the isotopic ratio anomaly $\delta^{18}O$ is defined by means of the isotope ratios of the sample and a reference standard as

$$\delta^{18}O = \left(\frac{(^{18}O/^{16}O)_{\text{sample}}}{(^{18}O/^{16}O)_{\text{standard}}} - 1 \right) \times 1000\,‰ \quad . \tag{12.1}$$

For water generally Standard Mean Ocean Water (SMOW) is used as standard, which also was used as a reference for the development of the EOS-90 of seawater. A sample with a $\delta^{18}O$ of –23.4 ‰ has 2.34% less heavy oxygen (^{18}O) than the SMOW standard.

When water evaporates at the ocean surface, it is easier for the lighter $H_2{}^{16}O$ molecules to leave the water surface than for the heavier $H_2{}^{18}O$ molecules. The latter take more energy than the former. Therefore, the water vapor in the atmosphere will be depleted in ^{18}O compared to the ocean. Any precipitation formed from that water vapor therefore will contain relatively less $H_2{}^{18}O$ than the surface ocean, although it is enriched with this heavier water relative to the water vapor due to the fractionation by the condensation process. With progressive cooling of the air mass and the accompanying condensation, the remaining water vapor will contain less and less $H_2{}^{18}O$. Therefore, precipitation formed in the colder high latitudes will contain less of the ^{18}O isotope than precipitation, formed in the warmer lower latitudes. Thus, as the temperature of an air mass decreases further when it moves polewards, $\delta^{18}O$ in the precipitation decreases monotonically, as in a kind of distillation cycle. In first order the $\delta^{18}O$ decrease is estimated to be proportional to the temperature fall in the air mass. In the present climate this results globally in a linear relationship between the *SAT* and $\delta^{18}O$ of the precipitation (Fig. 12.3). The deviations of the data points from the regression line are related to altitude and to the different source regions of the air mass and the water vapor. Locally, the regression coefficients for the linear relation between *SAT* end $\delta^{18}O$ in precipitation may differ from the global result, shown in Fig. 12.3. The direction coefficient stays of the order of 0.5 ‰ per °C with a tendency to higher values at high latitudes. Assuming that a similar linear relation holds for past climates, we can use the $\delta^{18}O$ values of snow, stored as ice in glaciers or ice sheets, as a surface air temperature proxy for those past climates. In order to use such a relation it is necessary that we can date the ice accurately. Ideal locations for a near-continuous series of $\delta^{18}O$ values are the thick ice sheets of Antarctica and Greenland.

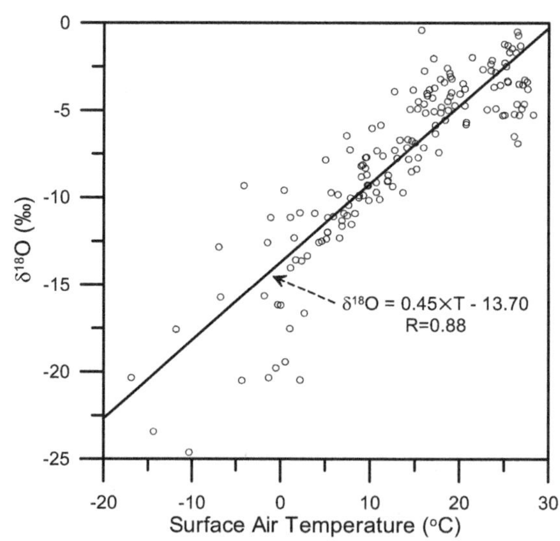

Fig. 12.3. Illustration of the relation between the annual mean surface air temperature and the annual mean $\delta^{18}O$ in precipitation from 177 meteorological stations in the tropics, midlatitudes, and polar regions. The data points have been derived from the seasonal values presented by Fricke and O'Neill (1999).

In 1992, the Greenland Ice Core Project (GRIP) of the European Science Foundation successfully completed the recovery of a 3028 m long ice core in the Summit region of central Greenland (~73°N, 38°W) (Dansgaard et al. 1993). One year later the American Greenland Ice Sheet Project Two (GISP2) completed drilling through the Greenland ice sheet at a site in the Summit region of central Greenland about 30 km further west (Mayewski et al. 1994). These projects have now provided the longest ice-core-derived paleoenvironmental records, >110,000 years, available from the Northern Hemisphere. The similarity between the two cores separated by 30 km provides strong support of climatic origin for even the minor features of the records and implies that investigations of subtle environmental signals (e.g., rapid climate change events with a few years onset and termination) can be rigorously pursued (Grootes et al. 1993).

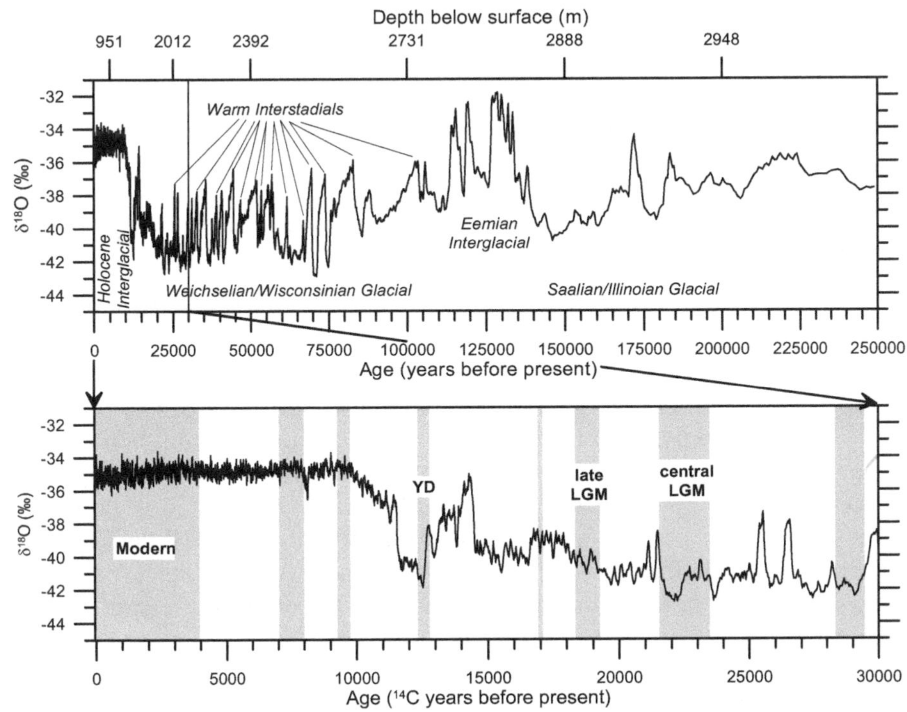

Fig. 12.4. The $\delta^{18}O$ values of the upper 2983 m GRIP ice core sampled every 50 cm (upper panel). This graph shows the cold period (low $\delta^{18}O$) of the last two glacials and the warm Eemian and Holocene interglacials. In the last glacial many short-period (~1500 years) warm interstadials or Dansgaard–Oeschger events are resolved. The lower panel zooms in on the last 30,000 years. The gray bands show intervals for which paleoceanographic evidence on the THC is discussed in this chapter.

It was possible to recognize annual layers in the GRIP ice core to obtain an excellent dating, particularly back to the cold Younger Dryas period (YD, ~12 500 to 11 500 years before present). Parameters to date the core included electrical conductivity as a measure for acidity, dust, nitrate, and ammonium, which all give clear annual layers, particularly in the recent Holocene period. Deeper ice was dated using ice flow models and ^{14}C analysis of the CO_2 in air inclusions in the ice.

The $\delta^{18}O$ values of the ice were determined every 0.5 m. The results of this analysis for the upper 2983 m are shown in Fig. 12.4. The last two ice ages or glacials and the warm interglacial in between appeared to be well resolved. Measurements of borehole temperatures have allowed a recalibration of the oxygen isotope–temperature relation for the GRIP ice core. They indicated that the temperature change at the end of the last glacial period was more than 20 degrees, a result found independently in the GISP2 borehole. This temperature change was significantly higher than the change that follows from the linear regression shown in Fig. 12.3.

The most dramatic finding from GRIP was that of the rapid climate changes (interstadials or Dansgaard–Oeschger events) in the last glacial period. These also had been observed in previous cores, but GRIP and GISP2 confirmed their existence, their number, and the extreme rapidity of their onset in unprecedented detail. In the widespread view, Dansgaard–Oeschger events are at least hemispheric in extent, if not global, and are related to rapid changes in the ocean circulation, in particular changes in the THC (e.g. Broecker et al. 1990; Stocker et al. 1992; Ganapolski and Rahmstorf 2001; Clark et al. 2002).

The possibility of rapid climate change induced by changes in the THC which are caused by global warming and the resulting environmental, economic and political consequences are presently attracting great attention (Broecker 1997, 2003). It has led to multimillion dollar field programs to monitor signs of a collapsing THC, like the British Rapid Climate Change program and the German Meridional Overturning Variability Experiment. However, Wunsch (2005) states that any relation of abrupt changes in the paleoceanographic record with the abrupt $\delta^{18}O$ changes in the Greenland ice cores may occur just by accident because of the similar spectral nature of these changes. He also argues that a reduction of the THC as a cause of an abrupt cooling event is highly improbable, given the relatively weak contribution of the high-latitude ocean to the total meridional heat transport (Fig. 10.5).

12.3. The past THC from oxygen isotopes in marine sediments

The fractionation of the stable oxygen isotopes, which occurs when $H_2{}^{16}O$ preferentially evaporates at the sea surface, will cause an increase of $\delta^{18}O$ in the surface water ($\delta^{18}O_w$) of areas with an evaporation excess. In those regions also the salinity of the surface waters will increase. This suggests a positive correlation of S and $\delta^{18}O_w$ in the surface waters of the ocean, as is illustrated for the modern Atlantic Ocean in Fig. 12.5. For the North Atlantic regions the relation is approximately linear (the straight line in Fig. 12.5). At higher latitudes, where lower surface salinities are found in the Atlantic, there seems to be a bifurcation in the S–$\delta^{18}O_w$ relation, possibly caused by freshwater which dilutes the surface seawater originating from local precipitation, river runoff, or melting land ice. Freezing of surface seawater also causes some fractionation, but its effects are generally assumed to be negligible compared to the effects of evaporation and precipitation. Different linear regressions are found for other ocean basins like the Mediterranean Sea (squares in Fig. 12.5) and the Indian and Pacific oceans. Apparently, the modern S–$\delta^{18}O_w$ relation of the ocean surface layer is determined by the hydrological cycle in the present-day climate.

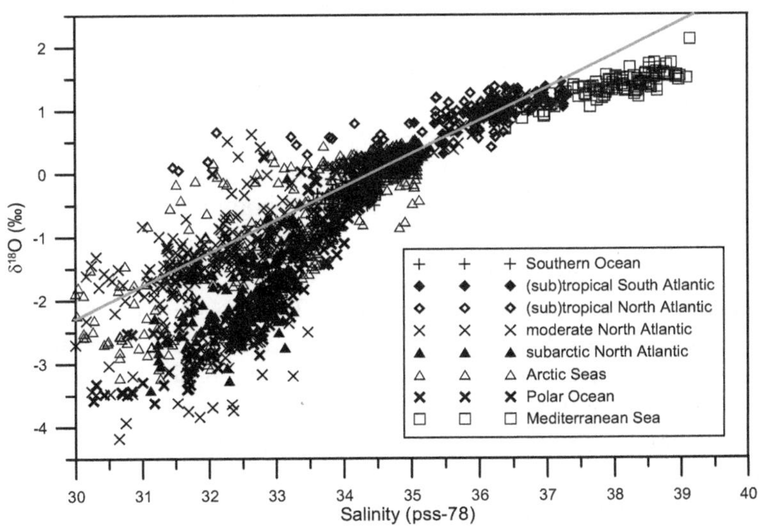

Fig. 12.5. The salinity–$\delta^{18}O$ relation for the surface waters (upper 100 m) of several regions of the Atlantic Ocean, including the Atlantic sector of the Southern Ocean and the Mediterranean Sea. The data used for this plot are from the ^{18}O data base of Schmidt et al. (1999). The straight line is the linear regression line for the equatorial and North Atlantic Ocean.

In marine sediments, remains of calcareous ($CaCO_3$) shells of small marine organisms are found that also contain both oxygen isotopes. We can determine also an isotopic ratio anomaly $\delta^{18}O$ for such shells ($\delta^{18}O_c$). For these determinations the standard reference material is normally a fossil marine belemnite of Cretaceous age (100 Myr) from the Pee Dee formation in Tennessee; hence: PDB. When marine organisms secrete a hard shell, isotope fractionation also occurs and the shell has less ^{18}O than the seawater from which it was formed. The magnitude of the fractionation, although slightly species dependent, depends on the temperature of the water. This gives us the possibility to use $\delta^{18}O_c$ from shells as a proxy to get information on the paleoenvironment in which the organisms lived. Generalized the dependence on environmental factors of $\delta^{18}O_c$ in calcite shells from marine sediments can be written as (Wefer et al. 1999)

$$\delta^{18}O_c = f(T_{sur}, Ice, Sal, Sel, Vit, Dis) \quad . \tag{12.2}$$

In this expression, T_{sur} is the temperature of the surrounding water when the shell is formed. The parameter "*Ice*" represents the influence of the relative excess of ^{16}O, stored in large continental ice sheets (Fig. 12.4), while "*Sal*" represents the influence of regional $\delta^{18}O$ variations in the seawater, connected with salinity variations (Fig. 12.5). "*Sel*" stands for the effects of selective growth in time and space, "*Vit*" for vital factors, including species effects, and "*Dis*" represents isotope effects of post-depositional dissolution. Oxygen isotopes are a common proxy for the determination of past oceanic temperatures. As a rule of thumb, an increase in the $\delta^{18}O_c$ in the calcite of a shell by 1‰ corresponds to a temperature decrease of ~4 °C. Equation (12.2) indicates that $\delta^{18}O_c$ also depends on other, partially unknown factors. So it is not completely possible to invert an observed $\delta^{18}O_c$ value to a temperature according to Eq. (12.2) unless we have all additional information on the environmental conditions in which the shell-forming organism lived and died. The most important condition is whether the organism had a planktonic or benthic lifestyle. Planktonic organisms have the environmental conditions in the near-surface photic zone recorded in their shells, while shells from benthic organisms in the sediment represent the environmental conditions of the local bottom water.

Foraminifera are small shell-forming organisms. They occur as free floating zooplankton species in the upper layers of the ocean and as bottom-dwelling benthic species. Foraminifera are regularly used in paleoceanography since their shells contain the oxygen isotopes which can be used as a proxy. The signal we can discern from the $\delta^{18}O_c$ distribution from benthic foraminifera in sediment cores resembles the cold–warm alternation of the glacials and interglacials (Fig. 12.6, upper panel). For the early period of the Pleistocene from 1.8 My to 1.2 My BP, the typical glacial–interglacial cycle has a periodicity of about 41,000 year, while the late Pleistocene period 0.6 My BP till present the dominant periodicity is a 100,000 year glacial–interglacial cycle. The period from 1.2 My to 0.6 My BP can be seen as a transition period. The cause of this periodicity shift is unexplained yet.

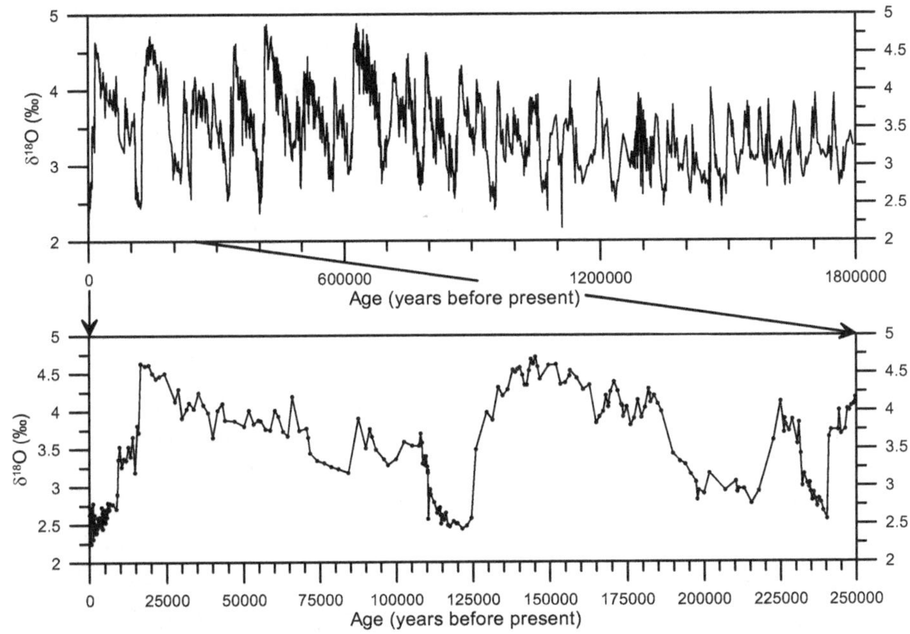

Fig. 12.6. The temporal variation of $\delta^{18}O$ determined from shells of benthic foraminifera sampled from the upper 109 m of the Deep Sea Drilling Project sediment cores 980/981 (upper panel) at a water depth of 2170 m in the Rockall Trough which covers a considerable part of the Pleistocene (Raymo et al. 2004). The lower panel shows a blowup of the top 32 m, coinciding with the time interval of the upper panel in Fig. 12.4. The data were downloaded from http://www.maureenraymo.com/climate_archives.php.

A blowup of the last 250,000 year (Fig. 12.6, lower panel), coinciding in time with the GRIP core, shows that $\delta^{18}O_c$ in the ocean sediments have a similar glacial–interglacial periodicity as the Greenland ice sheet, but with a smaller amplitude and opposite sign. If this $\delta^{18}O_c$ variation was completely determined by temperature changes only, the temperature in the north Atlantic at ~2000 m would have been about 8°C colder than at present. But since the present temperature at that depth is less than 3.5°C that implies temperatures well below the freezing point. Variations of $\delta^{18}O_w$

from the surrounding water, also a term in Eq. (12.2) via the "*Ice*" and the "*Sal*" effects will have played the major role in the observed variation. This is due to the fact that the continental ice sheets of the glacials contained water which went through the isotope fractionation cycle of evaporation and precipitation formation. The ice sheets, remnants from snow in the past, contained less of the heavy ^{18}O isotope, as still can be seen in the surviving Greenland ice sheet (Fig. 12.4). The oceans from which that water had evaporated not only became more saline than the modern Holocene ocean but also attained higher $\delta^{18}O_w$ values. This "*Ice*" effect in Eq. (12.2) has enabled paleoceanographers to use $\delta^{18}O_c$ as a timing device for sediment cores, since its record is found in ocean sediment cores all around the world.

Assuming that we can derive $\delta^{18}O_w$ from estimates of the magnitude of the size of continental ice sheets from the past, and ignoring the other factors influencing $\delta^{18}O_c$, an expression for the paleotemperature, derived from foraminifera shells is (Epstein et al. 1953)

$$T = 16.5 - 4.3(\delta^{18}O_c - \delta^{18}O_w) + 0.14(\delta^{18}O_c - \delta^{18}O_w)^2 \quad . \qquad (12.3)$$

The typical glacial–interglacial variation of the ocean's mean $\delta^{18}O_w$, connected with the formation and melting of high-latitude continental ice sheets amounts to 1.2‰ (Wefer et al. 1999; Bintanja et al. 2005).

Labeyrie et al. (1992) collected a series of $\delta^{18}O_c$ values from benthic foraminifera at a range of depths and latitudes in the North Atlantic Ocean. With these data they constructed characteristic $\delta^{18}O_c$ profiles for the (sub-)tropical (0–26°N) and moderate (29–54°N) North Atlantic, for the modern ocean as well as for the Last Glacial Maximum and Younger Dryas (Fig. 12.7).

For the modern Atlantic Ocean (Fig. 12.7, left panel) the data show nearly constant $\delta^{18}O_c$ values below 2000 m, indicative of the low vertical temperature gradients encountered there. At intermediate levels (750 to 2000 m) the stronger vertical $\delta^{18}O_c$ gradient agrees with the increased temperature gradient. The difference in $\delta^{18}O_c$ at intermediate levels between the (sub-)tropical and moderate Atlantic Ocean indicates an increased temperature at higher latitudes. This feature also is visible in the meridional temperature section, based on hydrographic observations (Fig. 4.2, upper panel), and is attributed to the presence of warm and saline MSOW between 30 and 50°N. The change in $\delta^{18}O_w$, associated with the northward salinity increase of the intermediate water, has an effect on the change in $\delta^{18}O_c$ opposite to the temperature effect but of smaller magnitude. Given the scatter of the data below 2000 m, the expected slope of 0.32 ‰/3000 m (the thin line in Fig. 12.7, left panel), derived from the modern temperature and salinity profiles, could not be rejected statistically. Apparently, the modern $\delta^{18}O_c$ distribution from the benthic foraminifera in the upper layers of the sediment agrees with the known hydrography of the North Atlantic.

The $\delta^{18}O_c$ profile for the Last Glacial Maximum (LGM) (Fig. 12.7, middle panel) shows an analogy with the modern profile, but shifted +1.8 ‰. With an estimated shift of the global mean $\delta^{18}O_w$ for the LGM of about +1.2 ‰ relative to the modern values, which leaves an additional shift of +0.6‰. If this shift is mainly determined by a change in temperature of the deep and intermediate water, the deep North Atlantic during the LGM was about 2.6°C cooler than the modern deep ocean. Another difference between the modern and the LGM North Atlantic is the absence of a meridional $\delta^{18}O_c$ gradient at intermediate levels during the LGM. Apparently the relatively warm MSOW at intermediate levels was absent during the LGM, with a ~100 m decreased sill depth near Gibraltar or the distribution and properties of that water type differed relative to the modern situation. The MSOW core was replaced by a colder intermediate water mass, probably with a northern origin.

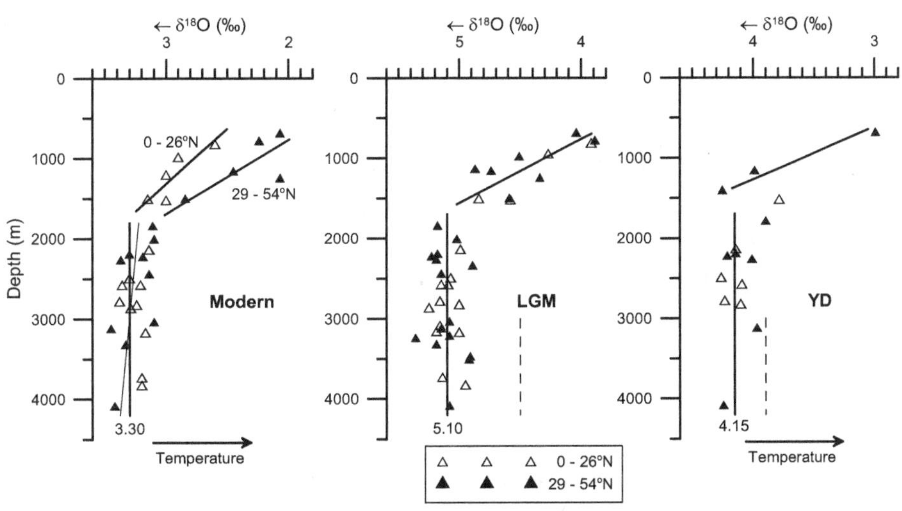

Fig. 12.7. Distributions with depth of $\delta^{18}O_c$ of benthic foraminifera for the (sub-)tropical (open symbols) and moderate North Atlantic Ocean (black symbols) from the modern ocean (left panel), the Last Glacial Maximum (middle panel) and the Younger Dryas (right panel). The straight lines give the approximate characteristic profiles. The dashed lines in the middle and right panels give the deep $\delta^{18}O_c$ values corrected for the estimated change due to water storage in continental ice sheets with +1.2 and +0.6‰, respectively. After Labeyrie et al. (1992).

For the Younger Dryas (YD) fewer data were available in the data set used by Labeyrie et al. (1992), especially for the intermediate levels above 2000 m (Fig. 12.7, right panel). At the levels of the deepwater, the $\delta^{18}O_c$ is shifted about +0.85 ‰ relative to the modern situation. The extent of the continental ice sheets during the YD and the accompanying shift of the global mean $\delta^{18}O_w$ are not precisely known, but a shift in $\delta^{18}O_w$ of +0.6,‰ half the shift from the modern to the LGM, seems to be a reasonable estimate. The remaining shift in $\delta^{18}O_c$ of 0.25‰ agrees, according to Eq. (12.3), with deepwater 1.1°C cooler than in the present situation.

The interpretation of Fig. 12.7 in terms of changed water mass properties is not straightforward, since the deep salinity is not known from proxies. The lower temperatures of the deep North Atlantic for the LGM and the YD do not directly imply that the NADW in those periods was colder than at present. A similar effect can be reached by an upward shift of the NADW and a stronger presence of the even colder AABW in the deeper layers. Additional proxies, reflecting the overturning of the THC, are required for the interpretation of the data in terms of the strength of the THC.

12.4. Stable carbon isotopes and the Atlantic paleo-THC

Nonconservative tracers like *AOU*, nutrients, *TIC*, and $\delta^{14}C$ are used to determine the aging of the deepwater in the ocean, and thereby enable oceanographers to determine the spreading path of the deep branch of the *THC* in the modern world ocean, as is described in Chapters 4 and 5. An important geochemical tracer for aging is the ratio of the stable isotopes ^{13}C and ^{12}C, expressed as

$$\delta^{13}C = \left(\frac{(^{13}C/^{12}C)_{\text{sample}}}{(^{13}C/^{12}C)_{\text{standard}}} - 1 \right) \times 1000 \text{ ‰} \quad . \tag{12.4}$$

As reference standard for $\delta^{13}C$ in calcite shells generally the PDB belemnite is used.

Connected with the aging of the deepwater is a change in the isotopic constitution of the *TIC*. In the present ocean values of $\delta^{13}C$ of the dissolved *TIC* decrease with increasing *TIC*, *AOU*, and nutrient concentrations, and with decreasing $\delta^{14}C$ (Fig. 12.8). These changes of the stable carbon isotope concentration are linked by deepwater circulation to the isotope fractionation which occurs during primary production in the surface layers of the ocean. The newly produced organic matter contains less of this isotope ($\delta^{13}C \approx -20$ ‰), while the surface water in contact with the atmosphere is slightly enriched in the heavy isotope ^{13}C ($\delta^{13}C \approx +1$ ‰). When this organic matter is remineralized in the deep ocean, the dissolved *TIC* will gain decreasing values of $\delta^{13}C$ with increasing age (Wefer et al. 1999). Since the ^{13}C depletion of deepwater is part of the oceanic carbon cycle, the best linear property–$\delta^{13}C$ fit is expected for *TIC*, which is confirmed by the data (Fig. 12.8, upper left panel). The correlation with *TIC* amounts to 0.97. Notwithstanding the superiority of this $\delta^{13}C$–*TIC* connection, $\delta^{13}C$ is often presented as a proxy for the paleonutrient distribution.

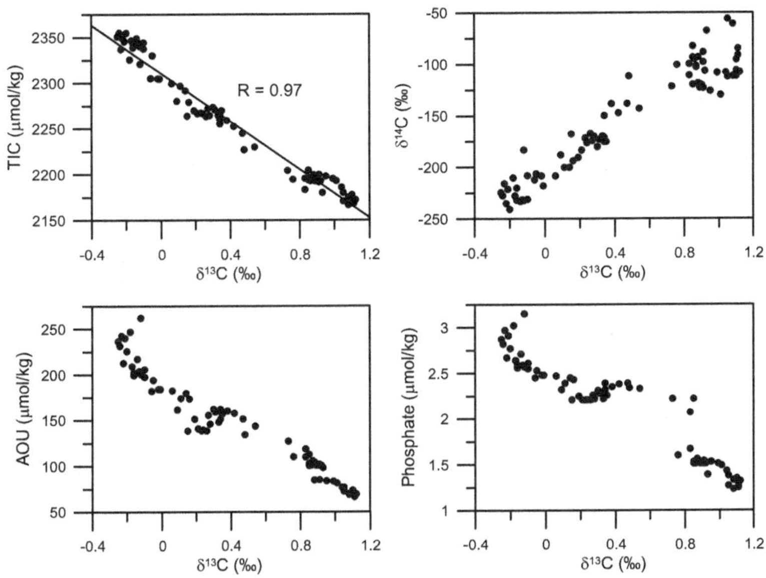

Fig. 12.8. $\delta^{13}C_w$–property plots for deepwater (>1000 m) of WOCE stations from the North and South Atlantic, Southern Ocean, and South and Equatorial Pacific Ocean.

Several reconstructions of the oceanic $\delta^{13}C$ structure of the ocean basins have been made (e.g. Duplessy et al. 1988; Bickert 1992; Sarnthein et al. 1994). From the $\delta^{13}C$ data from benthic foraminifera, published by Sarnthein et al. (1994), the meridional $\delta^{13}C$ structure in the eastern Atlantic Ocean for the modern period, and for the LGM can be drawn (Fig. 12.9). The $\delta^{13}C$ data were derived from sediment cores that were mainly located along the eastern margin of the Atlantic Ocean.

For the modern period of the last 4000 years, the $\delta^{13}C$ structure agrees with the *AOU* and nutrient structure derived from hydrographic observations (see, e.g., Fig. 4.10). The dominant deepwater type is the relatively young southward flowing NADW, centered near about 3000 m. In the bottom layer the north flowing AABW is encountered, characterized by relatively low $\delta^{13}C$ values. During the LGM the relatively young water (high $\delta^{13}C$) also seems to have a northern source near the Iceland–Scotland Ridge. However, this high-$\delta^{13}C$ Northern Component Water (NCW) or Glacial North Atlantic Intermediate Water (GNAIW) does not descend deep into the Atlantic Basin, contrary to the modern NADW. Near the equator its core is found at an approximate depth of about 1000 to 1500 m. In the modern THC this depth stratum is occupied by the aged low-$\delta^{13}C$ AAIW/UCDW. In the deepest layers a low-$\delta^{13}C$ Southern Component Water (SCW) is the dominant water in the Atlantic during the LGM. Its properties are reminiscent of those of AABW, but with definitely lower $\delta^{13}C$ values (< –0.8 ‰).

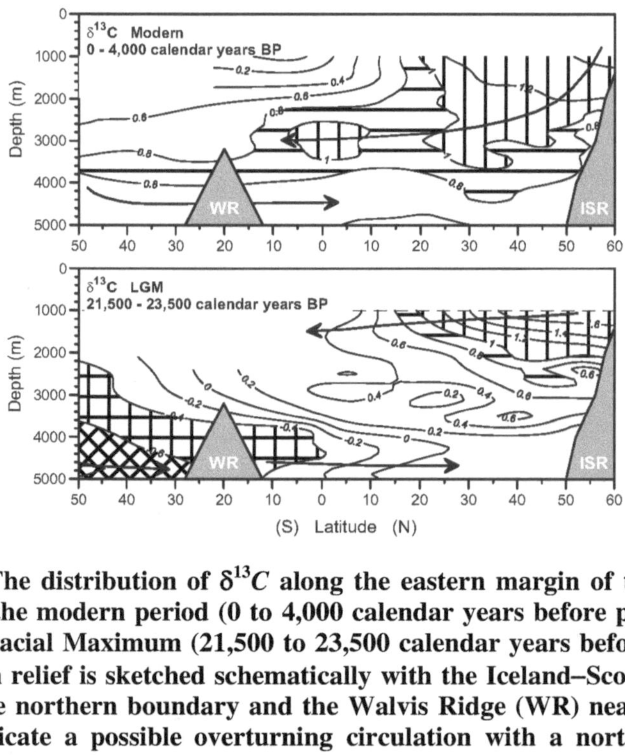

Fig. 12.9. The distribution of $\delta^{13}C$ along the eastern margin of the Atlantic Ocean for the modern period (0 to 4,000 calendar years before present) and the Last Glacial Maximum (21,500 to 23,500 calendar years before present). The bottom relief is sketched schematically with the Iceland–Scotland Ridge (ISR) at the northern boundary and the Walvis Ridge (WR) near 20°S. The arrows indicate a possible overturning circulation with a northern and a southern source. The data used for these sections are from the shells of the benthic foraminifera *Cibidoides wuellerstorfi* referenced to a PDB standard, published by Sarnthein et al. (1994).

Curry and Oppo (2005) analyzed a number of $\delta^{13}C$ profiles for the LGM from the western Atlantic Ocean basins based on sections of sediment cores across the North and South American continental slopes. These basins are the principal flow path of the major deep water masses today and during the LGM, because of the deep transport in a DWBC. This analysis confirmed that the deep water mass structure agreed with the structure in the eastern basins, shown in Fig. 12.9, with strongly ^{13}C, depleted SCW in the abyssal ocean and high-$\delta^{13}C$ GNAIW at intermediate levels. The GNAIW core reached to at least 30°S. Additionally they found evidence for the presence of a shallow LGM equivalent of an AAIW/UCDW core along the Brazilian continental slope at a depth of 1150 m. Curry and Oppo (2005) interpreted the large gradients in $\delta^{13}C$ between the GNAIW and the SCW compared with the modern $\delta^{13}C$ gradients between NADW and AABW as evidence of a vigorous overturning circulation in the Atlantic during the LGM. That vigorous overturning circulation is required to maintain the large gradients in the presence of probably enhanced turbulent mixing during the LGM (Wunsch 2003).

Between the shallow GNAIW core and the deep SCW core a near-homogeneous water mass ($0.2 < \delta^{13}C < 0.6$ ‰) with a thickness of 2 to 3 km was found in the deep eastern North Atlantic Ocean (Fig. 12.9). It is not clear whether the origin of this water mass was in the south or the north of the Atlantic, or whether it was a mixing product of SCW and GNAIW. Note that it is estimated that during the LGM the mean $\delta^{13}C$ in the world ocean decreased about 0.32 ‰ relative to the modern ocean, because of a change in the total amount of carbon stored on the continents as biomass and organic and inorganic carbon (Duplessy et al. 1988). The $\delta^{13}C$ difference between the modern AABW and the SCW during the LGM, however, is larger.

To study the variability of the $\delta^{13}C$ structure on shorter timescales we have constructed $\delta^{13}C$ profiles for a near-equatorial strip (0 to 10°N) in the North Atlantic Ocean (Fig. 12.10), based on data compiled by Sarnthein et al. (1994) for the eight time slices, indicated in Fig. 12.4. As we have seen from modern hydrographic data, the deepwater from the DWBC spreads eastward in this latitude band, and therefore the profiles are assumed to be characteristic for the Atlantic THC. The three modern profiles from the Holocene show a strong similarity with a high-$\delta^{13}C$ NADW core centered around 3000 m, an aged low-$\delta^{13}C$ intermediate core (probably AAIW/UCDW) near 1500 m, and aged low-$\delta^{13}C$ water (AABW) near 5000 m. The present THC seems to be representative for the Atlantic THC during the whole Holocene. The three profiles from the end of the last glacial (~17,000 to 24,000 BP) also show a common structure, with a core of relatively young (high $\delta^{13}C$) GNAIW near 1500 m located over a thick layer of aged water with a $\delta^{13}C$ maximum near ~4500 m. This agrees with SCW, spreading northward along the bottom while mixing slowly with the overlying NCW/GNAIW during upwelling from the bottom layer. The upwelling and mixing may lead to the observed $\delta^{13}C$ minimum above the bottom level, as described in Chapter 9. During the oldest available time slice from the glacial (28,300 to 29,500 years BP, open triangles in Fig. 12.10) the situation differed from the LGM. The $\delta^{13}C$ maximum, characteristic for NCW, was centered deeper, near about 3000 m, similar to the modern NADW. At intermediate levels between 1000 and 1500 m evidence for aged ^{13}C-depleted water was found. The near-bottom SCW had $\delta^{13}C$ values in between the modern and the LGM values at ~5000 m, while the $\delta^{13}C$ maximum was found at a depth of ~3900 m. Apparently the THC did vary also on (multi-) millennium timescales, shorter than the glacial–interglacial cycle. During the cold YD (open diamonds) the $\delta^{13}C$ profile had a shape in between the characteristic profiles of the modern interglacial and the LGM.

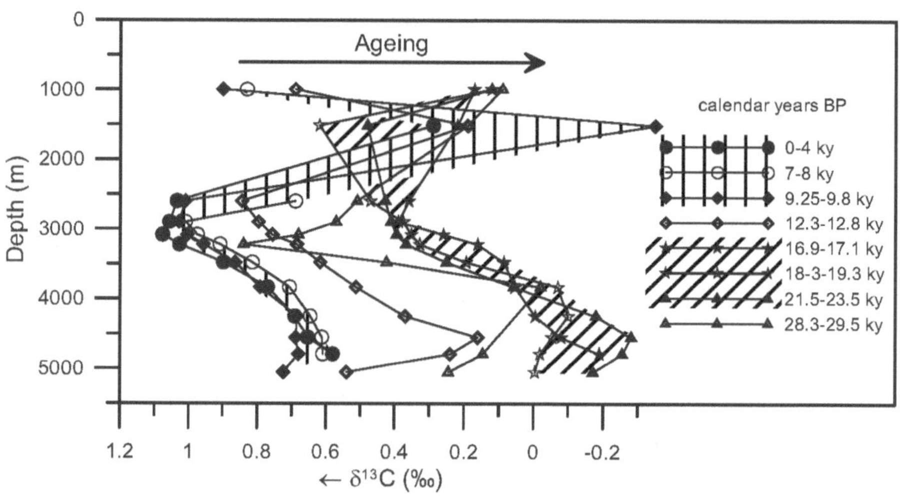

Fig. 12.10. Vertical profiles of $\delta^{13}C$ from the shells of benthic foraminifera in the near-equatorial North Atlantic (0–10°N, ~20°W) for the eight time slices shown by gray bands in Fig. 12.4. The data used for these profiles are from the shells of *Cibidoides wuellerstorfi* referenced to a PDB standard, published by Sarnthein et al. (1994).

It has to be noted here that the difference of the SCW between the Holocene and the LGM does not necessarily imply that the THC transport rates of the SCW or AABW and the NCW or NADW were slower during the LGM, although the characteristic deep $\delta^{13}C$ values were generally lower, interpreted as indicative for more aged water. Changes in the carbon cycle of the ocean also may have induced changes in the $\delta^{13}C$ characteristics of the Atlantic water types. The interpretation, therefore, of the $\delta^{13}C$ distribution during past climatic periods as characteristic for aging still has a strong qualitative character. A unique quantitative interpretation of the paleodata does not seem to be possible yet.

Above, we noted the possibility that within the last glacial the Atlantic THC may have shown some short-term variation, with different deep circulations before the LGM and during the YD. Raymo et al. (2004) analyzed low-frequency variations of reconstructed $\delta^{13}C$ profiles from the northern North Atlantic for the last 1.8 million years of the Pleistocene obtained from seven cores of the Deep Sea Drilling Project/Ocean Drilling Program (DSDP/ODP) in the northeastern North Atlantic Ocean to study long-term development of the glacial and interglacial characteristics. The water depths spanned the 1135 to 3427 m depth interval. From the $\delta^{18}O_c$ of benthic foraminifera the glacials and interglacials were easily recognizable (Fig. 12.6).

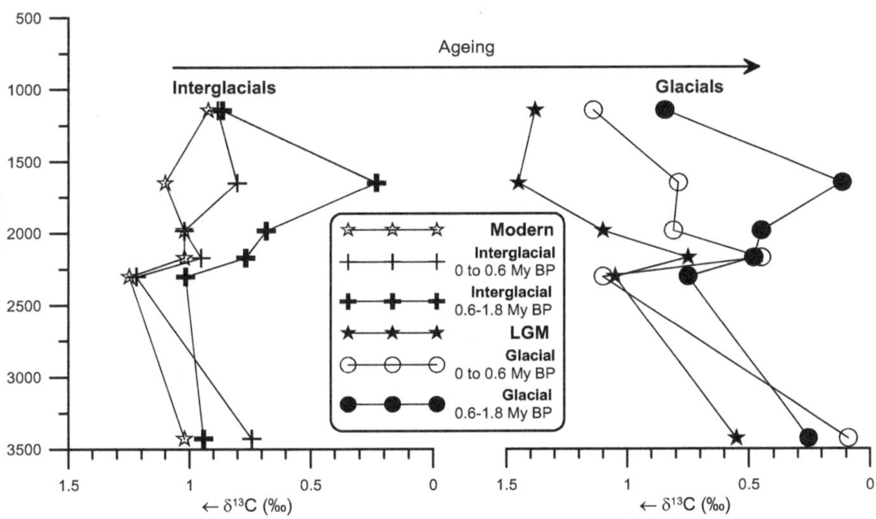

Fig. 12.11. Averaged $\delta^{13}C$ profiles of the shells of benthic foraminifera for interglacials (left panel) and glacials (right panel) in the northeastern North Atlantic Ocean over the periods 1.2 to 0.6 My and 0.6 My to present. For comparison the 10,000 years mean profiles of the present interglacial (open stars) and the LGM (black stars) were added. The data for these profiles were taken from Raymo et al. (2004).

For each glacial and interglacial period they determined the 10,000 year mean $\delta^{13}C$ value centered around the $\delta^{18}O_c$ maximum or minimum. It appeared that for the period 1.8 My to 0.6 My BP, covering the dominant

41,000-year periodicity and the transition period, characteristic glacial and interglacial $\delta^{13}C$ profiles emerged. For the last 0.6 My, the late Pleistocene, the mean glacial and interglacial profiles generally were also mutually coherent but different from the previous 1.2 My. The averaged $\delta^{13}C$ profiles for interglacials and glacials in both periods are shown in Fig. 12.11. In the depth interval where presently LSW is found in the northern North Atlantic (1000 to 2000 m) a layer with strong $\delta^{13}C$ depletion ($\delta^{13}C \approx 0.1$ to 0.2 ‰) was found in the period 1.8 to 0.6 My for the glacials as well as the interglacials, centered around 1640 m. During the last 0.6 My this ^{13}C-depleted core was still present but with higher $\delta^{13}C$ values (~0.8 ‰). Raymo et al. (2004) argued that the water at this depth level had an arctic origin, but with a high TIC concentration and low ^{13}C:^{12}C ratio due to a limited air–sea exchange of CO_2, because the arctic seas in those days were nearly permanently ice covered also during interglacials. This contrasts the present interglacial, where the Barents Sea, the Greenland Sea, and the Labrador Sea are largely ice free in summer. At a depth of 3427 m a considerable influence of AABW is found in the present northeastern North Atlantic Ocean (McCartney 1992; van Aken 2000). The strongest glacial–interglacial $\delta^{13}C$ shift of about 0.7‰ was found at this depth. This may reflect either shoaling of the SCW core during glacials or a change of the hydrographic source water properties of this core.

A more or less permanent feature was an intermediate $\delta^{13}C$ minimum near 2170 m derived from cores in the Iceland Basin. Presently the ISOW core is found near this depth in the Iceland Basin. The $\delta^{13}C$ minimum suggests that this reasonably well-ventilated water, possibly overflow water from the Norwegian Sea, was a permanent feature in the northeastern North Atlantic for most of the Pleistocene. We can compare the averaged $\delta^{13}C$ profiles with the profiles for the modern Holocene ocean and the LGM (stars in Fig. 12.11). The main differences are found in the depth interval of the present LSW core, 1000 to 2000 m. During the Holocene and the LGM a well-ventilated, high-$\delta^{13}C$ core was found in this depth interval ($\delta^{13}C = 1.45$ ‰ for the LGM, 1.10 ‰ for the Holocene), contrary to the $\delta^{13}C$ minima from the averaged profile for the last 0.6 My. Raymo et al. (2004) concluded that the glacial–interglacial signal may be more influenced by the production rate of deepwater around Antarctica rather than by changes in the NCW flux or the strength of the THC. Possibly NCW variability played a relatively passive role in global climate changes over much of the last 1.8 million years. The northern hemisphere high-latitude climate might have been forced by climate variability originating at low latitudes or from the region of the Southern Ocean, transmitted via the atmosphere. During the LGM and the present interglacial $\delta^{13}C$ profiles of the intermediate waters (1000 to 2000 m) in the North Atlantic apparently were atypical, possibly linked to recent changes in the arctic ice cover.

12.5. Cadmium and barium as paleoceanographic tracers of the THC

It has been known for some time that the toxic heavy metal cadmium (*Cd*), like other trace metals, tends to be present in the ocean proportional to essential macronutrients like phosphate and nitrate (Boyle and Keigwin 1982). Recently it has been shown that cadmium is used by diatoms and other phytoplankton species as well as bacteria to form enzymes that make CO_2 available for primary production from dissolved *TIC* (Cullen et al. 1999; Lane et al. 2000a, b). Consequently water loses dissolved inorganic cadmium and phosphate during primary production near the ocean surface. The newly formed organic matter, with a *Cd*:phosphate Redfield ratio of $3.5\ 10^{-4}$, settles out of the surface layer and is re-mineralized at depth when the deep water masses age. As a consequence, the aged deep water mass contains less oxygen and more *Cd* and phosphate. When forming their shells, benthic foraminifera also incorporate *Cd* from the bottom water into the calcium carbonate matrix. These processes make the *Cd* content of diatom shells a good proxy for the paleonutrient distribution, with a more direct link to nutrients than $\delta^{13}C$ which also depends on air–sea gas exchange of CO_2.

The relation of *Cd* enrichment of the calcite shells of foraminifera with the *Cd*/*Ca* ratio of the surrounding water shows no obvious offsets between species. In deepwater (depth > 3000 m) the empirical distribution coefficient *D*, defined as

$$D = \frac{(Cd/Ca)_{\text{foram}}}{(Cd/Ca)_{\text{water}}} \quad . \tag{12.5}$$

has a value of 2.9, while at depths shallower than about 1150 m *D* has a value of about 1.3. This vertical change of *D* mainly is due to pressure effects. Measured temperature effects on *D* are much smaller (Boyle 1992).

Boyle and Keigwin (1982) studied the geochemistry of a sediment core from the North Atlantic Ocean, collected over the western flank of the MAR at a depth of about 3200 m, close to the depth of the present-day core of NADW. The *Cd* content of the shells of benthic foraminifera was expressed as the *Cd*/*Ca* ratio in μmol/mol. While the distribution of $\delta^{18}O$ in the core reflected the last two glacials, the more or less similar vertical distributions of the *Cd*/*Ca* ratio and $\delta^{13}C$ in the sediment core showed that at the present NADW level the deepwater was more aged during the glacials than during the interglacials (the first three panels in Fig. 12.12). Apparently the *Cd*/*Ca* ratio as well as $\delta^{13}C$ in the shells of benthic foraminifera can be used as a proxy for the aging of deep water masses. Both proxies are involved in the primary production near the surface and remineralization of organic matter at depth, similar to dissolved oxygen, *TIC*, and nutrients like phosphate and nitrate.

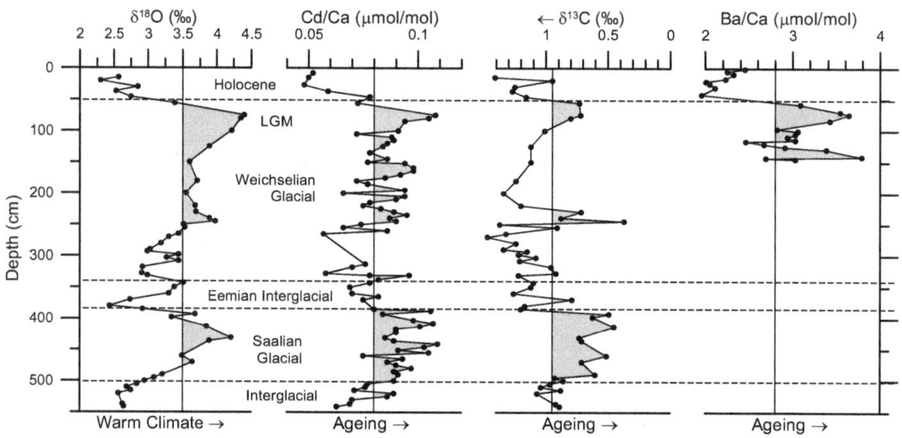

Fig. 12.12. The vertical variation of $\delta^{16}O$, *Cd*/*Ca*, $\delta^{13}C$, and *Ba*/*Ca* in a sediment core collected on the west flank of the Mid-Atlantic Ridge at 42°N, 32°W, at a depth of 3209 m. The $\delta^{16}O$ variation shows the alternation of interglacials and glacials (after Boyle and Keigwin 1982: Lea and Boyle 1990).

Another type of nutrient used in hydrographic research is dissolved silicate. This nutrient is not incorporated in organic matter, but in biogenic opal shells (SiO_2 nH_2O) in some types of plankton, e.g., diatoms. These hard parts are recycled less efficiently than the labile organic matter and a large part of the *Si*-containing particles end up in the sediment. Dissolution of biogenic opal in the sediment results in a flux of dissolved silicate from the sediment to the overlying deepwater. This difference of the *Si* cycle, compared to the cycles of oxygen, nitrate, and phosphate, is the cause of the differences in the aging signature of deepwater, as shown in, e.g., Fig. 4.9. No method exists for the direct reconstruction of the paleosilicate distribution, but Lea and Boyle (1989, 1990) discovered similarities between the cycling of barium (*Ba*) and silicate. *Ba* also is removed from surface waters by uptake onto biogenic particulate matter, apparently in the form of barite ($BaSO_4$). Unlike the vast portion of its organic matter host, barite is not remineralized in the upper water column by microbial activity but is transported into the deep ocean. There it is dissolved in the deepwater and/or the sediment, similar to the silicate shells. This similarity generates a *Ba* distribution in the present ocean coherent with the *Si* distribution (Fig. 12.13). The present concentrations of *Ba* in deepwater in the North Atlantic Ocean are about 55 nmol/kg and increase to about 150 nmol/kg in the North Pacific Ocean (Lea and Boyle 1990).

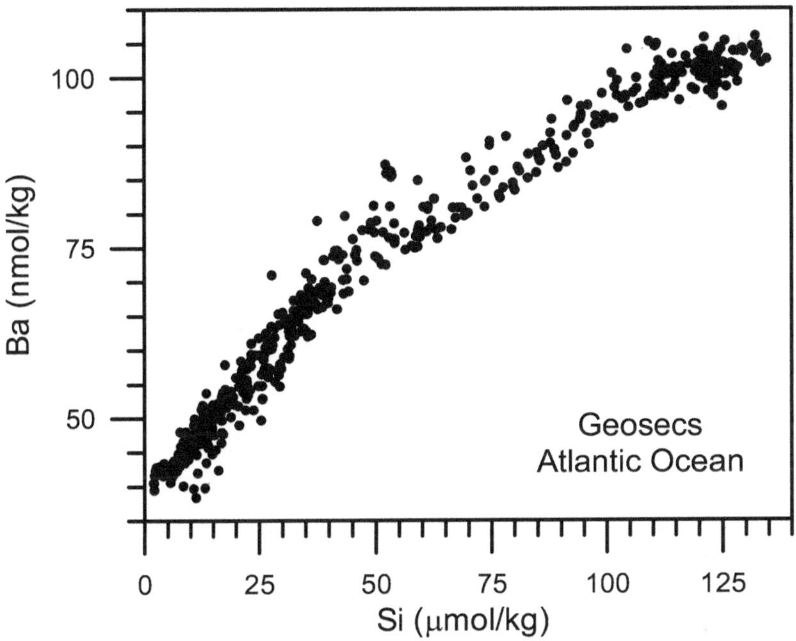

Fig. 12.13. The relation between dissolved barium and silicate in the Atlantic Ocean below 250 m depth. The data for this plot are from the GEOSECS Atlantic survey.

Like cadmium, barium also is incorporated into the calcite matrix of the shells of benthic foraminifera. The similar distribution of barium and silicate in the deep ocean makes the barium content of shells of benthic organisms a possible proxy for the reconstruction of the silicate distribution in past climatic periods. The barium content generally is expressed as a *Ba/Ca* ratio in μmol/mol. Lea and Boyle (1990) have presented the vertical *Ba/Ca* distribution in the top 145 cm of the North Atlantic sediment core (Fig. 12.12, right panel) which also was described by Boyle and Keigwin (1982). This proxy shows that the transition from the LGM to the Holocene at the level of the present NADW core in the North Atlantic was characterized by a decrease in the *Ba/Ca* ratio. This confirms the evidence for a relatively aged deep North Atlantic water mass during the LGM, compared with the modern ocean.

12.6. Stable carbon isotopes in the Southern Ocean

Charles and Fairbanks (1992) analyzed proxies from sediment core RC11-83, collected with a piston corer in the Southern Ocean near 42°S, 10°E, southwest of South Africa at a depth of 4718 m. The temporal development of the proxies reflected a sudden increase of $\delta^{13}C$ of benthic foraminifera from the last glacial to the Holocene (Fig. 12.14). Although the core came from a latitude and depth where in the modern ocean AABW is found, with the NADW/LCDW core about 2000 m higher up in the water column (Figs. 4.2 and 4.10), this sudden rise in $\delta^{13}C$ was interpreted as being associated with the "turn on" of NADW formation at the end of the last glacial.

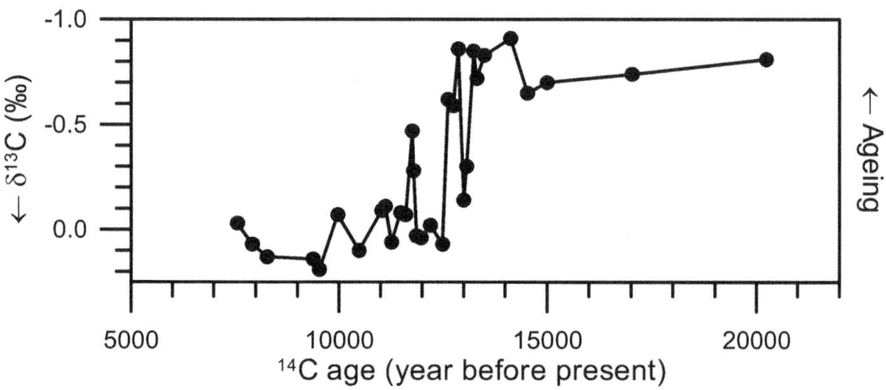

Fig. 12.14. The temporal change in $\delta^{13}C$ of benthic foraminifera from the LGM to the Holocene of piston core RC11-83 from a depth of 4718 m in the Southern Ocean, southwest of South Africa. After Charles and Fairbanks (1992).

Hodell et al. (2003) presented carbon and oxygen isotope data of sediment cores from ODP leg 177 in the same region in the Southern Ocean where piston core RC11-83 was collected. Their cores covered the depth range from 2082 to 4621 m. They found that the glacial–interglacial change of $\delta^{13}C$ of benthic foraminifera (low values during the LGM, high values in the Holocene), reported by Charles and Fairbanks (1992), was a common Southern Ocean feature during the last 1.1 million years in sediment cores collected at water depths below 2700 m. The typical glacial–interglacial difference in the late Pleistocene (0.6 My BP to present) amounted to 1.5 ‰. In contrast with the deeper layers of the Southern Ocean, the magnitude of the glacial–interglacial $\delta^{13}C$ signal near 2100 m was definitely smaller (typical glacial–interglacial difference of 0.5 ‰)

and largely could be explained by the global ocean mean glacial $\delta^{13}C$ shift of ~0.3 ‰.

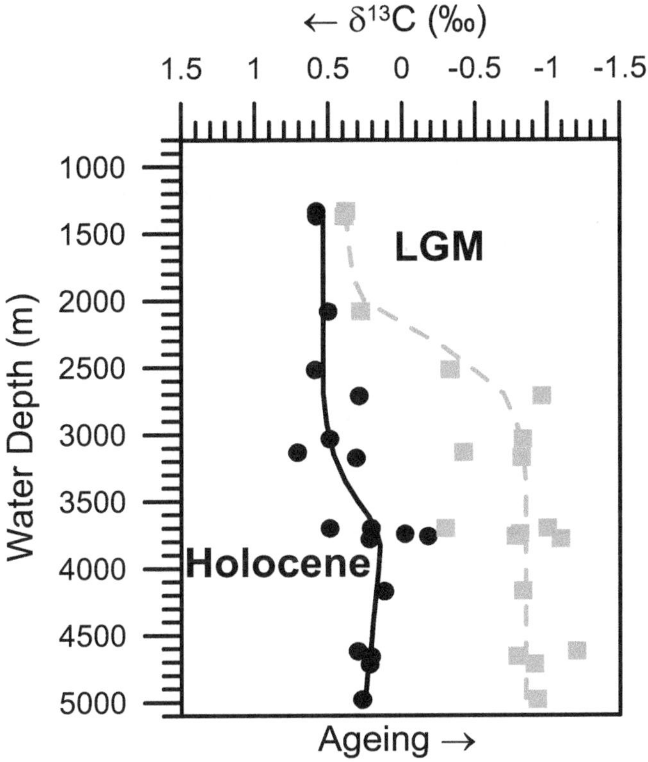

Fig. 12.15. Data points and reconstruction of the profiles of $\delta^{13}C$ of benthic foraminifera in the Southern Ocean, southwest of South Africa for the Holocene (black dots and solid black line) and the LGM (gray squares and dashed gray line). After Hodell et al. (2003).

Using additional data from other nearby ODP sites as well as piston core data, Hodell et al. (2003) were able to extend their Southern Ocean data for the LGM and the Holocene to the depth range of 1335 to 4981 m. With this extended data set they reconstructed the $\delta^{13}C$ profiles for these climatic periods (Fig. 12.15). Their results show that while during the Holocene only small vertical gradients in $\delta^{13}C$ were found; during the LGM a strong vertical gradient was concentrated in the 2100–2700 m interval where a sharp "chemocline" was found above an apparently strongly aged deep layer. A reasonable estimate for the characteristic $\delta^{13}C$ value of the SCW during the LGM is –0.85 ‰ (Figs. 12.14 and 12.15). This suggests that the SCW was strongly aged during the LGM.

12.7. Global water mass changes in the deep ocean

Boyle (1992) collected sediment cores from core repositories in order to analyze global changes of proxy distributions in the deep ocean basins of the world. These cores were sampled for analysis at or near the $\delta^{18}O$ maximum, representative for the LGM. The $\delta^{13}C$ data from the North Atlantic Ocean (Fig. 12.16, top left) show the aged water below 2500 m during the LGM. This is confirmed by the vertical distribution of the Cd_W obtained from Cd/Ca ratio of benthic foraminifera, which show a slight increase relative to the modern cadmium profile (Fig. 12.16, bottom left).

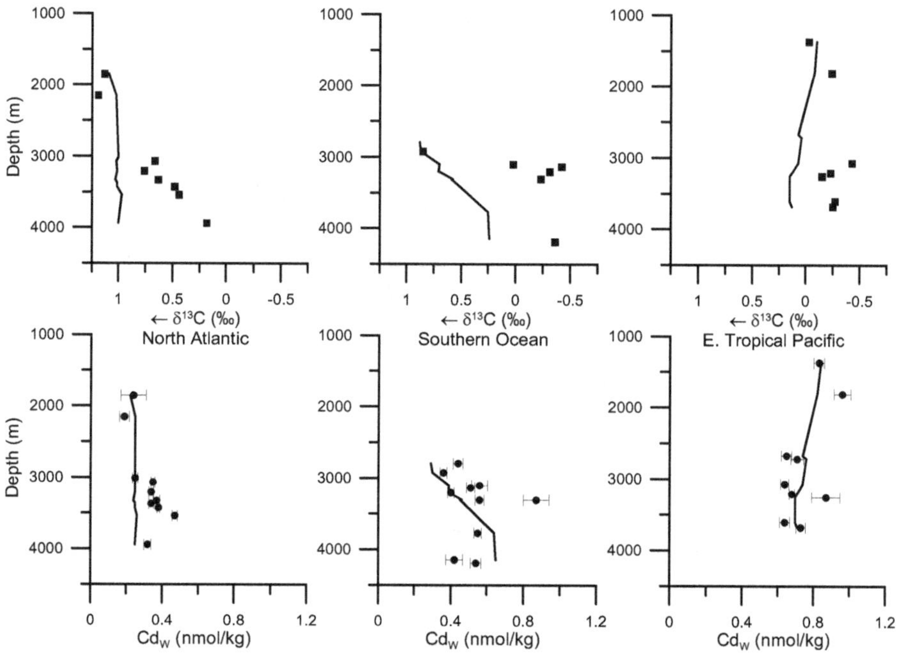

Fig. 12.16. Reconstructed profiles of $\delta^{13}C$ (top row) and dissolved cadmium (bottom row) for the North Atlantic Ocean (left panels), the Southern Ocean (central panels), and eastern Tropical Pacific (right panels). The full line shows the estimates of the modern bottom water properties near the sites, derived from GEOSECS water samples, while the dots show the values obtained for the LGM obtained from benthic foraminifera in sediment cores. The error bars indicate the standard analysis error of Cd_W derived from multiple analyses. A depth-dependent distribution coefficient was used to calculate Cd_W from the Cd/Ca ratio. After Boyle (1992).

The $\delta^{13}C$ profile for the Southern Ocean (Fig. 12.16, top central panel) confirms the strong ^{13}C depletion of the bottom water during the LGM, also shown in Fig. 12.15. However, the attribution of this depletion to aging, even when corrected for the global mean glacial–interglacial difference of 0.32 ‰, is not supported by the Cd_W profile for the Southern Ocean (Fig. 12.16, bottom central panel). The Cd_W hardly differs between the LGM and the modern ocean (0.52±0.05 versus 0.45±0.05 nnmol/kg). In the equatorial Pacific Ocean the deepwater below 2500 m also was depleted in ^{13}C during the LGM, compared with the modern ocean, with a mean $\delta^{13}C$ value of −0.3±0.1 ‰, similar to the Southern Ocean (Fig. 12.16, top right panel). But also in the deep tropical Pacific during the LGM the estimated mean value of Cd_W did not differ from the mean in the modern ocean (0.70±0.03 versus 0.72±0.01 nmol/kg). This value is about 0.2 nmol/kg higher than in the Southern Ocean.

In contrast to glacial aging, the proxies collected by Boyle (1992), shown in Fig. 12.16, do agree with an abyssal circulation below 2500 m during the LGM from the Southern Ocean toward the Atlantic and the Pacific oceans. In the Atlantic Ocean, this SCW is diluted by mixing with the overlying water (NCW/GNAIW), resulting in higher $\delta^{13}C$ and lower Cd_W values in the North Atlantic compared with the Southern Ocean. The mixing effects in the Atlantic during the LGM seem to be stronger than the aging effects. In the eastern tropical Pacific Ocean, aging and mixing with younger water apparently balanced each other with regard to $\delta^{13}C$, although the slight increase of Cd_W suggests some aging effects. Boyle (1992) also reported $\delta^{13}C$ and Cd_W values for the Northwest Pacific Ocean during the LGM that are similar to those of the Southern Ocean. The discrepancy between $\delta^{13}C$ and Cd_W profiles in the Southern Ocean during the LGM shows that a direct interpretation of these proxies in terms of aging is too simple an approach. Where the $\delta^{13}C$ showed a strong increase during the LGM below 2500 m, the Cd_W hardly differed between the LGM and the modern ocean. A generally accepted explanation for this discordance is not available yet. Boyle (1992) speculated that either changes in gas exchange of CO_2 or differences in ^{13}C depletion in plankton are responsible.

12.8. Ocean ventilation age from radiocarbon in sediment cores

The paleoceanographic examples presented in the previous section use proxies with a distribution similar to nutrients or *TIC*. Although these proxies document the changes of the water mass distribution in the deep ocean, they do not provide information on overall rates of water movements. In Chapters 4, 5, and 9 examples were given of the use of ^{14}C measurements for hydrographic research of the present ocean circulation. Radiocarbon from *TIC* dissolved in the surrounding ocean water also is built into the calcite shells of foraminifera. If the samples are younger than about 40,000 years, radiocarbon determinations on foraminifera in principle can be used to reconstruct the oceanic $\delta^{14}C$ distribution in past climatic periods. Because the sink function Q in Eq. (4.1) is precisely known for radiocarbon, a more quantitative reconstruction of the ocean circulation and its ventilation age come within reach. In paleoceanography the term "ventilation age" generally is used in quite a loose way. For the flushing of the deep sea it should be defined as the ratio of the deep sea volume divided by the production rate of new deepwater. For the study of the $\delta^{14}C$ distribution in the modern ocean often the ^{14}C age, corrected for the reservoir effect in the source region, is used as an approximation of the ventilation age. Since the reservoir effect of the source regions of deepwater during past climatic periods is not well known, in paleoceanography the ^{14}C age difference between deepwater and the overlying warm surface water is used as a crude or apparent ventilation age (Broecker et al. 2004).

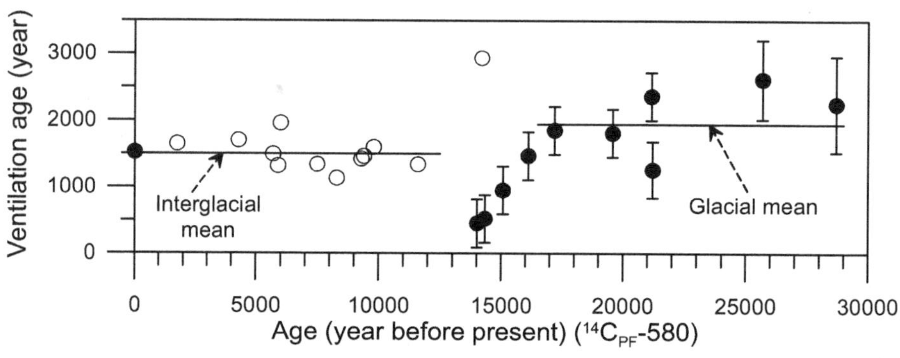

Fig. 12.17. Apparent ventilation age versus age from the western tropical Pacific Ocean at a depth of about 3200 m according to Andree et al. (1985) (open circles) and Shackleton et al. (1988) (black dots). The age of the sediment was determined from the $\delta^{14}C_{PF}$ of the planktonic foraminifera, corrected for a reservoir effect of 580 year the apparent ventilation age from the radiocarbon age difference between paired benthic and planktonic foraminifera.

The $\delta^{14}C$ value of the near-bottom water and the overlying surface water in the last 40,000 years can be estimated from the ^{14}C:C ratio in the calcite shells of paired samples of benthic and planktonic foraminifera, which are obtained from the same time slices of sediment cores. Since ^{14}C is subject to fractionation effects, similar to ^{13}C, corrections for the fractionation of ^{14}C are generally applied to marine samples by the ^{14}C laboratories. These corrections are based on the extrapolation of the observed fractionation of ^{13}C. Andree et al. (1985) and Shackleton et al. (1988) determined the paired ^{14}C ages of planktonic and benthic foraminifera in the deep Pacific Ocean sampled from sediment cores in the western tropical Pacific. These samples covered a time span from the present to nearly 30,000 years before present, according to the ^{14}C age of the planktonic foraminifera, assuming a reservoir effect for the tropical Pacific of 580 year (Fig. 12.17). The results suggest that while the apparent ventilation age in the modern ocean after ~12,000 years BP amounted to ~1500 years, the ventilation during the last glacial before 17,000 years BP was about 440 years slower. However, it has to be admitted that the spread of the glacial ventilation ages was quite large (standard deviation 480 years). During the transition period between 17,000 and 12,000 years BP the apparent ventilation age changed irregularly, with values between 450 and 2930 years. The slower ventilation of the Pacific deepwater during the last glacial agrees with the increased $\delta^{13}C$ during that period, but not with the hardly changed Cd_W presented by Boyle (1992) for the tropical Pacific (Fig. 12.16).

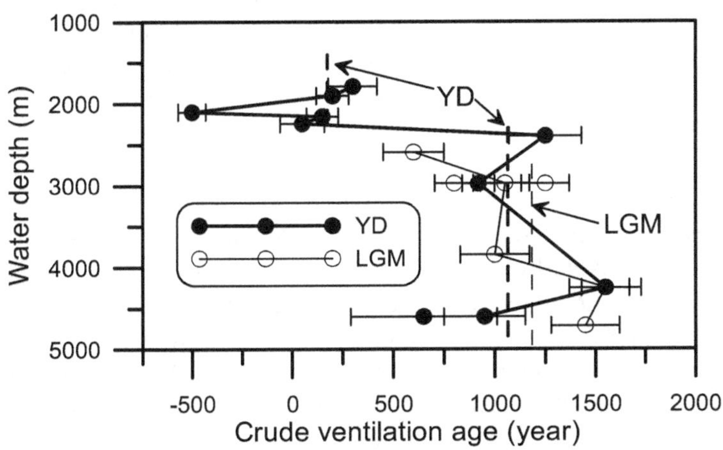

Fig. 12.18. Profiles of the apparent ventilation ages from the subtropical western North Atlantic Ocean for the YD and LGM derived from the ^{14}C age difference of benthic and planktonic foraminifera. The dashed lines show the mean values for the intermediate and deep water masses. After Keigwin (2004).

For the ventilation age of the Atlantic Ocean also several estimates exist. A recent example was presented by Keigwin (2004) who made a careful study of a series of sediment cores obtained from the continental slope region of the subtropical North Atlantic between the Caroline slope and the Bahama Outer Ridge, the region where the DWBC presently transports well-ventilated NADW along the American Continent. For the LGM, reliable estimates of the apparent ventilation age were available only for depths below 2500 m (the open symbols in Fig. 12.18). Their ^{14}C age, derived from the planktonic foraminifera in the sediment samples, came from a 2650 year time interval around 16,760 years BP. Below 2900 m the apparent ventilation age had a mean value of 1180±120 years, about 900 years older than in the present NADW core at the same latitude, derived from Geochemical Ocean Section Study (GEOSECS) ^{14}C data, and also about 450 years older than the present AABW in the northwestern Atlantic Ocean at 5000 m. During the YD the apparent ventilation age of the deepwater below 2300 m had a mean value of 1060±150 years, with a standard error similar to the deep LGM ventilation age. At intermediate levels the mean ventilation age above 2300 m was 175±50 year, ignoring the probably erroneous value of –500 year, which shows that during the YD the ventilation of the intermediate water was quite fast.

The use of the apparent ventilation age, derived from the radiocarbon age difference of paired benthic and planktonic foraminifera, as a quantitative proxy for the ventilation rate of deep ocean basins during past climatic eras requires several assumptions which are not always made explicit. It is assumed that, during the period under consideration the THC as well as the unknown reservoir effect in the source regions of the deepwater were stationary. Any variations of the THC, at the timescale of the 1000 to 2000 years long climatic interstadials, last too short a time to be resolved in the determination of the apparent ventilation age and are aliased to longer timescales. It also is assumed that the atmospheric $\delta^{14}C$ did not alter while the deepwater traveled from its source region to the studied basin. The atmospheric $\delta^{14}C$, however, has changed over the past 40,000 years (e.g., Hughen et al. 1998, 2004; Stuiver et al. 1998; Kitagawa and van der Plicht 1998). These changes can be attributed to changes of cosmic radiation due to variations in solar activity, as well as variations of the amount of carbon stored in the ocean and terrestrial biomass. The general decrease of the atmospheric $\delta^{14}C$ since about 17,000 BP will result in underestimating the oceanic apparent ventilation age. Adkins and Boyle (1997) showed that the apparent ventilation of the Pacific Ocean may be underestimated by 600 years due to this effect. They developed an alternative method to obtain a so-called projection age.

The determination of past apparent ventilation ages of deep ocean basins as a proxy for the ventilation rate by the THC is still hampered by other problems, presented by Broecker et al. (2004). They reviewed published ventilation rate estimates from ice-age Pacific sediments based on paired radiocarbon ages of coexisting benthic and planktonic foraminifera. They found a large scatter in the results, so large that the apparent ^{14}C age of the deepwater in the Pacific Ocean could lie anywhere between half and double of today's ventilation age. It did not become clear what was the cause of the wide scatter in radiocarbon ventilation ages. Before any useful conclusions can be drawn about how to cope with this uncertainty, the origin of this scatter has to be tracked.

12.9. A model interpretation of proxy data

Much of the interpretation of proxy data in terms of ocean circulation, as, e.g., presented in the examples from the previous sections, is verbal and intuitive. The interpretations are generally based on simple budgets calculations in box models that can be made for the integrated ocean properties. Boyle and Keigwin (1982) introduced a single-box model for a conservative tracer in the deep Atlantic Ocean to assess the quantitative significance of variations of *Cd/Ca* ratios and $\delta^{13}C$ values obtained from a North Atlantic sediment core. In a simplified form, that model is represented by the left panel of Fig. 12.19. The deep Atlantic water box is flushed from the north (NCW) with a volume flux V_N and from the south (SCW) with a volume flux V_S. Via deep upwelling a volume flux V_I leaves the deep ocean to shallower levels. The deep through-flow to other ocean basins is represented by a volume flux or flow rate V_T. Mass conservation (for an incompressible ocean) is given by

$$V_N + V_S = V_I + V_T \quad . \tag{12.6}$$

The tracer concentrations derived from proxy data have a characteristic value C_I in the deep box and source water values C_N and C_S for, respectively, the northern and southern source water types. This may be the dissolved phosphate concentration derived from the *Cd/Ca* ratio or the $\delta^{13}C$ proxy. We can add a source/sink term Q to deal with nonconservative behavior of the tracer. The mass conservation for the tracer C then is given by

$$V_N C_N + V_S C_S + Q = V_I C_I + V_T C_I \quad . \tag{12.7}$$

In this model the resulting tracer concentration in the deep ocean box will be the flux-weighted average of the concentrations in the source water types that combine to form the deep water mass in the box plus an aging term for relative to the mean residence time of the deepwater in the box:

$$C_I = \frac{V_N C_N + V_S C_S}{V_N + V_S} + \frac{Q}{V_N + V_S} = C_N + \frac{V_S (C_S - C_N)}{V_N + V_S} + \frac{Q}{V_N + V_S} \quad . \tag{12.8}$$

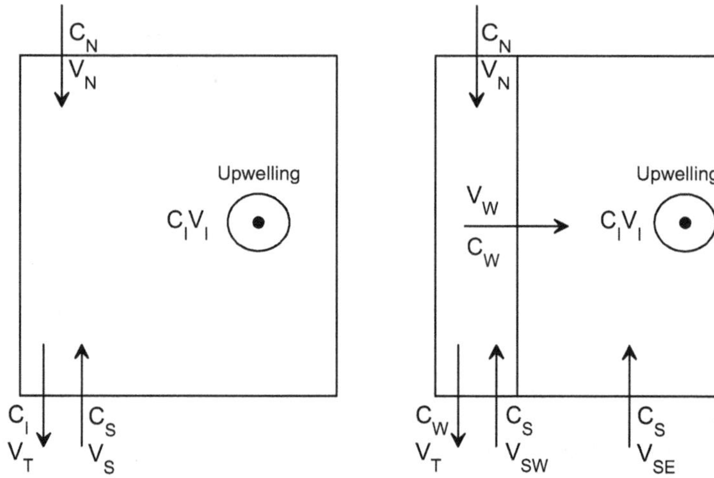

Fig. 12.19. Schematic single-box (left) and double-box (right) models of the deep ocean, given volume transports *V* and proxy concentrations *C*. In the double-box model the left box represents the DWBC. If the proxy represents a nonconservative parameter, a source term *Q* can be added to each box.

In other words, the value of C_I derived from the proxies then will change inversely with the NCW flow rate for conservative ($Q = 0$) tracers. Then determination of C_I effectively constrains the ratio V_S/V_N, according to

$$\frac{V_S}{V_N} = \frac{C_I - C_N}{C_S - C_I} \quad . \tag{12.9}$$

Boyle and Keigwin (1982) used this conservative diagnostic mode of the single box model as the basis for the interpretation of proxy data ($\delta^{18}O$, Cd/Ca, $\delta^{13}C$) in the North Atlantic Ocean. But Eq. (12.9) only constrains the ratio of the southern and northern flow rates, not their magnitude. Doubling or halving both V_N and V_S will result in the same conservative tracer concentration C_I in the deep ocean box.

The aging term for nonconservative tracers in Eq. (12.8) increases with the inverse of the sum of the flow rate of the SCW and NCW. Then it becomes difficult without additional information, to ascribe changes in C_I to either changes in the flow rate ratio V_S/V_N, or to a change in the residence time which is proportional to $1/(V_N+V_S)$. Even the flow rate ratio V_S/V_N no longer is constrained if non-conservative effects become important, since these add extra degrees of freedom to the system.

LeGrand and Wunsch (1995) extended the single-box model for the deepwater to a double-box model (Fig. 12.19, right panel). One box represents the slow abyssal flow in the ocean's interior, while a second box allows the existence of the DWBC with a relatively fast through-flow. The flow rate of NCW is again V_N, entering the DWBC from the NCW source region. The flow rate of the SCW is divided in a southwestern contribution to the DWBC, V_{SW}, and a southeastern contribution to the abyssal box, V_{SE}. Next to a through-flow rate in the DWBC, V_T, there is an export term from the DWBC to the abyssal box, V_W. In the abyssal box upwelling of deepwater into the overlying water is allowed, with a flow rate V_I. Upwelling in the narrow DWBC is ignored. Mass conservation for both boxes can be written as

$$\begin{aligned} V_N + V_{SW} &= V_T + V_W \\ V_W + V_{SE} &= V_I \end{aligned} \quad . \tag{12.10}$$

With characteristic tracer values C_N and C_S for the NCW and SCW and C_W and C_I for the DWBC box and the abyssal box, respectively, the equations for the conservative tracer now become

$$\begin{aligned} V_N C_N + V_{SW} C_S &= V_T C_W + V_W C_W \\ V_W C_W + V_{SE} C_S &= V_I C_I \end{aligned} \quad . \tag{12.11}$$

For both boxes a relation similar to Eq. (12.9) for the flow rate ratios V_{SW}/V_N and V_{SE}/V_W can be derived from these equations

$$\frac{V_{SW}}{V_N} = \frac{C_W - C_N}{C_S - C_W} \quad \text{and} \quad \frac{V_{SE}}{V_W} = \frac{C_I - C_W}{C_S - C_I} \quad . \tag{12.12}$$

By adding these results, the ratio of the flow rate for SCW and NCW, $V_S/V_N = (V_{SW} + V_{SE})/V_N$, can be derived to be

$$\frac{V_S}{V_N} = \frac{V_{SW} + V_{SE}}{V_N} = \frac{C_W - C_N}{C_S - C_W} + \frac{V_W}{V_N}\frac{C_I - C_W}{C_S - C_I} \quad . \tag{12.13}$$

Equation (12.13) shows that in a double-box model the flow rate ratio V_S/V_N no longer is constrained by the distribution of the conservative proxy data, since the ratio V_W/V_N, related to the divergence of the DWBC, also is unknown. Conservative tracers provide only local constraints on the ocean circulation. The tracer concentration in the abyssal box does not directly constrain the ventilation of the DWBC box.

The limited ability of conservative tracers to constrain the deep ocean circulation also was demonstrated by LeGrand and Wunsch (1995) with a low-resolution inverse model for the North Atlantic Ocean, with realistic physics: geostrophic flow, conservation vorticity, mass, heat, salinity, and phosphate. The latter tracer was treated as being conservative on the scale of the North Atlantic. The model only contained grid points below 1 km depth. The input of NADW into the DWBC at 50ºN was prescribed as the boundary condition, while the total fluxes of water and tracers across the boundaries at 1 km depth and 10ºN and 50ºN were treated as variables to be estimated with the model. The model with data from the modern ocean performed quite well when it was constrained by observations of potential temperature, salinity, and phosphate. With T, S, and ρ accurately specified at each grid point, the geostrophic shear and resulting abyssal circulation could be determined in the ocean interior. For the LGM the circulation in this model was constrained by the $\delta^{18}O_c$ and $\delta^{13}C$ data from 125 sediment cores. Information on the density structure in the ocean interior could be derived from the $\delta^{18}O_c$ data which were converted to temperature variables similar to Eq. (12.3). For simplicity it was assumed that gradients in salinity could be ignored in the calculation of the density gradients and the geostrophic shear. In the inverse calculations it appeared that the LGM $\delta^{13}C$ and $\delta^{18}O$ data were fully consistent with the circulation rates for the THC estimated for the modern ocean. However, the proxies also were consistent with a circulation rate of one-half the modern flux of NADW. Effectively there were an infinite number of circulation rates possible within the constraints of the available paleodata. Contrary to the experiment for the present ocean, density estimates derived from the $\delta^{18}O$ data were too scarce and inaccurate to enforce a reliable meridional geostrophic transport in the ocean interior, and thereby a flushing time for the deep ocean basins. This is similar to the lack of information on V_W in Eq. (12.13) which makes the double-box model nonconstrained. Wunsch (2003) showed that in principle ^{14}C derived ventilation ages can replace the density information in an inverse model to reconstruct the THC in past climatic periods. But as was stated by Broecker et al. (2004) there are still some serious problems to overcome before a reliable large-enough database with such data becomes available for inverse modeling of the past THC.

References

Adkins, J.F., and E.A. Boyle (1997) Changing atmospheric $\Delta^{14}C$ and the record of deep water paleoventilation ages. Paleoceanography, 12, 337–344

Anderson, L.A., and J.L. Sarmiento (1994) Redfield ratios of remineralization determined by nutrient data analysis. Global Biogeochemical Cycles, 8, 65–80

Andree, M., J. Beer, H. Oeschger, A.C. Mix, W.S. Broecker, N. Ragano, P. O'Hara, G. Bonani, H.J. Hofmann, E. Morenzoni, M. Nessi, M. Suter, and W. Woelfli (1985) Accelerator radio-carbon ages on foraminifera separated from deep-sea sediments. In: The Carbon Cycle and Atmospheric CO2: Natural Variations Archean to Present, E.T. Sundquist and W.S. Broecker (eds.), Geophysical Monographs, 32, American Geophysical Union, Washington, DC, pp. 143–153

Anonymous (1988) Data Announcement 88-MGG-02, Digital relief of the Surface of the Earth. NOAA, National Geophysical Data Center, Boulder, Colorado

Baumgartner, A., and E. Reichel (1975) The World Water Balance. Elsevier, Amsterdam, 179 pp.

Belkin, I.M. (2004) Propagation of the "Great Salinity Anomaly" of the 1990s around the northern North Atlantic. Geophysical Research Letters, 31, L08306, 10.1029/2003GL019334

Bickert, T. (1992) Rekonstruktion der Spätquartären Bodenwasserzirkulation im östlichen Süsatlantik über stabile Isotopen benthischer Foraminiferen. Berichte, Fachbereich Geowissenschaften, Universität Bremen, 27, 1–207

Bintanja, R., R.S.W. van de Wal and J. Oerlemans (2005) Modelled atmospheric temperatures and global sea level over the past million years. Nature, 437, 125–128

Böning, C.W., and A.J. Semtner (2001) High-resolution modelling of the thermohaline and wind-driven circulation. In: Ocean circulation & climate: observing and modelling the global ocean. G. Siedler, J. Church, and J. Gould eds. Academic Press, London, pp. 59–77

Bönisch, G., and P. Schlosser (1995) Deep water formation and exchange rates in the Greenland/Norwegian seas and the Eurasian Basin of the Arctic ocean derived from tracer balances. Progress in Oceanography, 35, 29–52

Bower, A.S., and H D. Hunt (2000a) Lagrangian observations of the deep western boundary current in the North Atlantic Ocean. Part 1: Large-scale pathways and spreading rates. Journal of Physical Oceanography, 30, 764–783

Bower, A. S., and H.D. Hunt (2000b) Lagrangian observations of the deep western boundary current in the North Atlantic Ocean. Part 2: The Gulf Stream-Deep Western Boundary Current crossover. Journal of Physical Oceanography, 30, 784–804

Boyer, T.P., S. Levitus, J.I. Antonov, R.A. Locarnini, and H.E. Garcia (2005) Linear trends in salinity for the world ocean, 1955–1998. Geophysical Research Letters, 32, L01604, doi:10.1029/2004GJ021791

Boyle, E.A. (1992) Cadmium and d13C paleochemical distributions during the stage 2 glacial maximum. Annual Reviews of Earth and Planetary Sciences, 20, 245–287

Boyle, E.A., and L.D. Keigwin (1982) Deep circulation of the North Atlantic over the last 200,000 years: geochemical evidence. Science, 218, 784–787

Broecker, W.S. (1987) The biggest chill. Natural History Magazine, 97, 74–82

Broecker, W.S. (1991) The great ocean conveyor. Oceanography, 4, 79–89

Broecker, W.S. (1997) Was a change in thermohaline circulation responsible for the Little Ice Age? Proceedings of the National Academy of Sciences, 97, 1339–1342

Broecker, W.S. (2003) Does the trigger for abrupt climate change reside in the ocean or in the atmosphere? Science, 300, 1519–1522

Broecker W.S., T. Takahashi, and T. Takahashi (1985) Sources and flow patterns of deep-ocean waters as deduced from potential temperature, salinity and initial phosphate concentration. Journal of Geophysical Research, 90, 6925–6939.

Broecker, W.S., G. Bond, M. Klas, G. Bonani, and W. Wolfi (1990) A salt oscillator in the glacial North Atlantic? The concept. Paleoceanography, 5, 469–477

Broecker, W.S., S. Sutherland, W. Smethie, T.-H. Peng, and G. Östlund (1995) Oceanic radiocarbon: Separation of the natural and bomb components. Global Biogeochemical Cycles, 9, 263–288

Broecker, W.S., E. Clarck, I. Hajdas, and G. Bonani (2004) Glacial ventilation rates for the deep Pacific Ocean. Paleoceanography, 19, doi: 10.1029/2003PA000974

Bryan, K. (1969) A numerical method for the study of the circulation of the world ocean. Journal of Computational Physics, 4, 347–376

Bryan, F. (1986) High-latitude salinity effects and interhemispheric thermohaline circulations. Nature, 323, 301–304

Bryan, K., and M.D. Cox (1968) A nonlinear model of an ocean driven by wind and differential heating. Parts I and II. Journal of Atmospheric Science, 25, 945–967

Bryden, H.L., H.R. Longworth, and S.A. Cunningham (2005) Slowing of the Atlantic meridional overturning circulation at 25° N. Nature, 438, 655–657

Bu, X., and M.J. Warner (1995) Solubility of chlorofluorocarbon 113 in water and seawater. Deep-Sea Research I, 42, 1151–1161

Buch, E., S-A. Malmberg, and S.S. Kristmannson (1996) Arctic Ocean deep water masses in the western Iceland Sea. Journal of Geophysical Research, 101, 11,965–11,973

Bullister, J.L., and S.P. Wisegarver (1998) The solubility of carbon tetrachloride in water and seawater. Seep-Sea Research I, 45, 1285–1302

Carpenter, W.B. (1871) On the Gibraltar Current, the Gulf Stream, and the general oceanic circulation. Proceedings of the Royal Geographical Society, XV, pp. 54–91

Charles, C.D., and R.G. Fairbanks (1992) Evidence from Southern Ocean sediments for the effect of North Atlantic deep-water flux on climate. Nature, 355, 416–419

Clark, P.U., N.G. Pisia, T.F. Stocker, and A.J. Weaver (2002) The role of the thermohaline circulation in abrupt climate change. Science, 415, 863–869

Clarke, R.A., and J.-C. Gascard (1983) The formation of Labrador Sea Water. Part I: large-scale processes. Journal of Physical Oceanography, 13, 1764–1778

Coachman, L.K., and K. Aagaard (1966) On the water exchange through Bering Strait. Limnology and Oceanography, 11, 44–59

Coachman, L.K., and K. Aagaard (1981) Re-evaluation of water transports in the vicinity of Bering Strait. in: The eastern Bering Sea shelf: Oceanography and resources, D.W. Wood and J.A. Calder (eds.), NOAA, Washington, pp. 95–110

Coachman, L.K., and K. Aagaard (1988) Transports through Bering Strait: Annual and interannual variability. Journal of Geophysical Research, 93, 15,535–15,539

Colin de Verdière, A. (1993) On the oceanic thermohaline circulation. In: Modelling Oceanic Climate Interactions, J. Willebrand and D.L.T. Anderson (eds.), NATO ASI Series, I-11, Springer Verlag, Berlin, pp. 151–183

Comiso, J.C., and A.L. Gordon (1987) Recurring polynyas over the Cosmonaut Sea and the Maud Rise. Journal of Geophysical Research, 92, 2819–2833

Cullen, J.T., T.W. Lane, F.M.M. Morel, and R.M. Sherrell (1999) Modulation of cadmium uptake in photoplankton by seawater CO2 concentration. Nature, 402, 165–167

Curry, W.B., and D.W. Oppo (2005) Glacial water mass geometry and the distribution of d13C of SCO2 in the western Atlantic Ocean. Paleoceanography, 20, PA1017, doi:10.1029/2004PA001021

Dansgaard, W., H.B. Clausen, N.S. Gundestrup, C.U. Hammer, S.F. Johnsen, P.M. Kristinsdottir, and N. Reeh (1982) A new Greenland ice core. Science, 218, 1273–1277

Dansgaard, W., S.J. Johnsen, H.B. Clausen, D. Dahl-Jensen, N.S. Gundestrup, C.U. Hammer, C.S. Hvidberg, J.P. Steffensen, A.E. Sveinbjörnsdóttir, J. Jouzel, and G.C. Bond (1993) Evidence for general instability of past climate from a 250 kyr ice-core record. Nature, 264, 218–220.

De Ruijter, W.P.M., A. Biastoch, S.S. Drijfhout, J.R.E. Lutjeharms, R.P. Matano, T. Pichevin, P.J. van Leeuwen, and W. Weijer (1999) Indian–Atlantic interocean exchange: dynamics, estimation and impact. Journal of Geophysical Research, 104, 20,885–20,910

Dickson, R.R., and J. Brown (1994) The production of North Atlantic Deep Water: Sources, rates, and pathways. Journal of Geophysical Research, 99, 12,31–12,341

Dickson, R.R., I. Yashayev, J. Meincke, B. Turrell, S. Dye, and J. Holfort (2002) Rapid freshening of the deep North Atlantic Ocean over the past four decades. Nature, 416, 832–837

Donners, J. (2005) The role of the South Atlantic in the upper branch of the global thermohaline circulation. PhD thesis, Utrecht University, 114 pp.

Duplessy, J.-C., N.J. Shackleton, R.G. Fairbanks, L.D. Labeyrie, D. Oppo, and L. Kallel (1988) Deep water source variation during the last climatic cycle and their impact on the global deep water circulation. Paleoceanography, 3, 343–360

Ellis, H. (1751) A letter to the Rev. Dr. Hales, F.R.S. from Captain Henry Ellis, F.R.S. dated Jan. 7, 1750–51, at Cape Monte Africa, ship Earl of Halifax. Philosophical Transactions, 47, 211–214

Emery, W.J., and J. Meincke (1986) Global water masses: summary and review. Oceanologica Acta, 9, 383–391

Epstein, S., R. Buchsbaum, H.A. Lowenstam, and H.C. Urey (1953) Revised carbonate–water isotopic temperature scale. Bulletin of the Geological Society of America, 64, 1315–1325

Fairbanks, R.G. (1989) A 17,000 glacio-eustatic sea level record: influence of glacial melting rates on the Younger Dryas event and deep ocean circulation. Nature, 342, 637–642

Ffield, A., and A.L. Gordon (1992) Vertical mixing in the Indonesian thermocline. Journal of Physical Oceanography, 22, 184–195

Fiadeiro, M. E., and G.Veronis (1984), Obtaining velocities from tracer distributions. Journal of Physical Oceanography, 14, 1734–1746.

Fine, R.A., and R.L. Molinari (1988) A continuous deep western boundary current between Abaco (26.5°N) and Barbados (13°N). Deep-Sea Research, 35, 1441–1450

Fofonoff, N.P. (1962) Physical properties of sea-water. In: The Sea: Ideas and observations on progress in the study of the seas, I, M.N. Hill (ed.), Interscience Publishers, New York, pp. 3–30

Fofonoff, N.P., and R.C. Millard Jr. (1983) Algorithms for computation of fundamental properties of seawater. UNESCO Technical Papers in Marine Science, 44, 53 pp.

Fomin, L.M. (1964) The dynamic method in oceanography. Elsevier Oceanography Series, 2, Elsevier, Amsterdam, 212 pp.

Frankignoul, C., and K. Hasselmann (1977) Stochastic climate models, part II, application to sea surface temperature, temperature anomalies, and thermocline variability. Tellus, 28, 289–305

Fricke H.C., and J.R. O'Neil (1999) The correlation between 18O/16O ratios of meteoric water and surface temperature: its use in investigating terrestrial climate change over geologic time. Earth and Planetary Science Letters, 170, 181–196

Gammelsrød, T., A. Foldvik, A.O. Nøst, Ø, Skagseth, L.G. Anderson, E. Fogelkvist, and T. Tanhua (1994) Distribution of water masses on the continental shelf in the southern Weddell Sea. In: The Polar oceans and their role in shaping the global environment, O.M. Johannessen, T.D. Muench and J.E. Overland (eds.) Geophysical Monographs, 85, pp. 159–176, American Geophysical Union, Washington DC.

Ganachaud, A., and C. Wunsch (2000) Improved estimates of global ocean circulation, heat transport and mixing from hydrographic data. Nature, 408, 453–457

Ganopolski, A., and S. Rahmstorf (2001) Simulation of rapid glacial climate changes in a coupled climate model. Nature, 409, 153–158

Garrett, C. (1979) Mixing in the ocean interior. Dynamics of Atmospheres and Oceans, 3, 239–265

Garwoord, R.W. Jr., S.M. Isakari, and P.C. Gallacher (1994) Thermobaric convection. In: The Polar Oceans and their role in shaping the global environment, O.M. Johannessen, T.D. Muench and J.E. Overland (eds.) Geophysical Monographs, 85, pp. 199–209, American Geophysical Union, Washington DC.

Gascard, J.-C., and A.C. Clarke (1983) The formation of Labrador Sea water, II, Mesoscale and smaller scale processes. Journal of Physical oceanography, 13, 1779–1797

Gill, A.E. (1982) Atmosphere and Ocean Dynamics. Academic Press, New York, 662 pp.

Gille, S.T. (2003) Float observations of the Southern Ocean. Part I: Estimating mean fields, bottom velocities, and topographic steering. Journal of Physical Oceanography, 33, 1167–1181

Gordon, A.L. (1986) Inter-ocean exchange of thermocline water. Journal of Geophysical Research, 91, 5037–5046

Gordon, A.L. (2001) Interocean exchange. In: Ocean circulation & climate: observing and modelling the global ocean. G. Siedler, J. Church, and J. Gould (eds.) Academic Press, London, pp. 303–314

Gordon, A.L., and J.C. Comiso (1988) Polynyas in the Southern Ocean, Scientific American, 258, 70–77

Gordon, A.L., R.D. Susanto, and A. Ffield (1999) Throughflow within Makassar Strait. Geophysical Research Letters, 26, 3325–3328

Gordon, A.L., R.D. Susanto, and K. Vranes (2003) Cool Indonesian throughflow as a consequence of restricted surface layer flow. Nature, 425, 824–828

Grootes, P.M., M. Stuiver, J.W.C. White, S.J. Johnsen, and J. Jouzel (1993) Comparison of oxygen isotope records from the GISP2 and GRIP Greenland ice cores. Nature, 366, 552–554

Hales, S. (1751) A letter to the president, from Stephen Hales, D.D. & F.R.S., Philosophical Transactions, 47, 214–216

Hallock, Z.R., and W.J. Teague (1996) Evidence for a North Pacific deep western boundary current. Journal of Geophysical Research, 101, 6617–6624

Haney, R.L., (1971) Surface thermal boundary condition for ocean circulation models. Journal of Physical Oceanography, 1, 241–248.

Hansen, B., W.R. Turrell, and S. Østerhus (2001) Decreasing overflow from the Nordic seas into the Atlantic Ocean through the Faroe Bank channel since 1950. Science, 411, 927–930

Hasselmann, K. (1976) Stochastic climate models. I: Theory. Tellus, 28, 473–485

Hátún, H., A.B. Sandø, H. Drange, B. Hansen, and H. Valdimarson (2005) Influence of the Atlantic subpolar gyre on the thermohaline circulation. Science, 309, 1841–1844

Helland-Hansen, B., and F. Nansen (1909) The Norwegian Sea, Its physical oceanography based upon the Norwegian researches 1900–1904. Rep. Nor. Fish. Mar, Invest. 2(1), 390 pp.

Hellerman, S. (1967) an updated estimate of the wind stress in the World Ocean. Monthly Weather Review, 95, 62–74

Hellerman, S., and M. Rosenstein (1983) Normal monthly winds stress over the world ocean with error estimates. Journal of Physical Oceanography, 13, 1093–1104

Hodell, D.A., K.A. Venz, C.D. Charles, and U.S. Ninnemann (2003) Pleistocene vertical carbon isotope and carbonate gradients in the South Atlantic sector of the Southern Ocean. Geochemistry, Geophysics, Geosytems, 4, doi:1004, 10.1029/202GC000367

Hogg, N.G. (2001) Quantification of the deep circulation. In: Ocean circulation & climate: observing and modelling the global ocean. G. Siedler, J. Church, and J. Gould (eds.) Academic Press, London, pp. 259–270

Hopkins, T.S. (1991) The GIN Sea — a synthesis of its physical oceanography and literature review 1972–1985. Earth Science Reviews, 30, 175–318

Huang, R.X. (1999) Mixing and energetics of the oceanic thermohaline circulation. Journal of Physical Oceanography, 39, 727–746

Hughen, K., J.T. Overpeck, S.J. Lehman, M. Kashgarian, J.R. Southon, and L.C. Peterson (1998) A new ^{14}C calibration data set for the last deglaciation based on marine varves. Radiocarbon, 40, 483–494

Hughen, K., S.J. Lehman, J.R. Southon, J.T. Overpeck, O. Marchal, C. Herring, and J. Turnbull (2004) ^{14}C activity and global carbon cycle changes over the past 50,000 years. Science, 33, 22–27

Hut, G. (1987) Stable isotope reference samples for geochemical and hydrological investigations. Consultant Group Meeting IAEA, Vienna 16–18 September 1985. Report to the Director General. International Atomic Energy Agency, Vienna

IPCC (2001) Climate change 2001: the scientific basis. contributions of working group I to the third assessment of the Intergovernmental Panel on Climate Change. J.T. Houghton, Y. Ding, D.J. Griggs, M. Noguer, P.J. van der Linden, X. Dai, K. Maskell and C.A. Johnson (eds.). Cambridge University Press, Cambridge, U.K., 881 pp.

Jakobsson, M., N.Z. Cherkis, J. Woodward, R. Macnab, and B. Coakley (1999) New grid of Arctic bathymetry aids scientists and mapmakers; Eos, Transactions, American Geophysical Union, 81, 89,93,96.

Jones, E.P., J.H. Swift, L.G. Anderson, M. Lipzier, G. Civitarese, K.K. Falkner, G. Kattner, and F. McLaughlin (2003) Tracing Pacific water in the North Atlantic Ocean. Journal of Geophysical Research, 108, doi:10.1029/2001JC001141

Jonsson, S., and H. Valdimarsson (2004) A new path for the Denmark Strait overflow water from the Iceland Sea to Denmark Strait. Geophysical Research Letters, 31, doi:10.1029/2003GL019214

Keigwin, L.D. (2004) Radiocarbon and stable isotope constraints on Last Glacial Maximum and Younger Dryas ventilation in the western North Atlantic. Paleoceanography, 19, doi:10.1029/2004PA001029

Kitagaw, H., and J. van der Plicht (1998) Atmospheric radiocarbon calibration to 45,000 yr. BP: Late Glacial fluctuations and cosmogenic isotope production. Science, 279, 1187–1190

Klinger, B.A., S. Drijfhout, J. Marotzke, and J.R. Scott (2004) Remote wind-driven overturning in the absence of the Drake Passage effect. Journal of Physical Oceanography, 34, 1036–2004

KNMI (1861) Onderzoekingen met den zeethermometer, als uitkomsten van wetenschap en ervaring, aangaande de winden en zeestroomingen in sommige gedeelten van den oceaan. Koninklijk Nederlandsch Meteorologisch Instituut, Utrecht, 183 pp. (in Dutch)

Knudsen, M. (1901) Hydrographic Tables, G.E.C. Gad, Copenhagen, 63 pp.

Krümmel, O. (1907) Handbuch der Ozeanographie, Band I. Verlag von J. Engelhorn, Stuttgart, 526 pp.

Labeyrie, L.L., J.-C. Duplessy, J. Duprat, A. Juillet-Leclerc, J. Moyes, E. Michel, N. Kallel, and N.J. Shackleton (1992) Changes in the vertical structure of the North Atlantic Ocean between glacial and modern times. Quarternary Science Reviews, 11, 401–413

Lane, T.W., and F.M.M. Morel (2000a) Regulation of carbonic anhydrase expression by zinc, cobalt, and carbon dioxide in the marine diatom thalassiosira weissflogii. Plant Physiology, 123, 345–352

Lane, T.W., and F.M.M. Morel (2000b) A biological function for cadmium in marine diatoms. Proceedings of the National Academy of Sciences, 97, 4627–4631

Latif, M., E. Roeckner, U. Mickolajewicz, and R. Voss (2000) Tropical stabilisation of the thermohaline circulation in a greenhouse warming simulation. Journal of Climate, 13, 1809–1813

Lazier, J.R.N. (1973) The renewal of Labrador Sea Water. Deep-Sea Research, 20, 341–353

Lea, D.W., and E.A. Boyle (1989) Barium content of benthic foraminifera controlled by bottom water composition. Nature, 338, 751–753

Lea, D.W., and E.A. Boyle (1990) Foraminiferal reconstruction of barium distributions in water masses of the glacial oceans. Paleoceanography, 5, 719–742

Leaman, K.D., and F. Schott (1991) Hydrographic structure of the convection regime in the Golfe du Lion. Journal of Physical Oceanography, 21, 575–598

Ledwell, J.R., E.T. Montgomery, K.L. Polzin, L.C. St. Laurent, R.W. Schmitt, and J.M. Toole (2000) Evidence for enhanced mixing over rough topography in the abyssal ocean. Nature, 403, 179–181

Lee, T.N., W.E. Johns, R.J. Zantopp, and E.R. Fillenbaum (1996) Moored observations of western boundary current variability and thermohaline circulation at 26.5° in the subtropical North Atlantic. Journal of Physical Oceanography, 26, 962–983

LeGrand, P., and C. Wunsch (1995) Constraints from paleotracer data on the Atlantic circulation during the last glacial maximum. Paleoceanography, 10, 1011–1045

Libby, W.F. (1952) Radiocarbon Dating. University of Chicago Press, Chicago

Macdonald, A.M., and C. Wunsch (1996) An estimate of global ocean circulation and heat fluxes. Nature, 382, 436–439

Mackensen, A., and T. Bickert (1999) Stable carbon isotopes in benthic foraminifera: proxies for deep and bottom water circulation and new production. In: Use of proxies in paleoceanography: examples from the South Atlantic. G. Fischer and G. Wefer (eds.) Springer Verlag, Berlin, pp. 229–254

Mamayev, O.I. (1975) Temperature–salinity analysis of world ocean waters. Elsevier Oceanography Series, 11, Elsevier Scientific Publishing Company, Amsterdam, 374 pp.

Manabe, S., and R.J. Stouffer (1999) Are two modes of thermohaline circulation stable? Tellus, 51A, 400–411

Marotzke, J. (1990) Instabilities and multiple equilibria of the thermohaline circulation. PhD thesis Berichte des Instituts für Meereskunde and der Universität Kiel, 194, 126 pp.

Marotzke, J. (2000) Abrupt climate change and thermohaline circulation: mechanisms and predictability. Proceedings of the National Academy of Sciences, 97, 1347–1350

Marotzke, J., and J. Willebrand (1991) Multiple equilibria of the global thermohaline circulation. Journal of Physical Oceanography, 21, 1372–1385

Marotzke, J., P. Welander, and J. Willebrand (1988) Instability and multiple steady states in a meridional-plane model of the thermohaline circulation. Tellus, 40A, 162–172

Marshall, J., and F. Schott (1999) Open-ocean convection: observations, theory, and models. Reviews of Geophysics, 37, 1–64

Mayewski, P. A., et al. (1994) Record drilling depth struck in Greenland, Eos Transactions of the American Geophysical Union, 75, 113–124

McCartney, M. S. (1992) Recirculating components to the deep boundary current of the Northern North Atlantic. Progress in Oceanography, 29, 283–383

McDougall, T.J. (2003) Potential enthalpy: A conservative oceanic variable for evaluating heat content and heat fluxes. Journal of Oceanography, 33, 945–963

Midtun, L. (1985) Formation of dense bottom water in the Barents Sea. Deep-Sea Research, 32, 1233–1241

Munk, W.H. (1966) Abyssal recipes. Deep-Sea Research, 13, 707–730

Munk, W.H., and C. Wunsch (1998) Abyssal recipes II: Energetics of tidal and wind mixing. Deep-Sea Research I, 45, 1977–2010

Nansen, F. (1912) Das Bodenwasser und die Abkühlung des Meeres. Internationale Revue der gesamten Hydrobiologie und Hydrographie, 5, 1–42

Nauw, J.J. (2003) Low-frequency variability of the wind-driven ocean circulation. PhD thesis, Utrecht University, Utrecht, 160 pp.

Nilsson, J., and G. Walin (2001) Freshwater forcing as a booster of thermohaline circulation. Tellus, 53A, 629–641

Nilsson, J., G. Broström, and G. Walin (2003) The thermohaline circulation and vertical mixing: does weaker density stratification give stronger overturning? Journal of Physical Oceanography, 33, 2781–2795

Onishi, H., and K. Ohtani (1999) On seasonal and year to year variation in flow of the Alaskan Stream in the Central North Pacific. Journal of Oceanography, 55, 597–608

O'Neill, J.R., R.N. Clayton, and T.K. Mayeda (1969) Oxygen isotope fractionation in divalent metal carbonates. Journal of Chemical Physics, 51, 5547–5558

Østerhus, S., and T. Gammelsrød (1999) The abyss of the Nordic seas is warming. Journal of Climate, 12, 3297–3304

Östlund, H.G., C. Craig, W.S. Broecker, and S. Spencer (eds.) (1987) GEOSECS Atlantic, Pacific, and Indian Ocean Expeditions, Shorebased data and graphics. GEOSECS Atlas Series, volume 7, U.S. Government Printing Office, Washington D.C. 200 pp.

Owens, W.B., and B.A. Warren (2001) Deep circulation in the northwest corner of the Pacific Ocean. Deep-Sea Research I, 48, 959–993

Pedlosky, J. (1996) Ocean Circulation Theory, Springer, Berlin, New York, 453 pp.

Peixoto, J.P., and A.H. Oort (1983) The atmospheric branch of the hydrological cycle and climate. In: Variations in the Global Water Budget, A. Street-Perrott, M Beran; and P Ratcliffe (eds.) Reidel, London, pp. 5–65

Peixoto, J.P., and A.H. Oort (1992) Physics of Climate. American Institute of Physics, New York. 520 pp.

Pickart, R.S. (1992) Water mass components of the North Atlantic deep western boundary current. Deep-Sea Research, 39, 1553–1572

Pickart, R.S., D.J. Torres, and R.A. Clarke (2002) Hydrography of the Labrador Sea during active convection. Journal of Physical Oceanography, 31, 428–457

Piola, A.R., and A.L. Gordon (1989) Intermediate waters in the southwest South Atlantic. Deep-Sea Research, 36, 1–16

Polzin, K.L., J.M. Toole, J.R. Ledwell, and R.W. Schmitt (1997) Spatial variability of turbulent mixing in the abyssal ocean. Science, 276, 93–96

Rahmstorf, S. (1995) Bifurcations of the Atlantic thermohaline circulation in response to changes in the hydrological cycle. Nature, 378, 145–149

Rahmstorf, S. (1996) On the freshwater forcing and transport of the Atlantic thermohaline circulation. Climate Dynamics, 12, 799–811

Raymo, M.E., D.W. Oppo, B.P. Flower, D.A. Hodell, J.F. McManus, K. A. Venz, K. F. Kleiven, and K. McIntyre (2004) Stability of North Atlantic water masses in face of pronounced climate variability during the Pleistocene. Paleoceanography, 19, PA2008, doi:10.1029/2003PA000921

Redfield, A.C. (1934) On the properties of organic derivatives in seawater and their relation to the composition of plankton. In: James Johnstone Memorial Volume, Liverpool University Press, Liverpool, pp. 176–192

Reid, J.L. (1981) On the mid-depth circulation of the world ocean. In: Evolution of Physical Oceanography, W. Warren and C. Wunsch (eds.), MIT Press, Cambridge MA, pp. 70–111

Richardson, P.L., and D.M. Fratantoni, (1999) Float trajectories in the deep western boundary current and deep equatorial jets of the tropical Atlantic. Deep-Sea Research, 46, 305–333

Roemmich, D., S. Hautala, and D. Rudnick (1996) Northward abyssal transport through the Samoan passage and adjacent regions. Journal of Geophysical Research, 101, 14,039–14,055

Rooth, C. (1982) Hydrology and ocean circulation. Progress in Oceanography, 11, 131–149

Rossby, T. (1999) Current switching as a mechanism for rapid climate change. ICES CM 1999/O:07

Rudnick, D.L. (1997) Direct velocity measurements in the Samoan Passage. Journal of Geophysical Research, 102, 3293–3302

Sandström, J.W. (1908) Dynamische Versuche mit Meereswasser. Annalen der Hydrographie und Maritime Meteorologie, 36, 6–23

Sandström, J.W. (1916) Meteorologische Studien im Schwedischen Hochgebirge. Göteborgs Kungliga Vetenskaps- og Vittershetsamhällets Handlingar, 49 pp.

Sarnthein, M., K. Winn, S.J.A. Jung, J.-C. Duplessy, L. Labeyrie, H. Erlenkeuser, and G. Ganssen (1994) Changes in east Atlantic deepwater circulation over the last 30,000 years: Eight time slice reconstructions. Paleoceanography, 9, 209–267

Saunders, P.M. (1994) The flux of overflow water through the Charlie–Gibbs Fracture Zone. Journal of Geophysical Research, 99, 12,343–12,355

Saunders, P.M. (1996) The flux of dense cold overflow water southeast of Iceland. Journal of physical Oceanography, 26, 85–95

Schiller, A., U. Mikolajewicz, and R. Voss (1997) The stability of the thermohaline circulation in a coupled ocean–atmosphere general circulation model. Climate Dynamics, 13, 325–347

Schlosser, P., B. Kromer, R. Weppernig, H.H. Loosli, R. Bayer, G. Nonani, and M. Suter (1994) The distribution of 14C and 39Ar in the Weddel Sea. Journal of Geophysical Research, 99, 10,275–10,287

Schmidt, G.A., G. R. Bigg, and E. J. Rohling (1999) Global Seawater Oxygen-18 Database, http://data.giss.nasa.gov/data/

Schmitz Jr., W.J. (1995) On the interbasin–scale thermohaline circulation. Reviews of Geophysics, 33, 151–173

Schott., F, and K.D. Leaman (1991) Observation with moored acoustic Doppler current profilers in the convection regime of the Golfe du Lion. Journal of Physical Oceanography, 21, 558–574

Schott, F., M. Visbeck, and J. Fischer (1993) Observations of vertical currents and convection in the central Greenland Sea during the winter of 1988/89. Journal of Geophysical Research, 98, 14,401–14,421

Schouten, M.W., W.P.M. de Ruijter, P.J. van Leeuwen, and J.R.E. Lutjeharms (2000) Translation, decay and splitting of Agulhas rings in the southeastern Atlantic Ocean. Journal of Geophysical Research, 105, 21,913–21,925

Scott, J.R., J. Marotzke, and P.H. Stone (1999) Interhemispheric thermohaline circulation in a coupled box model. Journal of Physical Oceanography, 29, 351–365

Shackleton, N.J. (1974) Attainment of isotopic equilibrium between ocean water and the benthic foraminifera genus Uvigerina: isotopic changes in the ocean during the last glacial. Gif sur Yvette: CNRS, Paris

Shackleton, N.J., J.-C. Duplessy, M. Arnold, P. Maurice, M.A. Hall, and J. Cartlidge (1988) Radiocarbon age of last glacial Pacific deep water. Nature, 335, 708–710

Sijp, W.P., and M.H. England (2004) Effect of the Drake Passage throughflow on global climate. Journal of Physical Oceanography, 34, 1254–1266

Sloyan, B.M., and S.R. Rintoul (2001) The Southern Ocean limb of the global deep overturning circulation. Journal of Physical Oceanography, 31, 143–173

Smethie Jr., W.M., R.A. Fine, A. Putzka, and E.P. Jones (2000) Tracing the flow of North Atlantic Deep Water using chlorofluorocarbons. Journal of Geophysical Research, 105, 14,297–14,323

Smith, W.H.F., and D.T. Sandwell (1997) Global sea floor topography from satellite altimetry and ship depth soundings. Science, 277, 1956–1962

Song, Q., A.L. Gordon, and M. Visbeck (2004) Spreading of the Indonesian throughflow in the Indian Ocean. Journal of Physical Oceanography, 34, 772–792

Spiess, F. (1928): Die Meteor Fahrt. Forschungen und Erlebnisse der Deutschen Atlantischen Expedition 1925–1927, Verlag Berlin, 376 pp.

Starr, V.P. (1968) The physics of negative viscosity phenomena. McGraw-Hill, New York, 256 pp.

Stephens, J.C., and D.P. Marshall (2000) Dynamical pathways of Antarctic Bottom water in the Atlantic. Journal of Physical Oceanography, 30, 622–640

Stocker, T.F., and D.G. Wright (1991) A zonally averaged ocean model for the thermohaline circulation. Part II: Interocean circulation in the Pacific–Atlantic Basin System. Journal of Physical Oceanography. 21, 1725–1739

Stocker T.F., D.G. Wright, and W.S. Broecker (1992) The influence of high-latitude surface forcing on the global thermohaline circulation. Paleoceanography, 7, 529–541

Stommel, H. (1958) The abyssal circulation. Deep-Sea Research, 5, 80–82

Stommel, H. (1961) Thermohaline convection with two stable regimes of flow. Tellus, 13, 224–230

Stommel, H., and A.B. Arons (1960) On the abyssal circulation of the world ocean–I. Stationary planetary flow patterns on a sphere. Deep-Sea Research, 6, 140–154

Stommel, H., and D.W. Moore (1989) An introduction to the Coriolis force. Columbia University Press, New York, 297 pp.
Stramma, L., M. Rhein, P. Brandt, M. Dengler, C. Böning, and M. Walter (2005) Upper ocean circulation in the western tropical Atlantic in boreal fall 2000. Deep-Sea Research I, 52, 221–240
Stuiver, M., P.J. Reimer, and T.F. Braziunas (1998) High precision radiocarbon age calibration for terrestrial and marine samples. Radiocarbon, 40, 1127–1151
SUN (1985) The International System of Units (SI) in Oceanography. Unesco Technical Papers in Science, 45, Unesco, Paris, 124 pp.
Sverdrup, H.U., M.W. Johnson, and R.H. Fleming (1942) The oceans: their physics, chemistry, and general biology. Prentice-Hall Inc., New York, 1087 pp.
Swallow, J.C., and L.V. Worthington (1957) Measurements of deep currents in the western North Atlantic. Nature, 179, 1183–1184
Swallow, J.C., and L.V. Worthington (1961) An observation of a deep countercurrent in the western North Atlantic. Deep-Sea Research, 8, 1–19
Swift, J.H. (1986) The arctic waters. In: B.G. Hurdle (ed.) The Nordic Seas. Springer-Verlag, pp. 129–153
Swift, J.H. and K.P. Koltermann (1988) The origin of Norwegian Sea Deep Water, Journal of Geophysical Research, 93, 3563–3569
Tarantola, A. (1987) Inverse problem theory. Elsevier Science Publishers B.V., Amsterdam, 613 pp.
Te Raa, L. (2003) Internal variability of the thermohaline ocean circulation. PhD thesis, Utrecht University, Utrecht, 158 pp.
Thompson, B. (1968) Collected works of count Rumford. The propagation of heat in fluids. S.C. Brown, (ed.) Harvard University Press, Cambridge, Massachusetts, pp. 117–284 (reprinted from 1798)
Thorpe, R. B., J. M. Gregory, T. C. Johns, R. A. Wood, and J. F. B Mitchell (2001) Mechanisms determining the Atlantic thermohaline circulation response to greenhouse gas forcing in a non-flux-adjusted coupled climate model. Journal of Climate, 14, 3102–3116
Toggweiler, J.R., and R.M. Key (2001) Thermohaline circulation. In: Encyclopedia of ocean sciences, J.H. Steele; S.A. Thorpe, and K.K. Turekian (eds.) pp. 2941–2947
Toggweiler, J.R., and B. Samuels (1995) Effect of Drake Passage on the global thermohaline circulation. Deep-Sea Research I, 42, 477–500
Tomczak, M. (1981) A multi-parameter extension of T–S diagram techniques for the analysis of non-isopycnal mixing. Progress in Oceanography, 10, 147–171
Tomczak, M. (1999) Some historical, theoretical and applied aspects of quantitative water mass analysis. Journal of Marine Research, 57, 275–303
Tomczak, M., and J.S. Godfrey (1994) Regional oceanography: an introduction. Pergamon, Oxford, U.K. 422 pp.
Trenberth, K.E., and J.M. Caron (2001) Estimates of meridional atmosphere and ocean heat transports. Journal of Climate, 14, 3433–3443

Tsuchiya, M., L.D. Talley, and M.S. McCartney (1992) An eastern Atlantic section from Iceland southward across the equator. Deep-Sea Research, 39, 1885–1917

Tziperman, E. (1997) Inherently unstable climate behaviour due to weak thermohaline ocean circulation. Nature, 386, 592–595

van Aken, H.M. (1995) Hydrographic variability in the bottom layer of the Iceland Basin. Journal of Physical Oceanography, 25, 1716–1722

van Aken, H.M. (2000) The hydrography of the mid-latitude Northeast Atlantic Ocean: I, The deep water masses. Deep-Sea Research I, 47, 757–788

van Aken, H.M. (2001) The hydrography of the mid-latitude Northeast Atlantic Ocean — Part III, The thermocline water masses. Deep-Sea Research I, 48, 237–267

van Aken, H.M. (2002) Surface currents in the Bay of Biscay as observed with drifters between 1995 and 1999. Deep-Sea Research I, 49, 1071–1086

van Aken, H.M., and G. Becker (1995) Hydrography and through-flow in the north-eastern North Atlantic Ocean: the NANSEN project. Progress in Oceanography, 38, 297–346

van Aken, H.M., A.K. van Veldhoven, C. Veth, W.P.M. de Ruijter, P.J. van Leeuwen, S.S. Drijfhout, C.P. Whittle, and M. Rouault (2003) Observations of a young Agulhas ring, Astrid, during MARE in March 2000. Deep-Sea Research II, 50, 167–195

van Aken, H.M., H. Ridderinkhof, and W.P.M. de Ruijter (2004) North Atlantic deep water in the south-western Indian Ocean. Deep-Sea Research I, 755–776

van der Schrier, G., and L.R.M. Maas (1998) Chaos in a simple model of the three-dimensional, salt-dominated ocean circulation. Climate Dynamics, 14, 489–502

Vellinga, M. (1997) Bifurcations and dynamics of the ocean's thermohaline circulation. PhD Thesis, Utrecht University, 129 pp.

Vellinga, M., and R.W. Wood (2002) Global climatic impacts of collapse of the Atlantic thermohaline circulation. Climate Change, 54, 251–267

Vellinga, M., R.W. Wood, and J.M. Gregory (2002) Processes governing the recovery of a perturbed thermohaline circulation in HadCM3. Journal of Climate, 15, 764–780

Vranes, K., and A.L. Gordon, and A. Ffield (2002) The heat transport of the Indonesian throughflow and implications for the Indian Ocean heat budget. Deep-Sea Research II, 49, 1391–1410

Wadhams, P., G. Budeus, J.P. Wilkinson, T. Løyning, and V. Pavlov (2004) The multi-year development of long-lived convective chimneys in the Greenland Sea. Geophysical Research Letters, 31, doi:10.1029/2003GL019017

Walker, S.J., R.F. Weiss, and P.K. Salamah (2000) Reconstructed histories of the annual mean atmospheric mole fractions for the halocarbons CFC-11, CFC-12, CFC-113, and carbon tetrachloride. Journal of Geophysical Research, 105, 14,285–14,296

Wang, X, P.H. Stone, and J. Marotzke (1999) Global thermohaline circulation. Part I: sensitivity to atmospheric moisture transport. Journal of Climate, 12, 71–82

Warner, M.J., and R.F. Weiss (1985) Solubilities of chlorofluorocarbons 11 and 12 in water and seawater. Deep-Sea Research, 32, 1485–1497

Warren, B.A. (1981) Deep circulation of the world ocean. In: Evolution of Physical Oceanography, W. Warren, and C. Wunsch (eds.) MIT Press, Cambridge MA, pp. 6–41

Warren, B.A. (1983) Why is no deep water formed in the North-Pacific? Journal of Marine Research, 41, 327–347

Warren, B.A., and W.B. Owens (1985) Some preliminary results concerning deep northern-boundary currents in the North Pacific. Progress in Oceanography, 14, 537–551

Warren, B.A., and W.B. Owens (1988) Deep currents in the central subarctic Pacific Ocean. Journal of Physical Oceanography, 18, 529–551

Weaver, A.J., and C. Hillaire-Marcel (2004) Global warming and the next ice age. Science, 304, 400–402

Weaver, A.J., C.M. Bitz, A.F. Fanning, and M.M. Holland (1999) Thermohaline ciculation: High-latitude phenomena and the difference between pacific and Atlantic. Annual Reviews of Earth and Planetary Sciences, 27, 231–285

Wefer, G., W.H. Berger, J. Bijma, and G. Fisher (1999) Clues to ocean history: a brief overview of proxies. In: Use of proxies in paleoceanography: examples from the South Atlantic. G. Fischer and G. Wefer (eds.) Springer Verlag, Berlin, pp. 1–68

Welander, P. (1986) Thermohaline effects in the ocean circulation and related simple models. In: Large-scale transport processes in oceans and atmosphere, J. Willebrand and D.L.T. Anderson (eds.) D. Reidel Publishing Company, pp. 163–200

Whitworth, T., B.A.Warren, W.D. Nowlin, S.B. Rutz, R. D. Pillsbury, and M.I. Moore (1999) On the deep western boundary current in the South-west Pacific basin. Progress in Oceanography, 43, 1–54.

Wijffels, S.E. (2001) Ocean transport of fresh water. In: Ocean circulation & climate: observing and modelling the global ocean. G. Siedler, J. Church, and J. Gould (eds.) Academic Press, London, 475–488

Wijffels, S.E., R.W. Schmitt, H.L. Bryden, and A. Stigebrandt (1992) Transport of freshwater by the oceans. Journal of Physical Oceanography, 22, 155–162

Wood, R.A., A.B. Keen, J.F.B. Mitchell, and J.M. Gregory (1999) Changing spatial structure of the thermohaline circulation in response to atmospheric CO_2 forcing in a climate model. Nature, 399, 572–575

Wright, D.G., and T.F. Stocker (1991) A zonally averaged model for the thermohaline circulation. Part I: Model development and flow dynamics. Journal of Physical Oceanography, 21, 1713–1724

Wright, D.G., C.B. Vreugdenhil, and T.M.C. Hughes (1995) Vorticity dynamics and zonally averaged ocean circulation models. Journal of Physical Oceanography, 25, 2141–2154

Wright, W.R., and L.V. Worthington (1970) The water masses of the North Atlantic Ocean: a volumetric census of temperature and salinity. Serial Atlas of the Marine Environment, Folio 19, American Geographic Society, New York, 9 pp.

Wunsch, C. (1985) Can a tracer field be inverted for velocity? Journal of Physical Oceanography, 15, 1521–1531

Wunsch, C. (1992) Decade to century changes in the ocean circulation. Oceanography, 5, 99–106

Wunsch, C. (1996) The ocean circulation inverse problem. Cambridge University Press, Cambridge, 442 pp.

Wunsch, C. (2002) What is the thermohaline circulation? Science, 298, 1179–1181

Wunsch, C. (2003) Determining paleoceanographic circulations, with emphasis on the Last Glacial Maximum. Quarternary Science Reviews, 22, 371–385

Wunsch, C. (2005) Abrupt climate change. An alternate view. Unpublished manuscript.

Wüst, G. (1933) Das Bodenwasser und die Gliederung de Atlantische Tiefsee. Wisschenschaftliche Ergebnisse der Deutsche Atlantische Expedition auf dem Forschungs- und Vermessungsschiff "Meteor" 1925–1927, VI, 1(1), 1–107

Wüst, G. (1935) Schichtung und Zirkulation des Atlantischen Ozeans. Die Stratosphäre. Wisschenschaftliche Ergebnisse der Deutsche Atlantische Expedition auf dem Forschungs- und Vermessungsschiff "Meteor" 1925–1927, VI, 1(2), 109–288

Wyrtki, K. (1961) The thermohaline circulation in relation to the general circulation in the ocean. Deep-Sea Research, 8, 39–64

Wyville Thomson, C. (1877) The voyage of the Challenger: The Atlantic. A preliminary account of the general results of the exploring voyage of H.M.S. challenger during the year 1873 and the early part of the year 1876. Volume I. Macmillan and Co. London

Index

ATMOSPHERIC AND OCEANOGRAPHIC SCIENCES LIBRARY

1. F.T.M. Nieuwstadt and H. van Dop (eds.): *Atmospheric Turbulence and Air Pollution Modelling*. 1982; rev. ed. 1984
ISBN 90-277-1365-6; Pb (1984) 90-277-1807-5
2. L.T. Matveev: *Cloud Dynamics*. Translated from Russian. 1984 ISBN 90-277-1737-0
3. H. Flohn and R. Fantechi (eds.): *The Climate of Europe: Past, Present and Future*. Natural and Man-Induced Climate Changes: A European Perspective. 1984
ISBN 90-277-1745-1
4. V.E. Zuev, A.A. Zemlyanov, Yu.D. Kopytin, and A.V. Kuzikovskii: *High-Power Laser Radiation in Atmospheric Aerosols*. Nonlinear Optics of Aerodispersed Media. Translated from Russian. 1985 ISBN 90-277-1736-2
5. G. Brasseur and S. Solomon: *Aeronomy of the Middle Atmosphere*. Chemistry and Physics of the Stratosphere and Mesosphere. 1984; rev. ed. 1986
ISBN (1986) 90-277-2343-5; Pb 90-277-2344-3
6. E.M. Feigelson (ed.): *Radiation in a Cloudy Atmosphere*. Translated from Russian. 1984
ISBN 90-277-1803-2
7. A.S. Monin: *An Introduction to the Theory of Climate*. Translated from Russian. 1986
ISBN 90-277-1935-7
8. S. Hastenrath: *Climate Dynamics of the Tropics*. Updated Edition from *Climate and Circulation of the Tropics*. 1985; rev. ed. 1991 ISBN 0-7923-1213-9; Pb 0-7923-1346-1
9. M.I. Budyko: *The Evolution of the Biosphere*. Translated from Russian. 1986
ISBN 90-277-2140-8
10. R.S. Bortkovskii: *Air-Sea Exchange of Heat and Moisture During Storms*. Translated from Russian, rev. ed. 1987 ISBN 90-277-2346-X
11. V.E. Zuev and V.S. Komarov: *Statistical Models of the Temperature and Gaseous Components of the Atmosphere*. Translated from Russian. 1987 ISBN 90-277-2466-0
12. H. Volland: *Atmospheric Tidal and Planetary Waves*. 1988 ISBN 90-277-2630-2
13. R.B. Stull: *An Introduction to Boundary Layer Meteorology*. 1988
ISBN 90-277-2768-6; Pb 90-277-2769-4
14. M.E. Berlyand: *Prediction and Regulation of Air Pollution*. Translated from Russian, rev. ed. 1991 ISBN 0-7923-1000-4
15. F. Baer, N.L. Canfield and J.M. Mitchell (eds.): *Climate in Human Perspective*. A tribute to Helmut E. Landsberg (1906–1985). 1991 ISBN 0-7923-1072-1
16. Ding Yihui: *Monsoons over China*. 1994 ISBN 0-7923-1757-2
17. A. Henderson-Sellers and A.-M. Hansen: *Climate Change Atlas*. Greenhouse Simulations from the Model Evaluation Consortium for Climate Assessment. 1995
ISBN 0-7923-3465-5
18. H.R. Pruppacher and J.D. Klett: *Microphysics of Clouds and Precipitation*, 2nd rev. ed. 1997 ISBN 0-7923-4211-9; Pb 0-7923-4409-X
19. R.L. Kagan: *Averaging of Meteorological Fields*. 1997 ISBN 0-7923-4801-X
20. G.L. Geernaert (ed.): *Air-Sea Exchange: Physics, Chemistry and Dynamics*. 1999
ISBN 0-7923-5937-2
21. G.L. Hammer, N. Nicholls and C. Mitchell (eds.): *Applications of Seasonal Climate Forecasting in Agricultural and Natural Ecosystems*. 2000 ISBN 0-7923-6270-5
22. H.A. Dijkstra: *Nonlinear Physical Oceanography*. A Dynamical Systems Approach to the Large Scale Ocean Circulation and El Niño. 2000 ISBN 0-7923-6522-4
23. Y. Shao: *Physics and Modelling of Wind Erosion*. 2000 ISBN 0-7923-6657-3
24. Yu.Z. Miropol'sky: *Dynamics of Internal Gravity Waves in the Ocean*. Edited by O.D. Shishkina. 2001 ISBN 0-7923-6935-1
25. R. Przybylak: *Variability of Air Temperature and Atmospheric Precipitation during a Period of Instrumental Observations in the Arctic*. 2002 ISBN 1-4020-0952-6
26. R. Przybylak: *The Climate of the Arctic*. 2003 ISBN 1-4020-1134-2
27. S. Raghavan: *Radar Meteorology*. 2003 ISBN 1-4020-1604-2
28. H.A. Dijkstra: *Nonlinear Physical Oceanography*. A Dynamical Systems Approach to the Large Scale Ocean Circulation and El Niño. 2nd Revised and Enlarged Edition. 2005
ISBN 1-4020-2272-7 Pb; 1-4020-2262-X
29. X. Lee, W. Massman and B. Law (eds.): *Handbook of Micrometeorology*. A Guide for Surface Flux Measurement and Analysis. 2004 ISBN 1-4020-2264-6
30. A. Gelencsér: *Carbonaceous Aerosol*. 2005 ISBN 1-4020-2886-5
31. A. Soloviev and L. Roger: *The Near-Surface Layer of the Ocean. Structure, Dynamics and Applications*. 2006 ISBN 1-4020-4052-0

32. G.P. Brasseur and S. Solomon: *Aeronomy of the Middle Atmosphere. Chemistry and Physics of the Stratosphere and Mesosphere.* 2005
ISBN 1-4020-3284-6; Pb 1-4020-3285-4
33. B. Wozniak and J. Dera: *Light Absorption and Absorbants in Sea Water.* 2006
ISBN 0-387-30753-2
34. A. Kokhanovsky: *Cloud Optics.* 2005 ISBN 1-4020-3955-7
35. T.N. Krishnamurti, H.S. Bedi, V.M. Hardiker, and L. Ramaswamy: *An Introduction to Global Spectral Modeling, 2nd Revised and Enlarged Edition.* 2006
ISBN 0-387-30254-9

SPRINGER NATURE

GPSR Compliance

The European Union's (EU) General Product Safety Regulation (GPSR) is a set of rules that requires consumer products to be safe and our obligations to ensure this.

If you have any concerns about our products, you can contact us on ProductSafety@springernature.com

In case Publisher is established outside the EU, the EU authorized representative is:

Springer Nature Customer Service Center GmbH
Europaplatz 3
69115 Heidelberg, Germany

FSC
www.fsc.org
MIX
Papier | Fördert
gute Waldnutzung
FSC® C083411

Kolibri 360 GmbH
Ferdinand-Jühlke-Straße 7
99095 Erfurt, Deutschland
produktsicherheit@kolibri360.de